Aligning Perspectives on Health, Safety and Well-Being

Series editors

Stavroula Leka, Institute of Work, Health & Organization, University of Nottingham, Nottingham, United Kingdom
Aditya Jain, Jubilee Campus, Nottingham University Business School, Nottingham, United Kingdom
Gerard Zwetsloot, University of Nottingham, TNO Innovation for Life, Hoofddorp, Noord-Holland, The Netherlands

Raising awareness of the interdisciplinary and complementary relationship of different research perspectives on health, safety and well-being is the main aim of the book series Aligning Perspectives on Health, Safety and Well-being. Combined research approaches on health, safety and well-being are becoming more and more popular in several research disciplines across and between the social, behavioural and medical sciences. Therefore, Aligning Perspectives on Health, Safety and Well-being stimulates the publication of interdisciplinary approaches to the promotion of health, safety and well-being. Recognizing a need within societies and workplaces for more integrated approaches to problem solving, the series caters to the notion that most innovation stems from combining knowledge and research results from related but so far separated areas. Volumes will be edited by expert authors and editors and will contain contributions from different disciplines. All authors, and especially volume editors are encouraged to engage in developing more robust theoretical models that can be applied in actual practice and lead to policy development.

More information about this series at http://www.springer.com/series/10757

Peter R.A. Oeij · Diana Rus
Frank D. Pot
Editors

Workplace Innovation

Theory, Research and Practice

Springer

Editors
Peter R.A. Oeij
TNO Innovation for Life
Leiden
The Netherlands

Frank D. Pot
Institute for Management Research
Radboud University
Nijmegen
The Netherlands

Diana Rus
Creative Peas
Amsterdam
The Netherlands

ISSN 2213-0497 ISSN 2213-0470 (electronic)
Aligning Perspectives on Health, Safety and Well-Being
ISBN 978-3-319-85887-6 ISBN 978-3-319-56333-6 (eBook)
DOI 10.1007/978-3-319-56333-6

Printed on acid-free paper

This Springer imprint is published by Springer Nature
The registered company is Springer International Publishing AG
The registered company address is: Gewerbestrasse 11, 6330 Cham, Switzerland

Foreword

The book '*Workplace Innovation: Theory, Research and Practice*' is targeted at a broad group of readers, namely policy makers, scientists and researchers, as well as workplace innovation practitioners such as employers, employees and employee representatives. This is why we asked four prominent individuals from these various fields to shed light on the question 'what is the meaning of workplace innovation for….?', respectively, science, policy makers, employers and employees.

Science and Research

Psychosocial Value in Work and Political Economy

This book offers important practical contributions to solving a major set of new challenges in our global economy relating to psychosocial wellbeing at work. We can see how this could be the case by placing both these problems, and the book's solutions, in a broad historical perspective.

These contributions—and in addition our recently developed Associationalist Demand–Control (ADC) model—both reflect a 'New Bargain' for the 21st century relating to the social structure of work and the wellbeing of individuals. They emphasize the 'psychosocial value of work activity', thereby outlining an alternative linkage between our most fundamental contemporary social institutions—work and the economy—and the wellbeing those are designed to create. This is a bargain to at least partially replace the material wellbeing Bargain #1 of the 17th–20th centuries that started in England with John Locke's 1690's formulations on property, and representative democracy. That bargain offered "everyman" actual control —in the form of "property rights" and new political processes—of the fruits of their labour, and was thus a major step forward for citizens whose property was otherwise always controlled by an absolute monarch with power theocratically legitimized in the western European countries of that era. It formed the platform for a broadly inclusive, albeit materialist, society.

Both this book and the new ADC model both go beyond that. "Bargain #2 offers—now—upon a social and psychological work organization platform—another pathway beyond Locke's original materialist formulation, to further positive social progress for our current uncertain times. The first half of the Bargain #2 is a creative goal for life: growth of capabilities (skills) for active, living beings (and in a social context: collaborative capabilities from creative coordination). These are based upon new social dynamics to link users' needs and workers' capabilities in smart, adaptive jobs. The second half of the Bargain #2 offers health—reduction of stress-related disease—based on a personal maintenance of stable internal self-regulation, which at the social level supports the welfare state's sustainability by reducing social costs." (Karasek, R., Dollard, M., Östergren, P-O, Formazin, M., Agbenyikey, W., Li, J., Cho, S-I, Houtman I. (submitted 2017). The Multi-level Job Content Questionnaire 2.0 (JCQ2) and the Associationalist Demand–Control (ADC) Theory)

This 'psychosocial value of work' was the platform for the original DC model, based on organizational sociology and psychology. Now, the recently developed Associationalist Demand–Control model goes further to address the political-economic challenges above, addressing three new important issues: explaining work stress effects in a multi-level work organization context; specifying multiple-level social relational processes that are allowing the needed organizational flexibility for stability-in-the-face-of-change; and explaining creative engagement and growth for both workers and organizations. Thus, the ADC model goes beyond the limits of single-discipline boundaries and describes multiple, linked levels of function in organizations where central control functions must coordinate the overall actions of sub-systems (for example: employees, departments) as they take integrated action in the environment. In summary, the ADC "model describes how systems can either organize themselves into higher levels of complexity (active hypothesis) or dissolve into systems with lower levels of complexity (strain hypothesis)—i.e. systems that grow and develop or systems no longer able to sustain their original complexity and capability. Thus the key issues are *coordination*, and the *association of parts*—rather than the physical reality of the parts themselves—moving beyond a purely materialist construction of reality." (Karasek et al., submitted 2017) The ADC model can be measured by either the User Version of the Job Content Questionnaire (JCQ) 2.0 (for practitioners), or the Researcher Version JCQ 2.0 (both to be published soon).

The ADC model further evolves the original DC model's 'active work' concept into the 'conducive economy', creating value for citizens in Bargain #2: producing the psychosocial wellbeing and social integration for healthy work that are otherwise missing in our market economy. On the 'production side' the 'conducive economy' contributes new ideas and innovation, and provides the new skills and training that are needed for future youth jobs in social policies for innovative economic development, also often missing in our conventional commodity-based economy.

Quite a number of authors in this book take the DC model as one of the foundations of their approach, developing it further or complementing it with other approaches—and hopefully the ADC model will encourage additional steps in this direction. The rich workplace innovation content of this book also supports, in my view, the perspective of a 'conducive economy'. Scientifically, I think the book is a valuable contribution to multi-level and multi-disciplinary theory and research in wellbeing and organizational performance. Moreover, the book shows a strong European research community that has been active for decades already and has been able to influence national and European policies several times. Those 'good practices' show the world how to take practical steps to insure a progressive and humane society in the future.

Robert Karasek is Director of Øresund Synergy and the JCQ (Job Content Questionnaire) Center and Emeritus Professor, Department of Work Environment, University of Massachusetts Lowell, USA and Emeritus Professor, Work and Organizational Psychology, Institute for Psychology, Copenhagen University, Denmark. He has received an American Psychological Association Lifetime Career Achievement Award for his work.

Policy Making

Fit for the Future

The nature of work has changed significantly during the course of human history. However, the pace of this change has accelerated in recent years, largely due to digital technologies. The future world of work will be different from the one we know. New technologies are already affecting job definitions and work patterns. They are transforming the relationship between employers and employees, the organization of work, and the types of business models used. Many of today's jobs and skill profiles did not exist a decade ago, while routine tasks are often vulnerable to automation.

Success in the new industrial revolution of course requires that our industry uses the best available technologies. But in parallel we also have to focus on the human factor. This is our main resource in Europe and we need to make the best use of it. Workers and employees need to find a new place in smart factories. Companies need to anticipate changes to ensure that both the workforce and the workplace are fit for the future.

European companies need to adapt to rapid change. Advances in automation, digitization and advanced manufacturing represent enormous opportunities for both employers and employees. But too few companies are actually rethinking the way people work and collaborate. Too few companies are remodelling their internal organization to tap into the capacities of all their employees—not only in their R&D departments. To be a leader of the new industrial revolution means to look beyond technologies. It requires having workplace innovation at the very DNA of the organization.

With this in mind, the European Commission proposed a series of initiatives. In 2013 we started the European Workplace Innovation Network, EUWIN. Under this project we produced practical materials, tools and know-how gathered in the EUWIN Knowledge Bank. We have stimulated change and adaptation in many firms, and hope that you will use these resources and connections to start your own workplace revolution. We also launched the Digital Agenda to help European companies to better exploit the potential of Information and Communication Technologies and the new industrial revolution. Finally, we are also implementing A Blueprint for Sectoral Cooperation on Skills to help Europeans find quality jobs in the offices and factories of the future.

Antti Peltomäki is Deputy Director-General of the Internal Market, Industry, Entrepreneurship and SMEs Directorate-General (DG GROW), European Commission.

Employer's Perspective

Social Innovation: "OUR" Future is "OUR" Responsibility
Entrepreneurship, creativity and passion …. Three human characteristics which have driven evolution and growth from the beginning of mankind are a source of energy abundantly available in every organization.

Competitiveness, flexibility, faster and faster time to market, continuous customized product innovation …. Are the characteristics of our organizations' challenges.

Working longer, urbanization, digitalization, mobility, … are some examples of the growing complexity of society.

Those organizations who have understood that product innovation and technology will not be sufficient to build the future AND who are able to bring "entrepreneurship, creativity, passion" of their employees together with the organizational and societal challenges are creating the most opportunities for growth, sustained bottom line results AND motivated people eager to contribute to the success of their teams and their companies.

I am very fortunate to live and lead a real-life example, proving that next to product- and technological innovation, social innovation (or workplace innovation) is an important pillar for success. This success is based on the conviction that: ***"our" future is "our" responsibility***, a conviction which is enabling every employee to be part of the whole, which is empowering each team and every team member to make the difference.

Is there a secret recipe? No, but there are important ingredients to be understood and to be supported by ALL involved people: leaders, social partners and employees.

- ***We get commitment by involvement***: Give time and room to employees to really participate and to contribute.
- ***Every process of change needs time***: The grass is not growing faster by pulling it! We need to take care by making the change process tangible and visible and by stimulating the behaviour which is helping us forward.
- ***Be willing to question ourselves as leaders***: Do we act and behave consistent with our messages, is our commitment "visible" on the shop floor, do we really give the others the ability to express their ideas and to implement them?
- ***Invest in training, coaching and continuous development***: Answering the need for competences and capabilities on "all" levels of the organization.
- ***Invest in continuous communication in two directions***: People cannot support what they do not know, what they do not understand. A "common" understanding of the objectives, the challenges, the need for help, the individual and team contribution is key next to celebrating success.

There is no secret recipe but sharing experiences and learning from each other has helped us to increase agility and to move faster. Therefore, it is an honour to write a foreword for this book and to thank the European Workplace Innovation Network for their successful approach in bringing organizations, politicians, unions and science together, supporting us in defining our "common" process of building "our" future.

Let's keep the rhythm, with enthusiasm.

Edwin Van Vlierberghe is Global Head of Supply Chain End to End, Bombardier Transportation, Belgium.

Employees' Perspective

We Need Work 4.0

Lately I've been studying, talking and even dreaming about the Fourth Industrial Revolution. There's no question that we are in the middle of a major shift in the world before the digitalization of everything. We have already seen the emergence of new business and money-making models, which significantly disrupt the old working life we're accustomed to. Technologies are not only revolutionizing one industry, but rather all industries as well as work tasks in a radical and exponential way.

The need for workplace innovation is obvious and enormous, but until now, it has mainly been excluded from the public, technology-driven debate. We have witnessed developments whereby industrial processes and the attitudes of people are changing more rapidly than ever. Employees are confused between the shining visions of new kinds of autonomous ways of working and the deep fear of losing their livelihood because of a lack of competencies or cheaper and noncomplaining robots.

It is evident that the Fourth Industrial Revolution also demands Work 4.0., where workplace innovations must play a major role. If even half of the companies would have the competencies for this, Europe would experience a higher level of wealth and prosperity.

Work 4.0 means a work community built on trust, partnership and open dialogue. Change-enthusiasm on the part of management is not enough for success; you need the whole company staff to back the change. Trust should be an important performance criterion for management. Workplace Innovation is doing exactly that.

Too few companies are rethinking the way people work and collaborate, or remodelling their internal organization to tap into the capacities of their employees. According to the European Company Survey, only 10 percent of companies achieve this kind of working community. That means 90 percent of companies constantly squander their possibilities to prosper!

What about the trade union strategies? There are factors that both block and support employees in being active development partners in companies.

One blocking factor is the rapidly changing labour market where unions are struggling for survival because of the declining membership rates. Unions are less visible in many workplaces and it is harder to get people behind cooperation initiatives. Secondly, the spreading polarization effect in the labour markets means that the traditional stronghold for the unions, the skilled worker with a permanent contract in industry and services, is shrinking. These developments may cause the unions to rethink or even abandon strategies based on trustful and cooperative relations with employers. On the other hand, unions are very well aware of these challenges and modify their strategies according to these pressures.

One supporting factor is that trade unions invest a lot in competence building, activity and involvement at work, improvement of work organization and quality of working life. Secondly, many trade unions realize that influencing via workplace innovations can be a sustainable part of union strategies, even if this has negative employment or wage effects in the short run. Usually, in the end, the benefits will be higher than potential losses. Workplace innovations are seen as a fair mechanism for managing change and securing the future.

Leila Kurki is senior adviser Finnish Confederation of Professionals STTK and member of the Advisory Board of European Workplace Innovation Network (EUWIN).

Contents

Editors and Contributors

About the Editors

Peter R.A. Oeij Senior Research Scientist, TNO, The Netherlands Organisation for Applied Scientific Research, Leiden, The Netherlands.

Diana Rus Director, Creative Peas, Amsterdam, The Netherlands.

Frank D. Pot Professor Emeritus, Radboud University, Nijmegen, The Netherlands.

Contributors

Tone Merethe Berg Aasen SINTEF Technology and Society, Trondheim, Norway

Tuomo Alasoini Tekes—The Finnish Funding Agency for Innovation, Helsinki, Finland

Oscar Amundsen Faculty of Social Sciences and Technology Management, Department of Education and Lifelong Learning, Norwegian University of Science and Technology, Trondheim, Norway

Isabella Biletta Eurofound, Dublin, Ireland

Sylvie Boermans TNO, The Netherlands Organisation for Applied Scientific Research, Leiden, The Netherlands

Andrea Ceschi Department of Human Sciences, Verona University, Verona, Italy

Antonio Corral IKEI Research & Consultancy, San Sebastián, Spain

Arianna Costantini Department of Human Sciences, Verona University, Verona, Italy

Vlad Dediu Centre for Organizational Health & Development, School of Medicine, University of Nottingham, Nottingham, United Kingdom

Steven Dhondt TNO, The Netherlands Organisation for Applied Scientific Research, Leiden, The Netherlands; KU Leuven, Leuven, Belgium

Rosemary Exton Workplace Innovation Europe CLG, Dublin, Ireland

Leif Jarle Gressgård Department of Social Sciences, International Research Institute of Stavanger, Stavanger, Norway

Kåre Hansen Department of Social Sciences, International Research Institute of Stavanger, Stavanger, Norway

Peter Hasle Center for Industrial Production, Aalborg University Copenhagen, Copenhagen, Denmark

Jürgen Howaldt Sozialforschungsstelle Dortmund (sfs), Central Scientific Institute of TU Dortmund University, Dortmund, Germany

Helge Hvid Roskilde University, Roskilde, Denmark

Aditya Jain Nottingham University Business School, Nottingham, United Kingdom

Maria Karanika-Murray Department of Psychology, Nottingham Trent University, Nottingham, United Kingdom

Ralf Kopp Sozialforschungsstelle Dortmund (sfs), Central Scientific Institute of TU Dortmund University, Dortmund, Germany

Stavroula Leka Centre for Organizational Health & Development, School of Medicine, University of Nottingham, Nottingham, United Kingdom

Chris Mathieu Department of Sociology, Lund University, Lund, Sweden

Zofia Mockałło Ergonomics Department, Laboratory of Occupational Psychology and Sociology, Central Institute for Labour Protection—National Research Institute, Warsaw, Poland

Peter R.A. Oeij TNO, The Netherlands Organisation for Applied Scientific Research, Leiden, The Netherlands

Agnès Parent-Thirion Eurofound, Dublin, Ireland

Frank D. Pot Radboud University, Nijmegen, The Netherlands

Paul Preenen Dutch Tax and Customs Administration, Utrecht, The Netherlands

Elise Ramstad Tekes—The Finnish Funding Agency for Innovation, Helsinki, Finland

Boleslaw Rok Kozminski University, Warsaw, Poland

Diana Rus Creative Peas, Amsterdam, The Netherlands

Riccardo Sartori Department of Human Sciences, Verona University, Verona, Italy

Vibeke Kristine Scheller Copenhagen Business School, Frederiksberg, Denmark

Jürgen Schultze Sozialforschungsstelle Dortmund (sfs), Central Scientific Institute of TU Dortmund University, Dortmund, Germany

Liv Starheim Implementations and Performance Management, Technical University of Denmark, Kongens Lyngby, Denmark

Marta Strumińska-Kutra VID Specialized University, Oslo, Norway; Kozminski University, Warsaw, Poland

Peter Totterdill Kingston University, London, United Kingdom; Workplace Innovation Europe CLG, Dublin, Ireland

Fietje Vaas TNO, The Netherlands Organisation for Applied Scientific Research, Leiden, The Netherlands

Pierre van Amelsvoort Catholic University Leuven, KU Leuven, Leuven, Belgium

Geert Van Hootegem Centre for Sociological Research, KU Leuven, Leuven, Belgium

Lander Vermeerbergen Centre for Sociological Research, KU Leuven, Leuven, Belgium

Greet Vermeylen Eurofound, Dublin, Ireland

Chris Warhurst Institute for Employment Research, University of Warwick, Coventry, United Kingdom

Mathijn Wilkens Eurofound, Dublin, Ireland

Sally Wright Institute for Employment Research, University of Warwick, Coventry, United Kingdom

Gerard Zwetsloot Centre for Organizational Health & Development, School of Medicine, University of Nottingham, Nottingham, United Kingdom

Rita Žiauberytė-Jakštienė TNO, The Netherlands Organisation for Applied Scientific Research, Leiden, The Netherlands

Chapter 1
Introduction: The Need to Uncover the Field of Workplace Innovation

Frank D. Pot, Diana Rus and Peter R.A. Oeij

1.1 Where Does Workplace Innovation Come from?

Since the 1990s there has been growing concern among companies on how they can remain productive and innovative in a globalised market and in the so-called knowledge economy. Public organisations share similar concerns because the population (i.e., citizens, customers, clients and patients), which is better educated than ever before, expects high-level services under circumstances of limited public budgets. Over time, the awareness grew that technological innovation is not enough to prepare private and public organisations to successfully face the future. Instead, the notion emerged that technological innovation should be complemented and integrated with non-technological innovation [in the terminology of the European Community Innovation Surveys (CIS)] or with social innovation in the workplace (as it is called in a number of countries). The financial and economic crisis of 2008 has only strengthened the urgency of this type of integration. Non-technological innovation in the CIS covers marketing and organisational innovation. Social innovation takes a slightly broader perspective and covers issues such as work organisation, human resources policies and labour relations in the workplace as well as marketing strategies and collaboration with suppliers, research institutes and social partners. Core elements of these non-technological approaches to

F.D. Pot (✉)
Radboud University, Nijmegen, The Netherlands
e-mail: frank.pot@ardan.demon.nl

D. Rus
Creative Peas, Amsterdam, The Netherlands
e-mail: d.rus@creative-peas.com

P.R.A. Oeij
TNO, The Netherlands Organisation for Applied Scientific Research, Leiden, The Netherlands
e-mail: peter.oeij@tno.nl

P.R.A. Oeij et al. (eds.), *Workplace Innovation*, Aligning Perspectives on Health, Safety and Well-Being, DOI 10.1007/978-3-319-56333-6_1

innovation are employee participation and better utilisation of human talents, including employees' as well as managers' talents. The objective of these approaches is to improve job quality and organisational performance simultaneously. Various names are being used for approaches which more or less relate to these issues and objectives. Some examples are: high performance workplaces, high involvement workplaces, innovative workplaces, innovative work organisation, workplace development, social innovation in the workplace, knowledge-based capital, relational coordination, employee-driven innovation and workplace innovation.

Since the beginning of this century, across Europe, researchers and consultants as well as national initiatives and programmes in English-speaking countries (UK, Ireland) have used the concept of workplace innovation (WPI) as an umbrella term for non-technological innovation. Programmes and initiatives in other countries have used concepts in their own languages, some of which are mentioned in an English translation among the examples above. More recently, the concept of workplace innovation has gained a high policy profile because in 2012 it was adopted by the European Commission in its industrial and innovation policy. The Commission established a European Workplace Innovation Network (EUWIN, to be pronounced as 'You Win') for 4 years, aiming at capacity building and dissemination of knowledge and practices. The initiative, led by TNO (NL) and Workplace Innovation (UK) began with 14 core partners and more than 30 ambassador organisations in 30 European countries encompassing companies, public policymakers, chambers of commerce, business federations, research organisations, and social partners. The contributing authors in this volume seem to share the idea that WPI is strongly related to employee involvement. However, they place a different focus regarding the objectives of quality of working life, organisational performance or a combination of both. They also differ in their treatment of preferred WPI-interventions such as organisational behaviour, HR-related measures, organisational design measures, or a combination of these.

In accordance with the purpose of the Springer book series 'Aligning Perspectives on Health, Safety and Well-being', this volume provides a number of interdisciplinary and complementary research, policy and practical perspectives on wellbeing at work and organisational performance. The book explores how different research perspectives regarding workplace innovation contribute to our understanding of simultaneously improving wellbeing at work and organisational performance. The included contributions are multi-disciplinary and stem from a wide variety of European scientists and practitioners.

1.2 Why Now a Book on Workplace Innovation?

Although sociotechnical forms of work organisation and dialogue-based forms of management have already been present for decades in multiple European countries, the link between these forms of organising and innovation has become more visible during the past two decades. During this period, national programmes and initiatives related to WPI as well as the European Network have flourished. EUWIN has managed to engage with over 5000 experts from all over the EU to join in the movement. In addition, new alliances of employers' associations, trade unions, research institutes and governments came into existence, even in some countries and regions where workplace innovation is a completely new concept. They took employee involvement as a constituting element of WPI and added topics such as better dialogue, improved work organisation, better labour relations, and enhanced organisational performance. Different approaches to organisational renewal have been practiced. Some of these approaches emphasise organisational design rules and focus on the output; others give priority to democratic dialogue and are primarily concerned with the process. The theoretical foundations of these approaches, definitions of concepts and practices also differ substantially. Sometimes the focus of new initiatives is primarily to improve organisational performance; at other times, wellbeing at work represents the focal goal. Hence, the associated strategies for change also differ markedly. We believe that after two decades of divergence, the time is ripe to take stock of what has happened in this respect over the past 20 years, identify commonalities and differences and forge a path towards a more unified body of knowledge regarding WPI. In this respect, EUWIN has facilitated the exchange of experiences, research and opinions and has contributed to forging tighter connections within the European workplace innovation community. EUWIN has laid bare the grounds to be explored; now it's Europe's turn to make WPI a reality. This book *"Workplace Innovation: Theory, Research and Practice"* presents the state-of-the art of WPI for policymakers, practitioners and researchers.

1.3 What Is the Book About?

The content of the book is organised in four parts: (I) Policy, (II) Theory and evidence, (III) Research and (IV) Practice. Each part contains a number of chapters, which will be introduced hereafter.

Part I—Policy

In Chap. 2 **Frank D. Pot, Peter Totterdill** and **Steven Dhondt** sketch an overview of the last two decades' European policy on the core topic of the book. The authors illustrate that workplace innovation is gaining a higher profile as an emerging European policy embedded in a broader economic and social EU policy to support organisational change in companies. However, Europe's policy is still divided into silos; there is hardly any integration of national policies regarding innovation, productivity, skills, employment, quality of jobs, occupational health and safety. The authors suggest that today's digitalisation and robotisation offer opportunities for a more integrated EU policy.

National programmes and initiatives are described in Chap. 3 by **Tuomo Alasoini, Elise Ramstad** and **Peter Totterdill**. The overview shows that workplace innovation programmes in Europe exclusively utilise soft regulation in its various forms. Activities are still unevenly distributed by geographical area, with most programmes being found in Northern and Western Europe. However, new promising developments are currently underway in many countries with no history of such programmes. Special attention is paid to methods for meeting the challenges of diffusion, an issue of crucial importance for the sake of improving the social effectiveness and legitimacy of programmes.

Chapter 4 deals with a particular case, namely the German Industry 4.0 policy to stimulate innovation, notably in smart and advanced manufacturing. **Jürgen Howaldt, Ralf Kopp** and **Jürgen Schultze**, in their essayist contribution, suggest that Germany has not learned much from its own history. Industry 4.0 is a technology-centred vision that underestimates risks and negative effects, as was the case with the completely computer-integrated and automated factory of the last century. The authors propose that workplace innovation (Work 4.0) should be given a more central place in the process of digitalisation.

Part II—Theory Matters and Evidence

The theory section starts with Chap. 5 by **Peter R.A. Oeij** and **Steven Dhondt**, who argue that the field of WPI is in dire need of an integrative theoretical underpinning. They propose a working definition of WPI, discuss a number of theoretical approaches that could be informative to the study and practice of WPI, and highlight some ways in which these approaches can inform a theory of WPI. Whereas the chapter does not develop a definitive theory of WPI, it presents the current state of the art in WPI research and suggests a path for aligning future research and theory development.

In Chap. 6, **Steven Dhondt, Lander Vermeerbergen** and **Geert Van Hootegem**, aim to show where the research on WPI as an organisational innovation is, and to outline why they think that it is not an issue that most of the proof is circumstantial. They first provide an overview of the current state of the art of the research on WPI. Next, they identify which kind of research approach would be the most promising for WPI. Finally, they discuss why some companies may not implement WPI, although WPI has been shown to deliver obvious results.

Given that workplace innovation promises to improve organisational performance, quality of working life and, consequently, wellbeing at work simultaneously, **Frank D. Pot's** Chap. 7 looks at wellbeing at work. He discusses how this promise can be theoretically backed up and shows the connection between wellbeing at work and workplace innovation in European policies. In some policies, however, there is an emphasis on improving wellbeing via individual coping behaviour. Pot argues that organisational interventions that improve the conditions for wellbeing (i.e., the 'conditional approach') represent a more successful strategy and that workplace innovation is a good example of such a conditional approach.

In Chap. 8, **Aditya Jain, Vlad Dediu, Gerard Zwetsloot** and **Stavroula Leka** present empirical evidence of determinants of wellbeing at work and relate this to the promotion of workplace innovation as such an antecedent. The chapter then proceeds to discuss commonalities in good practice interventions at the organisational level, specifically focusing on Vision Zero (zero accidents, zero harm, etc.) and PRIMA-EF, a model for the management of psychosocial risks. Considering the current evidence base, recommendations for future research and practice are made while also referring to the relevance of policy-level interventions to promote workplace innovation.

Part III—Research

The research section starts with Chap. 9 by **Arianna Costantini**, **Riccardo Sartori** and **Andrea Ceschi**, who present an overview of studies on WPI from a work and organisational psychology (WOP) perspective and discuss the advantages of integrating WOP and WPI research. Drawing on the definition of WPI as composed of two dimensions, i.e., a structural and a cultural orientation, they present evidence on how job autonomy, job flexibility, and participation in organisational life have an effect on quality of working life (QWL) and organisational performance (OP). The authors conclude that taking a systemic perspective in WPI implementation would be beneficial in promoting QWL and OP.

In Chap. 10 **Peter R.A. Oeij**, **Steven Dhondt**, **Rita Žiauberytė-Jakštienė**, **Antonio Corral** and **Paul Preenen** present evidence from the Eurofound study comprising 51 cases across 10 EU countries on why, how and with what effects, companies implement WPI. The authors show that successful WPI implementation results from an interplay between management-driven business goals and employee-driven quality of work goals. Moreover, they show that, whereas companies take different paths in implementing WPI, one key success factor is constructive cooperation between management, employees and employee representatives. They conclude that a whole-system approach focusing on the interplay between strategy, structure, and culture is most likely to lead to successful WPI implementation.

The chapter by **Helge Hvid** and **Vibeke Kristine Scheller** (Chap. 11), based on the study of six Danish organisations—part of the Eurofound study mentioned in Chap. 10—states that neither economic arguments nor arguments of humanising work are sufficient to get companies to implement WPI activities; workplace innovation also depends on institutional entrepreneurship. The article focuses particularly on institutional entrepreneurship exercised when WPI-related activities are implemented. Results suggest that institutional alliances and coalitions are not only an important part of institutional entrepreneurship to direct WPI in organisations; the sustainability of the introduced WPI activities also depends on the institutional alliances.

In Chap. 12, **Peter Totterdill** and **Rosemary Exton**, relying on data from the Eurofound study (see Chap. 10), present three UK case studies exemplifying the role of enterprise leadership for successful WPI implementation. Although the three case studies showcase journeys towards WPI from different starting points, they all demonstrate how a consistent approach to shared leadership can stimulate employee empowerment and bottom-up initiative, which, in turn, lead to successful WPI interventions. The authors highlight the importance of taking a systemic approach, a focus on long-term, small incremental changes as well as consistent values and leader behaviours for WPI success.

The contribution by **Liv Starheim** and **Peter Hasle** (Chap. 13) discusses the broader possibilities and challenges associated with a lean-inspired methodology as a tool for local WPI in hospitals. They present a case study in a Danish hospital where a lean-inspired intervention led to both improvements in the psychosocial work environment as well as to productivity improvements due to employee participation. The authors discuss some of the dilemmas associated with using lean for local WPI and conclude that lean can only be a successful tool for WPI if employees are actively involved in the process.

The contribution by **Marta Strumińska-Kutra, Boleslaw Rok** and **Zofia Mockałło** (Chap. 14) looks at the space for employee involvement in innovation-related activities in Poland's post-communist era. Using a qualitative approach, based on in-depth interviews and focus groups, they show that technological innovation often neglects the improvement of workplaces and state that this is associated with low work engagement. The authors conclude that the interaction between the individual level (employee) and the structural level (organisation) likely results in a gradual destruction of innovation, because innovation processes in Polish companies seem contrary to what WPI stands for. WPI stresses employee involvement to succeed in innovation, whereas Polish companies seem to do the opposite.

Despite the growing positive attention for workplace innovation it is threatened by new business models and labour relations such as in the platform economy. **Chris Warhurst, Chris Mathieu** and **Sally Wright** in Chap. 15 first place Uber and Uberisation amidst the various types of innovation. They then outline the model of work offered by Uber. Indeed, if digitalism can create new forms of work that diminish the quality of working life, 'vigilance' is needed to prevent a 'race to the bottom' in employment (work) standards. What constitutes employment and, with it, what constitutes decent, let alone good, job quality will need public and policy debate, a challenge to advocates of WPI.

In Chap. 16 **Agnès Parent-Thirion, Greet Vermeylen, Mathijn Wilkens, Isabella Biletta** and **Frank D. Pot** use data of the 6th European Working Conditions Survey (2015) to explore some key elements of workplace innovation: job autonomy (or discretion at work) and organisational participation and how a combination of these two dimensions relates to job quality, engagement and wellbeing. The analysis shows that employees in high involvement organisations (high discretion at work and high organisational participation) have the best scores on indices for job quality, engagement and wellbeing. The opposite is true for employees in organisations with low discretion at work and low organisational participation.

Part IV—Practice

This part on practice starts with Chap. 17 by **Pierre Van Amelsvoort** and **Geert Van Hootegem** with a coherent set of design approaches which together provide a framework for stakeholders interested in redesigning organisations for WPI. Their starting point is the sociotechnical design perspective for the design of core work processes, which is subsequently broadened by other approaches, such as Lean

Thinking, Total Productive Maintenance, HRM theories and ICT theories, to address issues related to the design of control, coordination and support systems. They conclude with outlining a path towards combining these approaches to develop a systemic concept of Total Workplace Innovation (TWIN).

Steven Dhondt, Peter Totterdill, Sylvie Boermans and **Rita Žiauberytė-Jakštienė** describe five steps to develop workplace innovation. Chapter 18 starts with explaining the building elements of WPI. Enriched with inspiring illustrations of the best practices from the Knowledge Bank of EUWIN and supplemented with easily applicable tips, it aims to be a practical tool for everyone interested in developing and implementing workplace innovation.

An important approach in the family of workplace innovation is employee-driven innovation. In this respect, the Norwegian cases are among the best documented cases. In Chap. 19, **Kåre Hansen, Oscar Amundsen, Tone Merethe Berg Aasen** and **Leif Jarle Gressgård** analyse why organisations differ in their abilities to obtain, explore and exploit their employees' ideas and knowledge in order to increase overall organisational innovation capacity. The findings indicate that all enterprises aiming to implement principles of employee-driven innovation should do a thorough assessment of their management styles and managers, in particular the middle managers (such as foremen, section heads, or shift bosses).

The expertise of work and organisational (W&O) psychologists is greatly underused in WPI practice. In their essay (Chap. 20), **Maria Karanika-Murray** and **Peter R.A. Oeij**, take an integrative, critical approach in discussing the potential role that W&O psychologists could play to strengthen the practice of WPI (see also Chap. 5, for a theoretical perspective). They propose a multidisciplinary, integrated perspective on WPI that takes into account the interplay between strategy, structure, and culture and outline avenues in which WPI practice can benefit from W&O psychology as well as how W&O psychologists can broaden their focus to provide a unique contribution to WPI practice.

In Chap. 21, **Fietje Vaas** and **Rita Žiauberytė-Jakštienė** argue that case examples of WPI best practices can supplement more abstract definitions of WPI by providing practitioners with more hands-on advice on how to implement WPI. Based on cases from the Eurofound study (see Chaps. 10–12), the authors develop three criteria for a case to be a good example for practical purposes: (1) the company should have implemented substantial WPI practices, (2) the case description should provide actionable information, and (3) the narrative should be inspiring. The authors underscore their argument by providing extensive and lively case descriptions.

Peter R.A. Oeij introduces in Chap. 22 the Innovation Resilience Behaviour tool (IRB-tool), aimed at improving teamwork in innovation teams, as an example of a WPI intervention. The tool is premised on the notion that innovation requires employee participation. The IRB-tool focuses on team processes, such as resilience, psychological safety, learning, voice, and leadership. By uncovering team defensiveness, a team's mindful infrastructure, and innovation resilience behaviour, the tool helps teams become more aware of issues that could hamper the success of innovation projects and provides ways to overcome these hurdles. The author concludes that the tool can be used to boost intrapreneurship and innovation.

1.4 What Do We Intend to Achieve?

We hope that our efforts will help to inform the growing workplace innovation community about the state of the art on workplace innovation in the four areas of policy, theory, research, and practice. We have tried to provide an overview of current approaches and thereby, uncover similarities and differences among them. By learning from different perspectives and by looking for convergence, we hope to contribute to mutually reinforcing workplace innovation approaches. Of course, the intention is not to promote any one best approach, as local circumstances differ and every approach, every project has to adapt to local circumstances. Nonetheless, we hope that the book will teach us about the core theories and practices that can guide organisations to better jobs and performance.

1.5 What Are the Plans from Here?

In the final Chap. 23 **Peter R.A. Oeij, Diana Rus** and **Frank D. Pot** analyse and discuss what has been achieved regarding convergence, core theories and practices. Finally, they share their thoughts about possible future activities/pointers in the four areas of policy, theory, research and practice. Their main message is that an integral, systemic approach of WPI is preferred in all these four realms. This seems to be the best guarantee for simultaneously arriving at better organisational performance and better quality of work. To enable this to happen there are consequently a number of tasks for various agents. The authors give suggestions on how policy can strengthen the case for WPI, how research could provide evidence and help practice, and on how employers and employees can utilise the opportunities that are unfolded in this book.

Author Biographies

Frank D. Pot ia an Emeritus professor of Social Innovation of Work and Employment, Radboud University Nijmegen (NL) and former director of TNO Work and Employment.

Diana Rus is a managing director at Creative Peas and a senior lecturer at the University of Groningen in The Netherlands. Her field of work is innovation management, open innovation and leadership development.

Peter R. A. Oeij is a senior research scientist and consultant affiliated with TNO, The Netherlands Organisation for Applied Scientific Research in Leiden, The Netherlands. His field of work is innovation management, workplace innovation, social innovation and team dynamics.

Part I
Policy

Chapter 2
European Policy on Workplace Innovation

Frank D. Pot, Peter Totterdill and Steven Dhondt

2.1 Introduction[1]

Workplace innovation is gaining prominence as an emerging European policy agenda for creating organisational performance and quality jobs. The evidence underpinning this is convincing (see Chap. 6), its theoretical foundations are sound (see Chap. 5) and quite a number of countries have developed supportive programmes (see Chap. 3). For policy makers, the principal objective is to encourage companies to adopt workplace innovation. The European Workplace Innovation Network (EUWIN), that started in 2013 describes workplace innovation as follows: "Workplace innovations designate new and combined interventions in work organisation, human resource management, labour relations and supportive technologies. It is important to recognise both process and outcomes. The term workplace innovation describes the participatory and inclusive nature of innovations that

[1]Most of this chapter was published earlier in Pot, Totterdill and Dhondt (2016).

F.D. Pot (✉)
Radboud University, Nijmegen, The Netherlands
e-mail: frank.pot@ardan.demon.nl

P. Totterdill
Kingston University, London, United Kingdom
e-mail: peter.totterdill@workplaceinnovation.eu

S. Dhondt
TNO, The Netherlands Organisation for Applied Scientific Research, Leiden, The Netherlands
e-mail: steven.dhondt@tno.nl

S. Dhondt
KU Leuven, Leuven, Belgium

P. Totterdill
Workplace Innovation Europe CLG, Dublin, Ireland

P.R.A. Oeij et al. (eds.), *Workplace Innovation*, Aligning Perspectives on Health, Safety and Well-Being, DOI 10.1007/978-3-319-56333-6_2

embed workplace practices grounded in continuing reflection, learning and improvements in the way in which organisations manage their employees, organise work and deploy technologies. It champions workplace cultures and processes in which productive reflection is a part of everyday working life. It builds bridges between the strategic knowledge of the leadership, the professional and tacit knowledge of frontline employees and the organisational design knowledge of experts. It seeks to engage all stakeholders in dialogue in which the force of the better argument prevails. It works towards 'win-win' outcomes in which a creative convergence (rather than a trade-off) is forged between enhanced organisational performance and enhanced quality of working life" (Dhondt 2012, p. 2).

This chapter describes the development of European policies regarding work organisation and workplace innovation over the last 20 years and its societal context. Three periods are being distinguished: the nineties, the Lisbon Agenda and the EU 2020 Strategy. There is quite some evidence for the positive effects of workplace innovation (Alasoini 2016; Dhondt et al. 2014; Fricke 1997; Naschold 1994; Oeij et al. 2015; Pot 2011; Ramstad 2009; see Chap. 6 as well). How this can be understood theoretically is explained in Chap. 5.

2.2 Recent Updates of Productivity and Industrial Democracy Policies

Workplace innovation, as it developed from the beginning of this century, is a member of the Sociotechnical Systems Design (STSD)-family (Mohr and Van Amelsvoort 2016), going back to the restructuring of Europe after the Second World War, starting more or less the same policies for productivity and industrial democracy in several Western European countries. Although consensus about the use of the concept of workplace innovation is growing and its policy profile is getting stronger, other different policy concepts are being used for more or less the same approach (Kesselring et al. 2014). Examples are 'innovative workplaces' (OECD 2010; EESC 2011) and 'sustainable work systems' (Docherty et al. 2002). And, as can be expected, in national programmes and initiatives (Totterdill et al. 2009; Pot et al. 2012) concepts in each country's language are being used. 'Workplace innovation' is also being used in the USA, Canada and Australia besides concepts such as 'high involvement workplaces' and 'relational coordination' (Gittell et al. 2010).

The emergence of interest in workplace innovation can be understood by looking at developments in the economy, technology and the labour market. The broader context is that in the early 1990s a significant shift in Europe's economy and businesses could be observed, fueled by information technology. This shift reversed the historical pattern where tangible capital was considered to be the main asset in companies. Around 1990 investments in intangible capital (in percentage of adjusted GNP), such as patents, R&D, marketing, organisational competences became higher than investments in tangible capital (Corrado and Hulten 2010). Regarding innovation the conviction grew in Europe that 'social innovation' (work organisation, competence development, employee participation, etc.) is probably more important than 'technological innovation' to explain the company's performance (Bolwijn 1986). This context explains the need to develop and utilise the skills and competences of the present and potential workforce to increase added value as part of a competitive and knowledge-based global economy (European Commission 2014). One more reason for 'workplace innovation' is that organisations can only fully benefit from technological innovation if it is embedded in a proper work organisation (Brynjolfsson and McAfee 2014, p. 138). Finally, there is a need to enhance labour productivity to maintain our level of welfare and social security in a near future with fewer people in the workforce due to the ageing population. This growing awareness of the need for new forms of work organisation led to a number of activities at the European level.

2.3 The '90s, the 'Green Paper', EWON and the European Work and Technology Consortium

During the mid-1990s the Employment Directorate General (DG EMPL) of the European Commission established ACTEUR, a policy advisory group which brought together representatives from national programmes and initiatives as well as officials from EU Member States where no national programmes had been established. At the same time individual lobbyists mobilised an influential coalition of researchers and policymakers, resulting in the publication in 1995 of 'Europe's next step: organisational innovation, competition and employment', a manifesto for the future of work organisation (Andreasen et al. 1995). Also in 1995, a different part of DG EMPL established the European Work and Technology Consortium. The Consortium brought together sixteen public policy and research organisations from ten EU Member States to create a 'Medium Term Plan for Collaborative Action for the Modernisation of Work Organisation' (Totterdill 2003). A seminal moment for those advocating the recognition of workplace innovation as a key dimension in the EU strategy came in 1997 with the publication of the Commission's Green (consultation) Paper 'Partnership for a new organisation of work': "The Green Paper invites the social partners and public authorities to seek to build a partnership for the development of a new framework for the modernisation of work. Such a partnership could make a significant contribution to achieving the objective of a productive, learning and participative organisation of work" (pp. 5–6). Interest in work organisation as a driver for European competitiveness and quality of working life had been growing, partly fueled by national initiatives such as those in Norway, Sweden, Ireland, France, Germany and the Netherlands. The Green Paper provided a rallying point for those who had been advocating recognition of workplace innovation, and there was high expectation that specific policy interventions would follow (Ennals 1998). Based on the responses to this consultation, a policy document '*Modernising the organisation of work—A positive approach to change*' was published by the European Commission in 1998.

A substantial volume of evidence for the positive effects of new forms of work organisation was provided by the European Work and Technology Consortium (1998). By 1998, it had become clear that, despite enthusiasm from some trade unions, there was little appetite amongst European social partners for intervention in the workplace whether regulatory or otherwise. Likewise several EU Member States and some senior officials within DG EMPL remained not enthusiastic, considering workplace innovation to be no more than a 'Nordic obsession' (Totterdill et al. 2012a). However, those officials at DG EMPL who believed in 'Modernising the organisation of work' won the debate.

ACTEUR was re-launched in 1997 by DG EMPL as the European Work Organisation Network (EWON) to support the policy of 'a new organisation of work' and instigated a series of policy dialogues, conferences and research projects until 2002, accompanied by a news bulletin. Meanwhile, Eurofound conducted a large scale research project into 'employee participation in organisational change' which provided again evidence for the positive relation between employee participation and organisational performance (EPOC: Eurofound 1997). EWON summarised for DG EMPL the positive research results in different countries (Savage 2001) and so did other researchers (Brödner and Latniak 2002). DG Research commissioned research into successful cases. In that report the concept of workplace innovation was used (Totterdill et al. 2002). EWON was discontinued by DG EMPL itself. This action was never explained to the participants. Most of the attention for organisational innovation was later assigned to EU OSHA, the European Agency for Safety and Health at Work (related to stress prevention and wellbeing at work) and to Eurofound.

In this first period, work organisation became a clear topic with support from the European Commission, in particular DG EMPL albeit this support was not translated into a clear policy vision towards companies and national governments.

2.4 The Lisbon Agenda and Work-In-Net and EDI: Fragmentation of a Message

In 2002, a European Social Fund (ESF)-programme (DG EMPL) focused on the realisation of a more flexible labour market and work organisation. One of the main areas proposed for investment was the 'design and dissemination of innovative and productive methods of work organisation'. EWON prepared a report on this subject for the Commission (EWON 2002). There are instances in Finland, Netherlands, Belgium, Germany and Sweden where it has been used as a foundation for national programmes or initiatives (see Chap. 3).

In 2004, facilitated by the 6th Framework Programme ERA-NET the establishment of the 'Work-In-Net' (WIN 2004–2010) consortium was one of the initiatives redeveloping the work organisation networks. (Alasoini 2005; WIN 2010), coordinating research in the field of 'Innovation of Work Organisation'. In the same period the Employee-Driven Innovation (EDI) Network was established, in particular by the Norwegian and Danish trade union confederations and researchers in the field of work organisation (Høyrup et al. 2012). This network was connected to the European programme 'Lifelong Learning in Europe (LLinE)'. EDI became part of the Norwegian government policy in 2008. A handbook for EDI was made, commissioned by the trade union confederation (LO) and the Confederation of Norwegian Enterprise (NHO) together (IRIS and NTNU 2011, see Chap. 19 in this book as well).

Since the Lisbon Growth and Jobs Strategy had been launched in 2000, the European Employment Strategy's overarching objectives have encompassed not only full employment, but also the promotion of quality and productivity at work. In the Commission the 'design and dissemination of innovative and sustainable forms of work organisation' (European Commission 2003) continued to be cited as a means of enhancing productivity, responsiveness and quality, as well as improving working life and the retention on the labour market of older employees. But by the middle of the decade, EU policy outputs relevant to the workplace read like a checklist of fashionable ideas of good practice, for example Corporate Social Responsibility, Financial Participation, Anticipating and Managing Change and Work-Related Stress. Each of these policy interventions made a potentially significant contribution in its own right to European economic and social policy objectives, but collectively offered an insufficiently integrated vision of the sustainable workplace (Totterdill et al. 2012a).

The renewed Lisbon strategy, agreed upon in March 2005, put growth and better jobs in the top of Europe's political priorities again and implied fresh commitment to a comprehensive approach. In the 'Guidelines for the employment policies of the Member States' we find the following text in proposed guideline 7: "Work-life balance policies with the provision of affordable care and innovation in work organisation should be geared to raising employment rates, particularly among youth, older workers and women, in particular to retain highly-skilled women in scientific and technical fields" (European Commission 2010b, p. 8). The title of the EU strategy for occupational safety and health (OSH) in this period was: 'Healthy and productive jobs'.

However, according to Totterdill et al. (2012a) 'better organisation of work' remained largely undefined in this policy narrative and its status as a factor 'which should be analysed' is a characteristic Commission obfuscation. There is very little evidence to show that the so-called 'Integrated Guidelines' stimulated action at national level to support new forms of work organisation. Member States such as those in the Nordic countries, Netherlands, France and Germany with a tradition of policies and programmes focused on workplace innovation continued to deliver such programmes; but countries with no such tradition continued, by and large, to ignore workplace innovation (e.g., Greece: Ioannou 2006).

This lack of traditions was also the case in the 'new' EU Member States. For some of them, the former socialist countries in Central and Eastern Europe, implementing workplace innovation is even more difficult because they have another tradition in which concepts such as productivity, industrial democracy and social dialogue had different meanings than comparable concepts in Western Europe. In the eighties the Central and Eastern European countries became familiar with the Japanese style of management and work organisation as most of these countries established productivity centres with Japanese aid funds and Japanese consultants (namely Japanese ex-managers in their 'second career'). These centres were connected to Western European centres through their membership of the European Association of National Productivity Centres (EANPC). The EANPC (2005) promotes not only productivity, but an integrated approach with quality of working life and sustainability. After these countries had entered the European Union the Japanese aid was discontinued, after which the Japanese concepts only survived in a number of companies. A separate set of exchanges of views on work organisation were organised in the eighties by the European Coordination Centre for Research and Documentation in the Social Sciences ('the Vienna Centre'), a strong network of researchers from East and West Europe (Grootings et al. 1991). The Vienna Centre had been established in the sixties by UNESCO and the International Social Science Council (ISSC). Most of the members of these two networks were researchers. Some were consultants or government officials. They inspired each other to contribute to work organisation policies in their own countries and on the European level.

Even in the Nordic countries implementing workplace innovation was not a matter of course; in Sweden the programmes and research were partly discontinued by the new centre-right government in 2006 (Sandberg 2013).

The outcome of this second period is a European policy pattern that has remained largely fragmented: a series of separate EU policy fields that add up to less than the sum of the parts.

2.5 EU2020 Strategy and EUWIN

Since the demise of the 1997 Green Paper, workplace innovation has fallen through the gaps between several policy platforms including competitiveness, innovation, employment and social inclusion—even though it has profound implications for each. The formulation of the EU's Europe 2020 vision and strategy during 2009–2010 (European Commission 2010a) therefore provided an important opportunity for European policymakers to learn from evidence of how innovation in working practices can address economic and social priorities and translate this into policies. However, that opportunity was missed by the policy makers of the initial EU2020 Strategy (Dortmund-Brussels Position Paper 2012).

While the broad vision behind Europe 2020 may represent widely acceptable goals, it fell into the same traps as the previous Lisbon strategy. In particular, there was no concrete model of how convergence between quite different policy objectives such as competitiveness, innovation, employment, health and safety and social inclusion will be achieved in practice. However, the good news was that this time, after DG EMPL had been in the lead since the mid-1990s, the initiative was taken by DG Enterprise and Industry, related to industrial and innovation policy.

2.5.1 A New Start

In March 2011, the European Commission's DG Enterprise and Industry (DG ENTR) organised a workshop on workplace innovation within the launch of its Social Innovation Europe initiative (Dhondt et al. 2011; Pot et al. 2012a). This launch reflected a growing recognition that innovation, central to the EU's 2020 economic strategy, has a clear social dimension. The studies of Corrado and Hulten (2010) had given policy makers a platform to support a broader set of factors to develop innovation. Key influences on the European Commission included a 2011 Opinion of the European Economic and Social Committee (employers' associations, trade unions, NGOs) on 'Innovative workplaces as a source of productivity and quality jobs' (EESC 2011) and the 'Dortmund-Brussels Position Paper' (Dortmund-Brussels Position Paper 2012) signed by more than 30 experts and practitioners across the EU, both calling for more proactive interventions by the European Commission. DG ENTR reacted to this Position Paper by organising a set of meetings to understand what they could do.

In order to define concrete ways to move the policy agenda forward at EU level, DG ENTR subsequently put the issue of workplace innovation on the agenda of the European Policy Group in Brussels in June 2012 which brought together the directors for Industrial Policy of all member states and leading industrial companies. Following a discussion with this group, DG ENTR decided to support and fund a European Workplace Innovation Network (EUWIN) embracing all 27 EU Member States, candidate countries, Switzerland and Norway.

2.5.2 DG GROW in the Lead

According to DG GROW (the former DG ENTR) workplace innovation improves motivation and working conditions for employees, which leads to increased labour productivity, innovation capability, market resilience, and overall business competitiveness. All enterprises, no matter their size, can benefit from workplace innovation, states DG GROW. It improves performance and working lives, and encourages creativity of employees through positive organisational changes, combines leadership with hands-on, practical knowledge of frontline employees and engages all stakeholders in the process of change.

The main objectives of the DG GROW initiative are to foster the uptake of workplace innovation across European businesses and raise policy maker awareness, at all levels, of the benefits of these innovations (http://ec.europa.eu/growth/industry/innovation/policy/workplace/index_en.htm).

Through DG GROW, the European Commission prioritised workplace innovation with, for example, the reinforced 2020 EU Industrial Policy Communication and the innovation policy. Crucially it established the EUWIN in 2013 to support this priority, to exchange good practices and establish 'workplace innovation alliances' of employers' associations, trade unions, governments and knowledge institutes.

This policy is also part of the 'Advanced Manufacturing Programme' (ADMA): "Workplace innovation has to provide advanced solutions for manufacturing industry, based on the newest technologies. Virtual reality and side laboratories, where employees can perform extra research and experimentation, not connected with their daily tasks, are examples of combining advanced manufacturing technologies and advanced workplaces. Furthermore, workplace innovation can help companies to enhance competitiveness by using the innovativeness and creativity of all employees. (…) The Commission has included workplace innovation aspects in the R&D&I programmes for advanced manufacturing. Explicitly including R&D on human-centred manufacturing could enhance the active and innovative role of people in factories and could contribute to design the workplaces of the future" (European Commission 2014, pp. 27–28).

2.5.3 DG EMPL

As reported earlier, DG EMPL was supporting workplace innovation through the ESF-funding. European agencies supporting the policies of DG EMPL continued to supply new ideas and policy recommendations. Eurofound already organised the first seminar on workplace innovation in 2005 and developed the concept over the years in their European Working Conditions Survey (EWCS; Eurofound 2012) and the European Company Survey (ECS; Eurofound 2015). In 2010, a workshop was organised by DG EMPL and some researchers to explore the concept of workplace innovation again (Totterdill 2010). Since the launch of Social Innovation Europe in 2011 'social innovation' was applied to work as well, called 'workplace innovation' which became a topic in policies of DG ENTR (Enterprise and Industry, now DG GROW) as well as DG EMPL. "With the Europe 2020 Strategy it also became a priority to support workplace innovation aimed at improving staff motivation and working conditions with a view to enhancing the EU's innovation capability, labour productivity and organisational performance" (European Commission 2015, pp. 169–70).

EU-OSHA commissioned a literature review on the relation between workplace innovation and occupational safety and health (OSH) (Eeckelaert et al. 2012) because the claim of workplace innovation is to improve quality of working life and organisational performance simultaneously. Consequently workplace innovation was connected to 'wellbeing at work' in the research priorities of OSH (EU-OSHA 2013a) as well as in the policy to extend OSH to 'wellbeing at work' (EU-OSHA 2013b).

In 2015 DG EMPL published the document 'Employment and social developments in Europe 2014'. Chapter 3 is about 'the future of work in Europe: job quality and work organisation for a smart, sustainable and inclusive growth'. One of the paragraph titles is 'Complementing technological innovation with workplace innovation' (p. 164). Presenting much empirical research—among which are Eurofound's European Working Conditions Surveys—its conclusion is that "Better jobs and work organisation yield a more productive workforce" (p. 137). Having better jobs and work organisation reduces the risk of stress, enhances wellbeing and leads to a lower tendency to quit the job. Better work organisation implies in particular a balance between job demands (job intensity) and job control (job autonomy), wholeness of tasks and more open access to decision-making processes. These are a few of the indicators which, the report suggests, should inform EU policy making to support investments in work organisation.

Other topics in that chapter are wages, OSH, and work-life and gender balance. It explores the ways in which technological change and innovation will transform the job landscape of the future (for example, polarisation of required skills and wages is seen as a risk) and can lead to a possible industrial renaissance in the EU. In this context managing the transition into a new labour market where many jobs succumb to automation must become a key priority for policymakers, according to DG EMPL.

The chapter then explores how work organisation can be shaped to increase productivity and labour market participation under the continuous pressure of ongoing structural changes (technological progress, globalisation, demographic change and the greening of the economy). It looks at how stimulating creativity and fostering exchanges between workers can prevent stress and help maintain physical and mental health, while at the same time improving productivity and innovation capacity. It sees how special arrangements can be implemented to accommodate older workers, workers with disabilities or certain diseases, and workers with family responsibilities. The chapter then discusses future challenges with respect to workplace learning. It ends by examining how expanding global value chains will affect work organisation, focusing on risks related to the global restructuring of value chains, virtual collaboration across time zones and the absence of multi-layered social dialogue.

One of the conclusions of that DG EMPL document is that for the knowledge-based potential to materialise, the knowledge triangle (knowledge, education, innovation) has to be complemented by forms of work organisation that use workers' human capital to their fullest. It will be important actively to engage employees in identifying and developing solutions while allowing them to participate in the implementation of work innovations so that they become more receptive to change. "In this context, an important policy would be to facilitate the creation of EU-wide platforms that allow employees and employers to exchange experiences in developing and implementing solutions related to production and work organisation. The specific characteristics of such platforms will vary between production entities and may take place at European or national level. They can promote the exchange of experiences, help identify best practices, monitor their implementation, assess their impact on productivity and identify social implications" (European Commission 2015, p. 163). The proposal to facilitate the creation of EU-wide platforms looks very similar to the EUWIN network (2013–2016) which was commissioned by DG ENTR.

All these activities have helped the concept of workplace innovation to gain a higher policy profile with other EU institutions. The policy concept of workplace innovation was also used by the European Parliament (2013) and IndustriAll European Trade Union (IndustriAll European Trade Union 2014) in their programmes for an industrial renaissance, as well as in national initiatives in Ireland and the UK, and in the translations of this concept into national programmes in Finland, Netherlands, Germany, Flanders/Belgium and Basque Country/Spain.

2.5.4 An Emerging European Policy

As shown in the previous section, the policies of DG EMPL and DG GROW concerning workplace innovation in the third period have many topics in common. Hence, as we see it, an integrated European policy across DGs is lurking. This should include DG Regions as well because workplace innovation alliances can play a major role in regional development as well as DG Research to support and improve these policies by research. So far in the EU2020 programmes little attention is paid to research on workplace innovation. Although Eurofound has strengthened workplace innovation in its surveys and EU-OSHA has put the subject in their list of research priorities (EU OSHA 2013, pp. 52–53), only a few new research opportunities have been created so far in the context of 'advanced manufacturing' and 'social innovation'.

2.6 Conclusions, Discussion and Perspectives

The European Commission has been developing bits and pieces of policy relating to work organisation since about 1995 instead of a coherent policy. Although its aim was always to achieve higher productivity, more innovation capability, more employment and better jobs simultaneously, the emphasis in the nineties was on productivity, in the beginning of this century on employment and the last ten years on innovation. The message that organisational performance and quality of working life are two sides of the same coin came primarily from representatives of the network of 'occupational safety, health and well-being' in Acteur, EWON, WIN, EDI and EUWIN.

Policies on work organisation and workplace innovation, however, have remained fragmented. The 'workplace innovation protagonists' refer to productivity, innovation, competitiveness and employment, but the 'productivity people', the 'innovation people', the 'competitiveness people' and the 'employment people' never refer to workplace innovation, with only a few exceptions. This is not different from the silo-policies in national governments. Nonetheless, there is some progress. The policies of DG GROW and DG EMPL clearly overlap although they have not been integrated up to the current date. More contacts between the two DGs are being planned. It is helpful that there is agreement on the use of the concept of workplace innovation as using different concepts makes it very difficult to develop policies and common understanding.

Policies of work organisation and workplace innovation have never resulted in legislation or regulations at EU level. Mentioning the issue in the Employment Guidelines did not seem to help much, nor did the existence of national legislation in a small minority of countries. Probably workplace innovation is not suitable for a legislative approach, because its implementation depends very much on the social dialogue at European, national, sectoral and organisation level. Furthermore there is a strong feeling among policy makers that they should not interfere in company policies. But EU- and national-authorities can stimulate that dialogue and develop campaigns for knowledge dissemination and capacity building. Alasoini (2016, pp. 20–23) calls that 'soft regulation' compared to 'hard regulation' (legislation). Some of these authorities stimulated workplace innovation, but unfortunately only for a short period of time as in Ireland, the Netherlands and Portugal. France, Germany and Finland (Alasoini 2016) are the exceptions with programmes that have been renewed several times over the past decades.

In the lobbying for and development of these policies an important role has always been played by researchers and their networks. Policies were developed bottom-up by coalitions of European Commission officials and researchers who organised seminars, meetings, workshops, etc., to convince the Commission's directors, directors-general and finally EU Commissioners. Sometimes also a few representatives of trade unions and/or employers' associations were active in these networks. These coalitions have appeared to be successful in helping to put workplace innovation higher on the political agenda.

Part of the policy to enhance workplace innovation should include more research into the obstacles and mechanisms that contribute to the failure to implement workplace innovation as a matter of course and into the mechanisms that support implementation. A couple of countries have experiences with national campaigns (see Chap. 3) but so far there is little evaluation research available.

On the one hand, the financial and economic crises after 2008 did not seem to have much influence on the attention paid to workplace innovation. Important steps in EU policy were taken during the crisis as well as in some countries. To give some examples: in the Netherlands the general employers' association (AWVN) argued in 2009 that because of the crisis workplace innovation had become even more urgent. In Ireland, the tripartite programme on workplace innovation ended according to plan just before the crisis, but the unions, in particular the Services, Industrial, Professional and Technical Union (SIPTU), continued to organise seminars and develop projects. Finland and Germany renewed their programmes during the crises as they had done for decades and Belgium started the 'Flanders Synergy' programme on workplace innovation in 2009. On the other hand, increased competition as a consequence of crisis and globalisation attracts opposite reactions such as cost cutting and a stronger command-and-control style of leadership, sometimes called 'the low road' as compared to 'the high road' of investing in people (Totterdill et al. 2002). This is more likely to happen in organisations and countries characterised by a liberal market economy (LME) rather than a coordinated market economy (CME) because there is already some trust between stakeholders and institutions such as social dialogue to solve industrial relations issues in collective agreements (Totterdill et al. 2016). However, as neoliberal ideology became more fashionable, social dialogue was also weakened in CMEs and hostility entered industrial relations again. Examples are the discontinuation of the Swedish workplace programmes after the government change in 2006 and the many strikes in 2015–2016 in the Netherlands in the 'metalektro' sector, not only for working conditions but for workplace innovation as well. The 'low road' is also more prevalent in countries in East and South Europe which are not yet familiar with ideas and examples of workplace innovation and where social dialogue is not well developed.

The idea of workplace innovation is nowadays actively disseminated in those countries, sometimes supported by EUWIN, sometimes by national programmes such as in Finland's support for Hungary and Romania (Makó et al. 2015). As in Western European countries researchers in those Member States play an important role in disseminating innovative ideas.

There is room for improvement. In the European Working Conditions Survey 2010, one question was: "Are you involved in improving the work organisation or work processes of the department or organisation?" Of the responding employees in the EU-27 countries 46.7% answered 'always' or 'most of the time' (Eurofound 2012).

For a number of reasons many enterprises, hospitals and government departments, for example, do not implement workplace innovation as a matter of course, in spite of the obvious benefits for employees and employers. That is why a better coordinated policy and more action is needed by governments, social partners and research institutes.

A good opportunity for more action of public authorities, including the European, is the debate on digitalisation and robotisation. We should remember the many failures in the past. Digitalisation and automation were implemented before optimising processes and work organisation. End users were not involved sufficiently. In their book 'The second machine age' Brynjolfsson and McAfee (2014) of MIT observe that in big companies with big ICT projects it takes five to seven years before the organisation has been redesigned and consequently before full benefits can be taken from the new technology. They say "Creativity and organizational redesign is crucial to investments in digital technologies" (p. 138). The concept developed by Brynjolfsson and McAfee is 'co-invention of organisation and technology'. This co-invention requires the creativity and collaboration on the part of the entrepreneurs, managers and workers at all levels. In our words: There can be no effective and sustainable returns on automation and digitalisation without workplace innovation. It seems that the Finnish and the German governments in particular are aware of this. They commissioned substantial programmes to support and investigate processes of digitalisation and robotisation including the design of 'good work' (Germany) or 'meaningful work' (Finland).

For the European Commission this is an opportunity to continue the policy of supporting workplace innovation. Workplace innovation could be easily integrated in EU policy agendas such as Innovation, New Skills, 'More and Better Jobs' and 'Social Dialogue'. On its innovation web page DG GROW refers to "key enabling technologies, such as workplace innovation" (DG GROW 2016). DG EMPL was in the lead of the 'New Start for Social Dialogue' in June 2016. The research programme Horizon 2020 prepares under the heading of 'societal challenges' calls for research on 'The human and social dynamics of the Fourth Industrial Revolution'. Let's hope there is more to come to support workplace innovation.

References

Alasoini, T. (2016). *Workplace development programmes as institutional entrepreneurs. Why they produce change and why they do not* (Doctoral Dissertations, Aalto University, 12/2016).

Alasoini, T., Ramstad, E., Hanhike, T., & Lahtonen, M. (2005). *European programmes on work and labour innovation—A benchmarking approach*. Helsinki/Bonn: Work-In-Net.

Andreasen, L. E., Coriat, B., den Hertog, J. F., & Kaplinsky, R. (1995). *Europe's next step: Organisational innovation, competition and employment*. Ilford: Frank Cass.

Bolwijn, P. T., van Breukelen, Q. H., Brinkman, S., & Kumpe, T. (1986). *Flexible manufacturing: Integrating technological and social innovation*. Amsterdam: Elsevier.

Brödner, P., & Latniak, E. (2002). *Sources of innovation and competitiveness: National programmes supporting the development of work organisation. Report to DG employment and social affairs*. Gelsenkirchen: Institute for Work and Technology.

Brynjolfsson, E., & McAfee, A. (2014). *The second machine age: Work, progress, and prosperity in a time of brilliant technologies*. New York/London: W.W. Norton.

Corrado, C., & Hulten, C. (2010). How do you measure a technological revolution? *American Economic Review, 100*(5), 99–104.

DG GROW. (2016). Retrieved September 18, 2016, from http://ec.europa.eu/growth/industry/innovation_en

Dhondt, S. (Ed.). (2012). *European learning network for workplace innovation (section 4 technical proposal, call for tender no 212/pp/ent/cip/12/c/n02c04)*. Hoofddorp: TNO.

Dhondt, S., Pot, F., & Kraan, K. (2014). The importance of organizational level decision latitude for wellbeing and organizational commitment. *Team Performance Management: An International Journal, 20*(7/8), 307–327.

Dhondt, S., van Gramberen, M., Keuken, F., Pot, F., Totterdill, P., & Vaas, F. (2011). *Workplace innovation, social innovation Europe launch event, Brussels*. Nottingham: UKWON.

Docherty, P., Forslin, J., & Dhani, A. B. (Eds.). (2002). *Creating sustainable work systems: Emerging perspectives and practice*. London/New York: Routledge.

Dortmund-Brussels Position Paper. (2012). *Workplace innovation as social innovation*. Online: http://www.kennisbanksocialeinnovatie.nl/nl/kennis/kennisbank/workplace-innovation-as-social-innovation/1009?q=dortmund. Accessed on May 6, 2016.

EANPC. (2005). *Productivity, the high road to wealth*. Brussels: European Association of National Productivity Centres.

Eeckelaert, L., Dhondt, S., Oeij, P., Pot, F., Nicolescu, G. I., Webster, J., et al. (2012). *Review of workplace innovation and its relation with occupational safety and health*. Bilbao: European Agency for Safety and Health at Work.

EESC. (2011). *Opinion of the European Economic and Social Committee on innovative work places as a source of productivity and quality jobs*. Online: http://www.eesc.europa.eu/resources/docs/ces543-2011_ac_en.doc. Accessed on May 6, 2016.

Ennals, R. (1998). Partnership for a new organisation of work and Europe as a development coalition: An interview with Allan Larsson, Director-General of DG-V, European Commission. *Concepts and Transformation, 3*(1/2), 143–152.

Eu, O. S. H. A. (2013). *Priorities for occupational safety and health research in Europe: 2013–2020*. Bilbao: EU OSHA.

Eurofound. (1997). *Employee participation and organisational change: EPOC survey of 6000 workplaces in Europe*. Dublin: European Foundation.

Eurofound. (2012). *Fifth European working conditions survey*. Luxembourg: Publications Office of the European Union.

Eurofound. (2015). *Third European company survey. Workplace practices: Patterns, performance and well-being*. Luxembourg: Publications Office of the European Union.

European Agency for Safety and Health at Work (EU-OSHA). (2013a). *Priorities for occupational safety and health research in Europe: 2013–2020*. Bilbao: EU OSHA.

European Agency for Safety and Health at Work (EU-OSHA). (2013b). *Well-being at work—Creating a positive work environment*. Bilbao: EU OSHA.

European Commission. (1997). Partnership for a new organisation of work. Green paper, Brussels.

European Commission. (1998). Modernising the organisation of work—A positive approach to change. Policy document, Brussels.

European Commission. (2003). *Council decision of 22nd July 2003 on guidelines for the employment policies of the member states*. OJEC, 5.8.2003, L 197/13, para 14.

European Commission. (2010a). *Europe 2020: A strategy for smart, sustainable and inclusive growth*. Brussels: COM(2010) 2020 final (March 3, 2010).

European Commission. (2010b). *Proposal for a council decision on guidelines for the employment policies of the member states: Part II of the Europe 2020 integrated guidelines*. Brussels: COM92010 193/3.

European Commission. (2014). *Advancing manufacturing—Advancing europe—Report of the task force on advanced manufacturing for clean production*. SWD(2014) 120 final, Commission staff working document, Brussels.

European Commission. (2015). *Employment and social developments in Europe 2014*. Brussels: DG Employment

European Parliament. (2013). *Report on reindustrializing Europe to promote competitiveness and sustainability*. Brussels: 2013/2006(INI).

European Work and Technology Consortium. (1998). *A medium term plan for collaborative action*. Nottingham: Centre for Work and Technology.

EWON. (2002). *The use of ESF funds in supporting the modernisation of work organisation*. Unpublished report for the European Commission. Online: http://uk.ukwon.eu/File%20Storage/5032375_7_EC-Work-Organisation-ESF-Final-Report.pdf. Accessed on August 13, 2015.

Fricke, W. (1997). Evaluation of the German work and technology programme from an action research point of view. In T. Alasoini, T. M. Kyllönen, & A. Kasvio (Eds.), *Workplace innovation: A way of promoting competitiveness, welfare and employment* (pp. 91–103). Helsinki: Ministry of Labour.

Gittell, J. H., Seidner, R., & Wimbush, J. (2010). A relational model of how high performance work systems work. *Organization Science, 21*(2), 490–506.

Grootings, P., Gustavsen, B., & Héthy, L. (Eds.). (1991). *New forms of work organization*. New Brunswick, NJ: Transaction Publishers.

Høyrup, S., Bonnafous-Boucher, M., Hasse, C., Lotz, M., & Møller, K. (Eds.). (2012). *Employee-driven innovation: A new approach*. London: Palgrave Macmillan.

IndustriAll European Trade Union. (2014). *Manifesto to put industry back to work*. Brussels: IndustriAll.

International Research Institute of Stavanger (IRIS) and NTNU Samfunnsforskning. (2011). *Handbøk I medarbeiderdrevet innovasjon (Manual employee driven innovation)*. Oslo: Ministry of Trade and Industry, in collaboration with LO and NHO.

Ioannou, C. A. (2006). Why is modern work organisation lacking from Southern European public policies? The case of Greece. *The International Journal of Comparative Labour Law and Industrial Relations, 22*(1), 19–37.

Kesselring, A., Blasy, C., & Scoppetta, A. (2014). *Workplace innovation: Concepts and indicators*. Brussels: European Commission.

Makó, C., Illéssy, M., Csizmadia, P., & Balázs, H. (2015). Mapping the context for transfer of Finnish workplace development practice: The case of Finland, Hungary and Romania. *Journal of Self-governance and Management Economics, 3*(2), 7–29.

Mohr, B., & Van Amelsvoort, P. (Eds.). (2016). *Co-creating humane and innovative organizations: Evolutions in the practice of socio-technical system design*. Sl: Global STS-D Network Press.

Naschold, F. (1994). The politics and economics of workplace development. In T. Kauppinen & M. Lahtonen (Eds.), *National action research programmes in the 1999* (pp. 109–155). Helsinki: Ministry of Labour (Labour Policy Studies 86).

OECD. (2010). *Innovative workplaces: Making better use of skills within organizations*. Paris: OECD.

Oeij, P., Žiauberytė-Jakštienė, R., Dhondt, S., Corral, A., Totterdill, P., & Preenen, P. (2015). *Workplace innovation in European companies*. Dublin: Eurofound.

Pot, F. D. (2011). Workplace innovation for better jobs and performance. *International Journal of Productivity and Performance Management, 64*(4), 405–415.

Pot, F., Dhondt, S., de Korte, E., Oeij, P., & Vaas, F. (2012a). Workplace innovation in the Netherlands. In I. Houtman (Ed.), *Work life in the Netherlands* (pp. 173–190). Hoofddorp: TNO Work and Employment.

Pot, F., Dhondt, S., & Oeij, P. (2012a). Social innovation of work and employment. In H.-W. Franz, J. Hochgerner, & J. Howaldt (Eds.), *Challenge social innovation: Potential for business, social entrepreneurship, welfare and civil society* (pp. 261–274). Berlin: Springer.

Pot, F., Totterdill, P., & Dhondt, S. (2016). Workplace innovation: European policy and theoretical foundation. *World Review of Entrepreneurship, Management and Sustainable Development, 12*(1), 13–32.

Ramstad, E. (2009). Promoting performance and the quality of working life simultaneously. *International Journal of Productivity and Performance Management, 58*(5), 423–436.

Sandberg, A. (Ed.). (2013). *Nordic lights: Work, management and welfare in Scandinavia*. Stockholm: SNS FÖRLAG.

Savage, P. (Ed.). (2001). *New forms of work organisation: The benefits and impact on performance*. Thematic Paper Presented to DG Employment and Social Affairs. Nottingham: European Work Organisation Network (EWON).

Totterdill, P. (2003). Europe's advantage? Work organisation, innovation and employment. In K. Cowling (Ed.), *Industrial policy in Europe: Theoretical perspectives and practical proposals* (pp. 74–87). Oxford: Routledge.

Totterdill, P. (2010). *Workplace innovation: Europe 2020's missing dimension*. Report of a workshop hosted by DG Employment, social affairs and equal opportunities, 23 June. Nottingham: UKWON.

Totterdill, P., Dhondt, S., & Milsome, S. (2002). *Partners at work? A report to Europe's policy makers and social partners*. Nottingham: The Work Institute.

Totterdill, P., Exton, R., Exton, O., & Gold, M. (2012). *Closing the gap between evidence-based and common practice*. Workplace innovation and public policy in Europe. *Lifelong Learning in Europe* (LLinE), No. 4. Online: http://www.elmmagazine.eu/articles/closing-the-gap-between-evidence-based-and-common-practice-workplace-innovation-and-public-policy-in-europe. Accessed on May 6, 2016.

Totterdill, P., Exton, O., Exton, R., & Gold, M. (2016). High-performance work practices in Europe: Challenges of diffusion. *European Journal of Workplace Innovation, 2*(1), 63–81.
Totterdill, P., Exton, O., Exton, R., & Sherrin, J. (2009). *Workplace innovation policies in European countries*. Nottingham: UKWON.
Work-In-Net (WIN). (2010). *The grand societal challenge: Sustainable European work to withstand global economic change and crisis*. Declaration March 11–12. Berlin: WIN.

Author Biographies

Frank D. Pot is an Emeritus professor of Social Innovation of Work and Employment, Radboud University Nijmegen (NL) and former director of TNO Work and Employment.

Peter Totterdill is a Director of Workplace Innovation Europe CLG, an international not-for-profit company committed to stimulating and resourcing better ways of working. He is also a Visiting Professor at Kingston University London (UK).

Steven Dhondt is a Senior research scientist at TNO (NL) and visiting professor at KU Leuven (B). Main research focuses: workplace innovation, quality of work, innovation policy, social innovation.

Chapter 3
National and Regional Policies to Promote and Sustain Workplace Innovation

Tuomo Alasoini, Elise Ramstad and Peter Totterdill

3.1 Introduction

In many European countries, programmes to reform working life originated in the 1960s and 1970s from the efforts of progressive researchers, trade unions, employers and policy-makers to make working conditions more humane, increase workplace democracy and develop methods of organizing and managing labour that provided alternatives to the Taylorist and Fordist doctrines. Key motives for reforming working life were the decline in the productivity potential of the mass production mind-set, workers' dissatisfaction with working conditions, and difficulties experienced by employers in recruiting labour for fragmented industrial work. Despite the fact that Europe moved into an era of slower economic growth and higher levels of unemployment in the 1980s, interest in programmes to improve work organization and working conditions was sustained in several countries. For example, in their summary of programmes in the field of "technology, work and organization" in the 1980s, Den Hertog and Schröder (1989) examined seven European countries with such programmes.

However, since the 1980s, targets related to work humanization and workplace democracy have been diluted and shifted into the background within many of the programmes. A new generation of programmes launched in various European countries in the 1990s were increasingly motivated by productivity- and competitiveness-related targets. Typical objects in the programmes included

T. Alasoini (✉) · E. Ramstad
Tekes—The Finnish Funding Agency for Innovation, Helsinki, Finland
e-mail: tuomo.alasoini@tekes.fi

E. Ramstad
e-mail: elise.ramstad@tekes.fi

P. Totterdill
Kingston University, London, United Kingdom
e-mail: peter.totterdill@workplaceinnovation.eu

P.R.A. Oeij et al. (eds.), *Workplace Innovation*, Aligning Perspectives on Health, Safety and Well-Being, DOI 10.1007/978-3-319-56333-6_3

team-based organizational structures, flexible working methods, continuous improvement tools and business practices based on trust and (direct) worker participation. A survey by Business Decisions Limited (2000) conducted in the late 1990s found 18 government supported programmes for "new forms of work organization" in 11 European countries. Another study by Brödner and Latniak (2003), using somewhat stricter criteria, found national programmes to support the development of work organization in seven EU-15 member states.

In a recent review commissioned by the European Agency for Safety and Health at Work, Eeckelaert et al. (2012) examined national programmes and strategies for social innovation in the workplace by monitoring the extent to which occupational safety and health issues are included. The analysis comprised 18 countries. It found examples of national programmes in 11 member states. Indications of programme evaluations were discovered in the case of eight countries.

This chapter takes a more detailed look at main national- and regional-level policy approaches in Europe for setting up workplace innovation during the last two decades. The results of the analysis are mirrored against the previous overviews on the subject (above) and the policy matrix presented below. Special attention is given to the challenges of diffusion of workplace innovation and different methods for meeting such challenges within programmes.

"A workplace development programme" is understood to mean three things: first, development is guided by a shared framework that applies to several companies simultaneously; second, the content of the framework has been accepted by management and staff of the companies and by other major stakeholder groups such as policy-makers, social partners, and researchers, consultants and other experts; third, the involved companies engage in exchange of information, interaction and cooperation (Alasoini 2008). This definition is quite broad, and programmes may differ significantly with regard to the conceptual framework, the number of stakeholders and participating companies, and inter-organizational communication.

3.2 The Policy Matrix

It is possible to identify two main forms of policy intervention to promote workplace innovations. *Hard regulation* refers to directives or rules that have the force of law and are designed to impose minimum standards of practice or define the specific rights of individuals or organizations. *Soft regulation*, or animation, refers to non-binding, persuasive and proactive policy intervention designed to drive economic or social changes beyond the scope of more passive regulatory mechanisms. We can make a further distinction between direct and indirect hard regulation, as well as between direct, meso-level and indirect soft regulation (Table 3.1).

Hard regulation has been rare in the promotion of workplace innovations. Organizational, management or work-related practices associated with workplace innovation are typically not regarded as targets for hard/direct public intervention. It is also difficult to find empirical support for the view that deregulation can

Table 3.1 Policy options in the promotion of workplace innovation

Hard/indirect regulation Directives or binding rules which focus indirectly on workplace innovation through some other policy area (e.g. product market, labour market or occupational safety and health)	*Hard/direct regulation* Directives or binding rules which focus directly on workplace innovation (e.g. work-related, organizational or management practices)	
Soft/indirect regulation General policy frameworks and recommendations, conferences, "good practice" guides, etc.	*Soft/meso-level regulation* Educational and training programmes, coaching, research, learning networks, etc.	*Soft/direct regulation* Subsidised consultancy, development and action-oriented research projects, tax credits, etc.

contribute to workplace innovation. Deregulation may decrease companies' willingness to invest in workplace and other innovation because it enables companies to improve their competitiveness through increased flexibility in the use of labour or in other terms and conditions of employment (De Spiegelaere et al. 2014; Vergeer et al. 2015). As shown by the previous overviews, the most common means of promoting workplace innovations in different countries has been soft regulation in its various forms.

3.3 The Problems of Slow Diffusion and Low Awareness

Numerous empirical studies and development projects observed that companies have been able to improve their performance by innovations in work organization and management practices (e.g. Howaldt et al. 2016; Pot 2011; Ramstad 2014). However, this knowledge has not inspired many other companies to follow suit. Large-scale European surveys (Eurofound 2012, 2015) show that while companies may have tested individual "good practices" to varying degrees and in varying forms, few have tried to apply them in a comprehensive or systematic way and to utilize workplace innovation as a strategic competitive asset.

Our main argument is that it is unrealistic to expect accessible "good practice" cases or validated research data to constitute a driving force for a massive, self-sustaining "wave of change" in working life or for the improvement of European competitiveness. Unlike simple off-the-shelf management techniques or tools, workplace innovations typically require tacit knowledge and local learning. This process is not at the level of an individual company merely a matter of "adoption" or "implementation" of some principle or practice but rather entails "reconciling", "fine-tuning" and "re-invention".

It is obvious that publicly funded workplace development programmes cannot have a significant effect on the progress of working life change in themselves. Programmes mainly offer a means to reinforce desirable trends and provide a fertile and sheltered ground in which emerging niche-innovations can develop. Programmes must be guided by visionary and insightful understanding of current trends and promising niche-innovations, and the ability to implement such promotional measures. Any programme must also have a more or less explicit "programme theory", a plausible and sensible model of how a programme is supposed to function and produce the desired processes, leading to the desired outcomes (Funnell and Rogers 2011).

Three elementary types of knowledge are needed for designing and implementing programmes. Programme designers and implementers must have knowledge on both new workable designs (*design knowledge*) and collaborative and participatory ways to construct them (*process knowledge*) in companies. In addition, *dissemination knowledge* is needed to support diffusion of experiences of new designs and participatory and collaborative processes of change for the benefit of a wider group of actors (Alasoini 2011). Dissemination knowledge is necessary, because the legitimacy of publicly supported intervention cannot be based primarily on the success of individual demonstration projects but must rather be justified by a variety of positive externalities such as knowledge spill-overs and network effects produced by the intervention.

Low activity amongst policy-makers in the promotion of workplace innovation in many countries can be explained, in addition to a knowledge gap, by the fact that the organization of work is seen a managerial prerogative. Public intervention in areas even remotely associated with management and work organization is a sensitive issue in all European countries. Therefore, publicly supported programmes to promote workplace innovation usually require the consent and support of social partners and their readiness for cooperation in this area.

3.4 Workplace Innovation in European Policy Contexts

European countries differ greatly in their history of programmes to promote workplace innovation. In order to get insight and up-to-date information about the programmes, a literature review was undertaken and experts and practitioners were contacted in each country. The following description provides a brief overview of main national and regional initiatives in recent years.

3.4.1 Norway

Norway is the pioneer in Europe; its history of experiments in job redesign and workplace democracy dates back to the early 1960s. The Norwegian Confederation of Trade Unions (LO) and the Confederation of Norwegian Enterprise (NHO) have been key players in workplace innovation activities, whereas the role of the government has been less direct. An important milestone was the agreement on enterprise development signed by LO and NHO in 1982. The parties agreed upon increasing value creation in companies through broad mutual cooperation. Their main instrument to achieve this goal is the Joint Action Programme (HF). HF encourages companies to develop arenas for cooperation, and funds research and development projects in companies and broader cooperation conferences. In addition, the social partners give advice, run seminars, spot new trends and create meeting places for regional coalitions and networks. In recent years, LO has actively promoted the idea of an "employee-driven innovation" as a means of democratising innovation processes in companies (see Chap. 19 of Hansen et al. in this volume).

The cooperation structures that have emerged and developed between the social partners have formed the framework also for longer-term R&D programmes such as Enterprise Development 2000 (ED 1994–2001) and Value Creation 2010 (VC 2001–07) ED 2000 was a path-breaking programme in that it was not built on projects in individual companies but on larger (mainly regional) modules comprising several enterprises and research organizations. The idea of networking between enterprises and enterprises and researchers was an integral part of VC 2010 too; programme funding focused exclusively on the work input of researchers in regional enterprise networks. Researcher education and training was a component of project activity in both programmes.

VC 2010 was discontinued after its mid-term evaluation and the Research Council of Norway launched the new Funding Programme for Regional R&D and Innovation (VRI 2007–17) to promote innovation and value creation through regional cooperation and strengthening R&D effort within and for the regions. Unlike ED 2000 and VC 2010 which focused exclusively on enterprises, VRI targets both private and public-sector organizations. The link to workplace innovation in VRI is less direct and the scope of issues that can be funded broader than in preceding programmes. In addition to action research in companies, VRI funds regional foresights, competence brokering between companies and research organizations, and mobility schemes among students, researchers and business people.

3.4.2 *Sweden*

Sweden also has a rich history of innovative working life experiments since the late 1960s both at company and national levels. However, Sweden has not implemented extensive workplace development programmes at the national level since the massive Work Life Fund programme ended (1990–95) after some 25,000(!) funded projects. In recent years, R&D on working life has undergone several reorganizations and institutional arrangements. The closure of the National Institute for Working Life after a government change in 2007 is considered the most significant symbolic event in the weakening position of working life research in Swedish research and R&D policy (Sandberg 2013).

The promotion of workplace innovation in Sweden is now characterized by increasing integration with regional industrial and innovation policy. An important milestone in such integration was the establishment of the Swedish Agency for Innovation Systems (VINNOVA) in 2001. Among other tasks, VINNOVA aims to create sustainable development in business and public sectors by highlighting the importance of human factors as a creative resource. For example, the programme Management and Work Organization Renewal (2007–16) focuses on strategic management and work organization for well-functioning workplaces and thereby the efficiency and long-term development of operations. The programme has had several research calls with separate funding for development activities.

In 2013, VINNOVA provided a grant for a Sustainable Work platform. Sustainable work is a generic policy concept that highlights the dynamic fit between employees and working conditions, promoting health, well-being, learning and influence as well as productivity, innovation and growth. The platform is a consortium of leading Swedish occupational safety and health researchers, the social partners, public authorities and representatives from private companies. Its aim is to strengthen synergy between research centres in Sweden and the rest of Europe, as well as dialogue with social partners and users of research, to increase the scope for research on sustainable and innovative workplaces within Horizon 2020 and to provide policy pointers for authorities and other actors in the field.

3.4.3 Finland

Finland is a latecomer compared to its Nordic neighbours in introducing workplace innovation onto the political agenda. In Finland, the first national programmes to reform work processes, work organization and management practices started in 1993 and 1996 in the aftermath of a severe economic depression, as the National Productivity Programme and the Workplace Development Programme TYKE were launched. The two programmes ran in parallel as part of government programmes until 2003 in close cooperation between the Ministry of Labour and social partners. In 2004, the Ministry launched the new 6-year Workplace Development Programme TYKES. With the adoption of a new "broad-based innovation strategy" for Finland in 2008, the implementation of TYKES was transferred to the innovation funding agency Tekes and the promotion of workplace innovation was included in the operation of Tekes on a permanent basis. During 1996–2010, more than 1800 development projects, applied research projects and network projects were funded through the TYKE and TYKES programmes.

In 2011, the government decided to draw up a National Working Life Development Strategy for Finland. The strategy was prepared by the Ministry of Employment and the Economy in cooperation with other ministries and social partners. The Working Life 2020 project was established in 2012 to implement the strategy. As part of this strategy, Tekes launched the 6-year programme "Liideri—Business, Productivity and Joy at Work" for the renewal of companies' business operations through developing management, ways of working and employees' participation in innovation as well as making innovative use of digital technologies. The aim is for at least 300 companies to launch programme-funded projects.

Promotion of workplace innovation has been on the Finnish policy agenda through various governments and activities of Tekes and social partners for more than 20 years. The policy rationale has increasingly shifted from the context of industrial relations to that of innovation policy. Unlike the TYKE and TYKES programmes which targeted at many kinds of companies and public work organizations on an equal footing, Liideri has its focus on growth- and export-oriented SMEs in accordance with the new Tekes strategy. To substitute for the narrowed focus of Tekes in the promotion of workplace innovation, many industry development programmes for the improvement of productivity and employee well-being have been launched by labour market organizations in the context of Working Life 2020 (Alasoini 2016).

3.4.4 Denmark

Denmark ranks at the top in European comparisons of learning-oriented forms of work organization, employee participation in innovation, work autonomy and job satisfaction. Although work organization development has attracted much attention in Denmark since the 1970s, Denmark differs from the other Nordic countries in the sense that there have not been as extensive government funded programmes at the national level. Improvements in quality of work have to some degree taken place through massive national campaigns focusing on promoting "developmental work" and reducing monotonous work, but, more importantly, through local joint efforts by the social partners and individual companies. Compared to its Nordic counterparts, the "Danish innovation model" has been based to a greater extent on the significant role played by SMEs and their employees, often supported by local trade unions, in the national innovation system.

Denmark's top position in both learning-oriented forms of work organization and broad participation of employees in innovation processes is an outcome of interplay between two institutional mechanisms. The dynamic and practically-oriented system of vocational training led by both labour market parties has equipped a large number of employees with improved ability to participate in innovation processes. Second, the collaborative and decentralized system of industrial relations has contributed to forming of labour-management partnerships for change within companies. Since 2006, also the Danish LO has taken an increasingly active role in the promotion of "employee-driven innovation", a new concept inspired by the idea of a "learning organization" (Høyrup et al. 2012), as part of its policy.

3.4.5 Germany

Germany has had a continuous chain of national programmes to develop working life since 1974 with the Humanization of Work programme. In the 1980s, the programme underwent many changes in content, and in 1989 it was further developed into the Work and Technology programme. In 2001, the Federal Ministry of Education and Research launched a five-year framework concept “Innovative Development of Work—The Future of Work”, to cope with new demands caused by demographic, technological and organizational changes on employees and companies. This was followed by a framework programme “Working–Learning–Developing Skills: Potential for Innovation in a Modern Working Environment” in 2007 as part of federal research funding policy, with the aim of developing new solutions for working life in a knowledge-based and service-oriented economy and by closing knowledge gaps, stimulating change processes and supporting political decisions for these goals. The framework comprised several smaller-scale programmes that funded research and action-oriented research projects. About 1500 companies were assisted.

The most recent German framework programme in this area, “Innovation for Production, Services and Work of Tomorrow”, was launched in 2014. The programme is established until 2020 with a total budget of approximately €1 billion. Research programme “The Future of Work—Innovations for Work of Tomorrow” forms a separate entity within the broader framework. The overall aim of the programme is to develop and disseminate new concepts and tools for designing and organizing work in an increasingly digitalized working environment with the help of research, practically-oriented development projects and transfer activities in nine thematic fields of action. SMEs form the special target group. The programme will be conducted in close cooperation with the social partners and many other German and international programmes and networks, including activities related to *Industrie 4.0.* “The Future of Work” is co-funded through the ESF.

In addition to national programmes, some federal states in Germany like Bremen, North Rhine-Westphalia and Saxony have run programmes of their own. As the most outstanding example, the Federal State of *North Rhine-Westphalia* launched a social-oriented technology policy programme (SoTech) in 1985 influenced by the national Humanization of Work programme and in which the emphasis was on developing socially acceptable ways of shaping new technologies. SoTech was followed in 1994 by the more development-oriented QUATRO programme which placed more emphasis on work organization, group work and employees’ social competencies.

Six years later, the Ministry of Economy and Labour of North Rhine-Westphalia launched a new programme framework under the heading of “Work-Oriented Modernization” funded through the ESF. Many of the instruments created under this framework are still in existence under the new labour policy framework “Staying Competitive and Securing the Future”. Led by a regional government agency Innovative Employment Promotion Company (G.I.B.), the framework promotes capacity building, harnessing diverse sub-regional agencies in promoting workplace innovation and recruiting companies. G.I.B.'s consulting services support SMEs in short-term workplace change projects as well as in the longer-term development of organizational strategy. Approximately 1500 companies are supported annually. In addition, the development and dissemination of new methods and tools for the use of companies are funded through the framework. As in the case of the German federal programme, digitalization as an enabler of workplace innovation has recently become an increasingly important new topic in labour policy for SMEs and their employees in North Rhine-Westphalia.

3.4.6 The Netherlands

The Netherlands is a European pioneer in experiments with sociotechnical work design starting from the early 1970s and has a strong tradition in research in work, organization, innovation and occupational safety and health. However, with certain exception like the "Technology, Work and Organization" (TAO) programme in the 1980s and 1990s, the country does not have a similar history of large-scale national programmes as, for example, Germany. The promotion of workplace innovation has taken place mainly in individual companies and at industry level through joint efforts by social partners and through the ESF.

Workplace innovation has been on the agenda of the government and social partners in the 2000s under the heading of "social innovation". In 2006, the Dutch National Centre for Social Innovation (NCSI) was established by employers' associations, trade unions and research organizations as a continuation of the former Innovation Platform, aiming to achieve better synergy between actions by separate players in this field. NCSI's mission was to put social innovation on the national agenda, disseminate knowledge and stimulate action among companies and company networks, trade unions and knowledge institutions. The Centre's activities were subsidised from many sources, including three ministries. In 2012, the founding organizations and the sponsors considered the mission of the Centre to be fulfilled and it was closed down. Some of the activities of NCSI were taken over by other organizations (Pot et al. 2012).

In 2015, the Dutch Ministry of Economic Affairs launched the national technology-oriented Smart Industry strategy. Smart Industry does not refer only to the intelligent use of new digital technologies, but is also about changing companies' attitudes to customer orientation, work organization and cooperation. With some pressure from trade unions and researchers, a "Field Lab for Social Innovation" was created as part of the strategy. The government funds the launch of the field lab to develop practices of skills development, work organization and employee involvement needed as part of technological development in companies. The objective is to have companies to co-fund the projects. In the collective agreement of 2016 in the metalworking and electrical engineering industries, the social partners agreed upon a joint social innovation project within the Smart Industry framework.

3.4.7 Flanders

The Flemish government, under the auspices of the Minister of Work, Education and Training, started a work organization development programme, entitled Flanders Synergy, in 2006. This 18-month programme was funded through the Flemish ESF Agency. Flanders Synergy was a rather small programme that comprised only 12 practically-oriented company projects. However, it signified an important breakthrough in policy thinking, indicating that for the first time organizational innovation was being accepted as a legitimate target of policy intervention and an integral part of the government's global innovation strategy.

The programme was inspired by the new consensual atmosphere between the social partners, embodied in the Pact of Vilvoorde of 2001. This pact contained a long-term socioeconomic vision for Flanders, including an ambitious target to improve workability through better organization of work. Improved workability rate means that a higher proportion of employees will have jobs with qualities that lead to positive social effects, such as diminished work-related health problems and sick absenteeism and postponement of retirement. Flanders Synergy was one means to help the government and social partners to reach the target set in the pact.

In 2008, the ESF Flanders launched a new call under the heading of "social innovation" as a 2-year follow-up programme with increased financial resources. At the same time, in 2009, a new pact between the Flemish government and social partners updated their engagements made in the previous pact. Flanders Synergy established its position as a competence pool. It operates now as a membership organization under a public-private cooperation structure, based on interaction between companies, social partners, knowledge institutions and certified consultants. The aim is to create innovative and adaptive organizations that improve quality of work through building star cases and prototypes, evidence-based standards and a broad ecosystem for diffusing these experiences. Between 2009 and 2015, about 300 organizations actively participated in the redesign of their structure within this framework.

3.4.8 The UK and Scotland

Successive governments in the UK have been resistant to ideas of intervention in support of workplace innovation. Neo-classical economic perspectives have tended to dominate policy-making, leading to the invisibility of the workplace despite its potential significance in addressing major national policy priorities such as productivity and skills utilisation.

Despite policy indifference the UK Work Organisation Network (UK WON), a coalition of researchers, social partner organizations and other stakeholders created in 1997, secured ESF support for seven consecutive years from the government's Department of Work and Pensions. ESF support triggered a diverse range of activities including action research, learning networks, workshops and the creation of online knowledge resources. UK WON was registered as a not-for-profit company in 2001. It subsequently diversified sources of income to include a wide range of European, regional and commercial funding. This has enabled UK WON to sustain and expand its portfolio of activities, all of which contribute to the stimulation and resourcing of workplace innovation.

Scotland has led the way in the UK in terms of how best to improve skills utilisation within workplaces through new forms of work organization. Building on this, in 2013, the Scottish Centre for Employment Research led the establishment of a Workplace Innovation Consortium which, with support from the European Regional Development Fund, developed the *Innovating Works ... improving work and workplaces* initiative to support SMEs in simultaneously improving business performance and job quality. The "Scottish approach" to workplace innovation is inextricably linked to initiatives to improve job and employment quality and in 2015 the government established the Fair Work Convention with an invited membership of employers, unions and academics. In 2016, the Convention delivered its framework and vision of a competitive and inclusive economy in which, by 2025, people in Scotland will have a world-leading working life where fair work drives success, well-being and prosperity for individuals, businesses, organizations and society. Fair work is defined as work that offers effective voice, opportunity, fulfilment, security and respect and workplace innovation is seen as a route to delivering fair work, tackling inequality and improving competitiveness. Together, workplace innovation and fair work are central to Scotland's Economic Strategy.

The government has also encouraged businesses to sign up the Scottish Business Pledge, indicating their support for and adoption of a suite of practices tackling low pay and promoting equality, opportunity and innovation in business. All public agencies are charged with promoting fair work and encouraging take up of the Pledge in their delivery of services to business. Scottish Enterprise, one of the country's economic development agencies, established a dedicated Workplace Innovation Service in 2016 and launched a series of pilot projects comprising support to enable companies to diagnose current working practices and implement appropriate workplace innovation measures; a programme of Masterclasses for raising awareness of workplace innovation amongst businesses; a leadership team programme for the support of senior managers; and a deeper Engagement Programme based on action-learning and in-company support for cluster of companies. Scottish Enterprise aims to support more than 3000 companies through these measures by 2019.

3.4.9 Ireland

In Ireland, the government, together with the Irish Productivity Centre and social partners, launched the first national programme to help companies modernize their work organization in the spirit of labour-management partnership in 1995. The 3-year New Work Organization in Ireland programme was a rather small effort, intended to create "best practices" for further dissemination. In 1997, the government established a special organization, re-established 4 years later as the National Centre for Partnership and Performance (NCPP), to support workplace change and innovation through partnership. In 2003, the government requested the NCPP to establish a special "Forum on the Workplace of the Future", resulting in the development of an ambitious National Workplace Strategy 2 years later.

The strategy was supported by a nationally unique social partnership framework and a strong institutional nexus of state agencies. The strategy aimed to improve innovation capability by increasing the role of workplace innovation in the national innovation system through the adoption of an integrated and coherent approach to workplace and workforce development (Flood et al. 2008). A set of nine inter-dependent and mutually reinforcing characteristics for Ireland's workplaces were defined. As part of the strategy, and delivered by the social partnership agreement *Towards 2016*, the Workplace Innovation Fund (WIF) was unveiled in 2007. The WIF was administrated by Enterprise Ireland, a state development agency focused on transforming Irish industry. The fund was organized into three strands, with the aim to support workplace innovation in SMEs based on partnership, finance initiatives by the social partners relating to the strategy and implement national campaigns to raise awareness among employers and employees of the strategy.

Favourable political climate for strengthening the role of workplace innovation on the political agenda was aborted by the economic and fiscal crisis post-2007. The NCPP was dissolved in 2010 as part of a wider process of consolidating public service agencies and as a consequence of fragmentation of national social partnership. Between 2007 and 2009 the fund had supported 36 company-based projects. After the abolition of the NCPP and WIF, the promotion of partnership-based workplace innovations has taken place through the efforts of individual trade unions, such as the Services, Industrial, Professional and Technical Union (SIPTU) which created the Ideas Institute, offering expert support to companies and their trade union representatives.

3.4.10 France

Anact (L'Agence nationale pour l'amélioration des conditions de travail) was formed in 1973 against a backdrop of industrial relations conflict, in part a result of the Taylorist forms of work organization that predominated in French enterprises.

Anact was created as a statutory national agency, involving social partners particularly through regional economic development strategy, but funded by the state with the aim of improving health and safety and reducing conflict through the introduction of a consistent policy framework for new forms of work organization.

Since 2008, Anact has run the Fund for the Improvement of Working Conditions (FACT) that provides short-term intervention in SMEs or groups of SMEs for projects adopting a comprehensive approach to improving working conditions. By 2014, 102 projects were underway, about 20% of them covering groups of SMEs.

ANACT's Social Innovation Fund (FISO), established in 2013 by the President, François Hollande, offers advances to finance socially innovative projects across the French regions. Two further programmes, aimed specifically at the co-operative and social enterprise sector respectively, provide financial support for eligible projects.

3.4.11 The Basque Country

During the last decade the Provincial Council of Gipuzkoa, representing one third of the Basque Country, has developed a system of programmes and policies aimed at promoting a knowledge-based economy. Policy measures to support the transition towards an innovation economy were incorporated in 2014, promoting the participation of workers in workplace innovation through the development of new, partnership-based models at company level. Improving job design, skills utilisation and development and "employee voice" and employee ownership all fall within the remit of this new policy framework.

Support is available to a wide range of actors for the stimulation and resourcing of workplace innovation including companies and business associations based in Gipuzkoa, trade unions, and regional bodies concerned with innovation, education, economic development and research. Activities or areas eligible for support at company level include the promotion of companies' intangible assets; developing legal and ethical frameworks for inclusive decision-making; supporting company strategies that guarantee the continuity of business activity and the handover to the next generation through succession planning; promoting worker participation; experimentation and intervention in advanced innovation formulas; and the assessment, valorisation and dissemination of "good practices" relating to participation and sustainability. The framework provides support also for creating sustainable intervention models in regional organizations; developing new ways of satisfying the needs, challenges and social and regional opportunities; and territorial and social innovation dynamics that integrate economic, social and educational agents for the competitive and balanced development of sectoral and regional spheres. In 2014–15, about 100 organizations participated in these activities.

3.4.12 Building Capacity Elsewhere in Europe

In 2013 the European Workplace Innovation Network (EUWIN) was established by the European Commission with a strong remit to stimulate workplace innovation especially in those EU member states which currently lack formal institutional support (Pot et al. 2016). A programme of conferences and workshops, as well as the creation of an online Knowledge Bank, enabled the sharing of "good practices" between companies and has engaged policy-makers, business support agencies, social partners and other stakeholders in countries throughout Europe. EUWIN interventions have led to the creation of ad hoc task groups which are exploring ways of stimulating and resourcing workplace innovation in countries such as Estonia, Lithuania, Poland and Portugal.

3.5 Summary of the Policy Overview

The overview shows that policies for the promotion of workplace innovation still resort exclusively to soft regulation, ranging from general policy frameworks and recommendations to more direct forms. These activities are still unevenly distributed by geographical area, highlighting the gap between the more active north and the more passive south, and the absence of programmes in most of the new EU member states. In countries like Norway and Germany, the promotion of workplace innovation has a well-established position that dates back to several decades. Finland and Flanders are examples of cases in which workplace innovation has in recent years gained increased foothold as part of mainstream labour and innovation policy thinking. There are also cases like Scotland and the Basque country in which new policy frameworks for the promotion of workplace innovations has only recently been established. Finally, in some countries exploration of ways of stimulating and resourcing workplace innovation are just starting.

3.6 Methods for Meeting the Challenges of Diffusion

The uneven distribution of programmes by geographical area highlights the importance of European-level initiatives such as EUWIN. One of the biggest challenges faced by all programmes has been to find efficient means of disseminating good practices from one context to another and bring about cumulative innovations among a larger group of companies than those directly participating in programme activities. Meeting the challenges of diffusion is an issue of crucial importance not only for the sake of improving the social effectiveness and legitimacy of programmes, but also for encouraging new countries and regions to take further steps.

There are different options to tackle this issue (Alasoini 2011). The *first* method available is to deploy various means of transfer (training, seminars, publications, data banks. etc.) more efficiently and selectively for clearly targeted groups of potential adopters. The progress of ICT and social media provides new opportunities to overcome the traditional trade-off between richness and reach of information sharing. The limitation of this method is that diffusion is still considered a mechanical process, without any radically new tools to support local learning and re-invention among potential adopters.

The *second* alternative is to shift programme resources from the "over-resourced" innovation creation stage (the stage at which "good practices" are created) to support the "under-resourced" group of "second-wave" adopters. This shift of emphasis can help to overcome problems inherent in the first method. However, it would require ample financial and expert resources on the part of the programme in the form of providing support to a relatively large group of follower companies.

Third, it is possible to utilize better empirical evidence-based research to enhance the knowledge provided by demonstration projects. A rigorous analysis of "what works and why" and of causal mechanisms can help to improve the overall trustworthiness of project results for wider use. As important as being able to produce evidence-based information is for any programme, a major limitation in practice is that achieving hard evidence with wide and off-the-shelf applicability in various kinds of work contexts is often difficult.

Storytelling, forum theatre and other narrative methods have proved an efficient way to share experiences. The *fourth* option is based on the idea of enriching the information provided by demonstration projects, data banks or "good practice" cases by rendering this information more interactive and easier to adopt by other companies. This can take place at interactive workshops and dialogue seminars in which it is possible for "first-wave" and "second-wave" adopters to embark upon intensive exchange of information on a face-to-face basis. It often requires ample resources to arrange and lead these forums.

The *fifth* method is to utilize development or learning networks. Using such networks corresponds to a blurring of the tendency to think in phases. The idea behind the use of networks is to bring together within programmes several actors (companies, consultants, researchers, etc.) who share an interest in similar issues but have a broad diversity of complementary expertise. The actors launch large-scale joint projects or, alternatively, several parallel experiments and learn from each other through intensive exchange of information. Many European countries have experience of utilizing regional, industry-wise or thematic networks in support of both innovation creation and the diffusion of "good practices" (e.g. Alasoini 2008; Bessant and Tsekouras 2001; Ekman et al. 2011). Building up such networks and achieving the confidential interaction relations required for networks to function requires mutual trust between participants and a long-term commitment on the part of the programme.

3.7 Conclusion

As this overview shows, national- and regional-level policies to promote workplace innovation in Europe differ widely by their approach and content. There is no universal solution to how to target programme resources successfully. The answer depends largely on the wider policy framework and the overall national or regional development infrastructure that provide tangible or intangible support to sustain workplace innovation.

According to the "best practice framework" for national workplace development strategies developed by Naschold (1994), the strategic justification for such strategies should arise primarily from industrial policy. It is easy to agree with his view that without a clear link to mainstream industrial or innovation policy there is a danger that the promotion of workplace innovation becomes side-tracked from challenges faced by companies themselves and excessively entangled in problem settings that arise, for example, from the industrial relations system or the academic world. However, subordination of workplace innovation only as a component of industrial or innovation policy includes another kind of risk. An important aspect of workplace innovation that gives it special status is the idea that employees, in addition to enterprise owners and management, have legitimate interests. The crucial role employees play in workplace innovation should be explicitly recognized.

Rather than considering workplace innovation only an instrument of owner- or management-driven enterprise development, a broader policy framework is necessary. Although the industrial relations system must not guide strategic goal setting, it is appropriate to involve social partners to strengthen the social legitimacy of programme activities. The engagement of trade unions and employers' associations with an advisory role is a common feature of all major European programmes. The same logic applies to the engagement of research organizations; their interests must not guide the goal setting, but their role is crucial in bringing new insights to programme activities by helping to explore and experiment with new solutions.

To meet the challenges of diffusion, part of programme resources should be used for strengthening the research and development infrastructure conducive to workplace innovation. This can take place through enhancement of knowledge and building of networks. Many programmes have included training, education, research or the development of development methods as part of their activities. Network building is of crucial importance too. In modern polycentric societies, no individual actor can claim to have superior knowledge on workplace innovation or monopolize such knowledge indefinitely. New useful knowledge is generated through dialogue between various innovation actors rather than by "trickling" information from the "top down" or from "the core" to "the periphery" (Fricke 1994; Ramstad 2008). At best, national and regional programmes are forums for such dialogue.

References

Alasoini, T. (2008). Building better programmes: Learning networks in the promotion of workplace innovation. *International Journal of Action Research, 4*(1–2), 62–89.

Alasoini, T. (2011). Workplace development as part of broad-based innovation policy: Exploiting and exploring three types of knowledge. *Nordic Journal of Working Life Studies, 1*(1), 23–43. http://rossy.ruc.dk/ojs/index.php/njwls/article/view/2334/651. January 2, 2013.

Alasoini, T. (2016). *Workplace development programmes as institutional entrepreneurs: Why they produce change and why they do not* (Doctoral Dissertations 12/2016, School of Science, Aalto University, Helsinki).

Bessant, J., & Tsekouras, G. (2001). Developing learning networks. *AI and Society, 15*(1–2), 82–98.

Brödner, P., & Latniak, E. (2003). Sources of innovation and competitiveness: National programmes supporting the development of work organisation. *Concepts and Transformation, 8*(2), 179–211.

Business Decisions Limited. (2000). *Government support programmes for new forms of work organisation: A report for DG Employment and Social Affairs*. Luxembourg: Office for Official Publications of the European Communities.

De Spiegelaere, S., Van Gyes, G., & Van Hootegem, G. (2014). Labour flexibility and innovation, complementary or concurrent strategies? A review of the literature. *Economic and Industrial Democracy, 35*(4), 653–666.

Den Hertog, J. F., & Schröder, P. (1989). *Social research for technological change: Lessons from national programmes in Europe and North America*. Maastricht: University of Limburg.

Eeckelaert, L., Dhondt, S., Oeij, P., Pot, D. F., Nicolescu, G. I., Trifu, A., et al. (2012). *Review of workplace innovation and its relation with occupational safety and health: Literature review*. Luxembourg: Publications Office of the European Union.

Ekman, M., Gustavsen, B., Asheim, B. T., & Pålshaugen, Ø. (Eds.). (2011). *Learning regional innovation: Scandinavian models*. Houndmills: Palgrave Macmillan.

Eurofound. (2012). *Fifth working conditions survey: Overview report*. Luxembourg: Publications Office of the European Union.

Eurofound. (2015). *Third European company survey—Overview report*. Luxembourg: Publications Office of the European Union.

Flood, P. C., Guthrie, J. P., & Liu, W. (2008). *New models of high performance work systems*. Dublin: National Centre for Partnership and Performance.

Fricke, W. (1994). Scientific knowledge, social change and action research. In T. Kauppinen & M. Lahtonen (Eds.), *National action research programmes in the 1990s* (pp. 47–69). Helsinki: Ministry of Labour.

Funnell, S., & Rogers, P. (2011). *Purposeful program theory: Effective use of theories of change and logic models*. San Francisco: Wiley.

Howaldt, J., Oeij, P. R. A., Dhondt, S., & Fruytier, B. (2016). Workplace innovation and social innovation: An introduction. *World Review of Entrepreneurship, Management and Sustainable Development, 12*(1), 1–12.

Høyrup, S., Bonnafous-Boucher, M., Hasse, C., Lotz, M., & Møller, K. (Eds.). (2012). *Employee-driven innovation: A new approach*. Houndmills: Palgrave Macmillan.

Naschold, F. (1994). *The politics and economics of workplace development: A review of national programmes*. Helsinki: Ministry of Labour.

Pot, F. (2011). Workplace innovation for better jobs and performance. *International Journal of Productivity and Performance Management, 60*(4), 404–415.

Pot, F., Dhondt, S., De Korte, E., Oeij, P., & Vaas, F. (2012). Workplace innovation in the Netherlands. In I. Houtman (Ed.), *Work life in the Netherlands* (pp. 173–190). Hoofddorp: TNO.

Pot, F., Totterdill, P., & Dhondt, S. (2016). Workplace innovation: European policy and theoretical foundations. *World Review of Entrepreneurship, Management and Sustainable Development, 12*(1), 13–32.

Ramstad, E. (2008). *Innovation generating model—Simultaneous development of work organization and knowledge infrastructure: Experimenting in the field of organizational development.* Helsinki: Tekes.
Ramstad, E. (2014). Can high-involvement innovation practices improve productivity and the quality of working-life simultaneously? Management and employee views on comparison. *Nordic Journal of Working Life Studies, 4*(4), 25–45. http://rossy.ruc.dk/ojs/index.php/njwls/article/view/4706/2393. Accessed January 7, 2015.
Sandberg, Å. (Ed.). (2013). *Nordic lights: Work, management and welfare in Scandinavia.* Stockholm: SNS förlag.
Vergeer, R., Dhondt, S., Kleinknecht, A., & Kraan, K. (2015). Will 'structural reforms' of labour markets reduce productivity growth? A firm-level investigation. *European Journal of Economics and Economic Policies, 12*(3), 300–317.

Author Biographies

Tuomo Alasoini is a Chief Adviser at Tekes—The Finnish Funding Agency for Innovation.

Elise Ramstad is a Senior Adviser at Tekes—The Finnish Funding Agency for Innovation.

Peter Totterdill is a Director of Workplace Innovation Europe CLG, an international not-for-profit company committed to stimulating and resourcing better ways of working. He is also a Visiting Professor at Kingston University London (UK).

Chapter 4
Why Industrie 4.0 Needs Workplace Innovation—A Critical Essay About the German Debate on Advanced Manufacturing

Jürgen Howaldt, Ralf Kopp and Jürgen Schultze

In the international debate, the digitalisation of industry and development of "advanced manufacturing" occupy an important position as significant factors for maintaining competitiveness and safeguarding jobs. In Germany, in recent years, the term "Industrie 4.0" has increasingly featured in this debate. Given that the capabilities of digital systems are unquestionably increasing, we argue that the debate surrounding "Industrie 4.0" is in need of considerable reorientation, in light of a new innovation paradigm, so that the real challenges in the transition to a knowledge-based society are not missed. Effective approaches should build on existing strengths. One suitable approach which could excellently integrate the various participative working cultures specific to Europe and the new requirements of "Industrie 4.0" is a consistent orientation to high involvement innovation practices. Similar approaches have been developed in many European countries, with names such as "high performance workplaces", "high involvement workplaces", "innovative workplaces", "sustainable work systems" and "employee driven innovation" (Pot et al. 2012, p. 262). These concepts, much in agreement to workplace innovation, are often about emphasising the close relationship between organisational performance (labour productivity, innovation capabilities) and better jobs (competence development, wellbeing at work). Furthermore, a sustainable approach should take into account the contribution to be made by future industrial structures to addressing the big social challenges (e.g. environment, health, demographic change). We briefly outline the vision of "Industrie 4.0" and its

J. Howaldt (✉) · R. Kopp · J. Schultze
Sozialforschungsstelle Dortmund (sfs), Central Scientific Institute of TU Dortmund University, Dortmund, Germany
e-mail: howaldt@sfs-dortmund.de

R. Kopp
e-mail: kopp@sfs-dortmund.de

J. Schultze
e-mail: schultze@sfs-dortmund.de

P.R.A. Oeij et al. (eds.), *Workplace Innovation*, Aligning Perspectives on Health, Safety and Well-Being, DOI 10.1007/978-3-319-56333-6_4

problems. We use "Halle 54" (a plant of Volkswagen car industry that became famous for its failed vision of an factory without people on basis of computer integrated manufacturing) and a more recent example ("Enterprise 2.0") to illustrate the dysfunctionalities and contradictions of technology-centred approaches, whose excessive automation ambitions, despite having failed repeatedly, are currently experiencing a resurgence in "Industrie 4.0". Against this background we define the contours of a new innovation paradigm, which focuses on the question of the conditions for developing and maintaining modern societies' capacity for innovation. This leads to the conclusion that the discussion about "Industrie 4.0" should follow a more comprehensive innovation strategy with social innovation and workplace innovation at the core. This means also to give more attention to the development of sustainable and integrated business models, and the enhancement of companies' ability to innovate through comprehensive utilisation of the potential of their employees and of society. In Germany this debate labelled by "Work 4.0" (Arbeiten 4.0) becomes stronger in recent time.

4.1 Vision, Core Objective Promoters and Problems of Industrie 4.0 in Germany

According to a current definition, "Industrie 4.0" is a term which refers to "the fourth industrial revolution, a new level of organisation and management of the entire value chain across the product life-cycle. This cycle is geared to increasingly individualised customer wishes, and extends from the idea, the development and production work, and the delivery of a product to the end customer, to recycling, including the associated services. It is based on the availability of all relevant information in real-time as a result of networking all the parties involved in value creation, and on the ability to infer from the data the optimal value stream at any time. Linking people, objects and systems creates dynamic, real-time optimised, self-organising and inter-enterprise value creation networks which can be optimised according to various criteria such as costs, availability and resource usage" (Plattform Industrie 4.0 2015, p. 3).

The term "Industrie 4.0" was introduced in 2011 at the Hanover Fair in Germany and covers the most important activities and projects of the German government to promote computerisation of the manufacturing industry (smart factory). In 2012, it became the focus of a working group on "Industrie 4.0" chaired by Robert Bosch GmbH and acatech.[1] The "Industrie 4.0-working group" represents the core of the promoters (Kagermann et al. 2013). Their implementation recommendations to the German Federal Government were presented in April 2013, again at the Hanover Fair in a final report. The three industry associations Bitkom, VDMA and ZVEI concluded from this to create an "Industrie 4.0-platform", "to put the

[1]Acatech is Germany's National Academy of Science and Engineering.

pre-competitive conditions in place for the economic implementation and realisation of the Industrie 4.0 vision […]. Through dialogue involving different sectors, the aim is to develop concepts for technologies, standards, business models and organisation models, and promote their practical implementation" (Plattform Industrie 4.0 2015, p. 3). According to its protagonists, the "Internet of Things" is increasingly penetrating all social aspects of production, services, trade and consumption. The authors of one of the key studies on the topic write that "Industrie 4.0" centres on "the real-time capable, intelligent, horizontal and vertical networking of humans, machines, objects and ICT systems for the dynamic management of complex systems" (Bauer et al. 2014, p. 18). Custom products should be manufacturable at bulk-product prices, as a result of humans, machines and artefacts communicating with each other, and the emergent product is so computerised that it can optimise its own production process. "Industrie 4.0" holds out the prospect of a completely new logic and quality of production management, which should make it possible for "intelligent products, machines and equipment to exchange information autonomously, initiate actions and control each other independently in real-time" (ibid). Networking does not end at the factory gate, rather it encompasses the relationship between factories and suppliers, with the result that it can extend to form widespread value creation networks. The concept is being driven "by computer scientists, engineers, innovation policy actors, influential business associations and larger technology-intensive enterprises" (Hirsch-Kreinsen 2014, p. 421).

The hype surrounding "Industrie 4.0" so far has occurred primarily at the discursive level and policy concepts following expectations in information science. Even though some big players (e.g. Siemens) started a number of dedicated "Industrie 4.0"-pilot projects and despite there is given a lot of (financial and practical) support by governmental institutions, it has received only a modest reception especially among broad swathes of Germany's *Mittelstand*.[2] This is not necessarily due to information deficits, a lack of innovative spirit or slowness to react. There are not only opportunities but also a number of risks involved (e.g. high investment costs, data safety, issues of compatibility within production structures, qualification requirements, work conditions), which might lead to a hesitant attitude towards "Industrie 4.0". It seems that, on a practical level, companies follow still a corporate concept that is based on lean production. New digital technologies are continuously being integrated into this context. This is the following of a path dependent development (see also Hirsch-Kreinsen 2016, p. 22).

"Industrie 4.0" sometimes seems to transport back to another era. Last century, in the 1950s and 1960s, there were widespread attempts to draw far-reaching conclusions from technological developments for the design of organisational structures and work. In their ground-breaking study for the subsequent debate on the relationship between technology and work, "Industrial labour and worker consciousness", Kern and Schumann (1970) note: "In the literature on the sociology

[2]Mittelstand stands for small and medium sized companies.

of industry, there is a concept that proved to be particularly appealing, which embeds the historical relationship between industrial technology and human labour in a three-phase model" (ibid, p. 27). The model is guided by the assumption that the respective technological conditions lead to workers being employed in particular ways, and determine the skill sets that are needed. One "assumed a rising line of development from skilled crafts and trades to mechanisation (assembly line production) and then to automation; to this corresponded, respectively, the worker-types of the autonomous craftsman, the heteronomous low-skilled worker on the production line, and finally the requalified worker now doing hardly any physical work" (Pfeiffer 2010, p. 234). In the 1970s, the three-phase model was increasingly surpassed in industrial sociology, and the "end of technological determinism" (Lutz 1987) became the new basic consensus. This was combined with an understanding of innovation in which technological and social innovations are mutually dependent. Nevertheless, at first these insights were slow to have any practical effect. Both in the popular idea of technological development and in the social-science (sociological) discourse of innovation research, the primacy of technology—even if in an enlightened version—remained dominant. This technology-centred view led to spectacular failures in the past, yet seems to be gaining new impetus in the "Industrie 4.0" debate. Thus we read today: "Industrie 4.0 is feasible, human 4.0 not so easily" (MTM aktuell 2014, p. 4).

The example of experiments with computer integrated manufacturing (CIM) in a plant of Volkswagen in Wolfsburg called "Halle 54" became famous for its principal failure and illustrates the problematic consequences of choosing a technocentric approach. In other words: "Industrie 4.0" is in danger to become CIM 2.0. The notion of "Industrie 4.0" is remarkably reminiscent of the disappointed hopes in the 1980s of a fully automated factory in the automotive industry. "Halle 54" was a production and final assembly hall at Volkswagen's Wolfsburg plant, which at the time of its commissioning in 1983 was considered to be an advanced computer-integrated manufacturing (CIM) concept and blueprint for a fully automated factory. Like today it was accompanied by promises of higher productivity and the elimination of monotonous activities in favour of the highly-skilled jobs that remained. Its failure became legendary. The reasons were seen in an "overengineering" trailed after the lack of quality, demotivation, lack of acceptance, rise of absentism. Thus the model ultimately failed because of its radically contra-anthropocentric rationalisation strategy. "The idea of a fully automated factory ran up against its material limits just as Taylorism reached its limits as a model for the organisation of work and production" (Pfeiffer 2010, p. 233). Instead, new concepts of production and organisation (e.g. lean production, teamwork, learning organisation) found their way into the day-to-day activities of many businesses. Whilst these concepts stay ambivalent according to working conditions, they were aiming to comprehensively exploit the potentials of human labour in sense of dezentralization of responsibility and widening of scopes for decision-making (see also Kern and Schumann 1984; Minssen et al. 1991).

Since then, a "variety of more or less innovative production concepts have emerged" (Heßler 2014, p. 16), which focus on the social and cultural aspects of business organisation and management. According to Heßler (2014), the 1990s are characterised by the coexistence and mixing of different concepts, in which the relationship between humans and machines is configured context-specifically. Nevertheless, robots continued to be developed, and work was indeed successfully done to "enable them to identify errors or deviations in the process themselves, and learn from this" (Heßler 2014, p. 16)—in other words, so that they gather experience-based knowledge. These old discussions have striking similarities to the current debate, with the result that in the context of the design of work as well, there are reflections on whether "in precisely the context of "Industrie 4.0", the time has come to implement a few 'old' ideas" (Hartmann 2014, p. 7). The experiences of Halle 54 can teach us not only that the social aspects need to be incorporated into the vision and architecture of technology design from the outset, but also that there is a need for a realistic assessment of the reach of the concepts. It can be assumed, for instance, that such advanced technologies can be usefully applied only in particular industries and areas of production (see also Ittermann and Niehaus 2015, p. 31). Even if "Industrie 4.0" is "treated from the outset as a socio-technical system, in which humans are to remain central as comprehensive decision-makers or as cognitive all-rounders" (Howaldt and Kopp 2015, p. V6) the beginning of the debate around "Industrie 4.0" was astonishingly close to the technology-centred logic of that time.

A more recent example than "Halle 54" for failed technological driven approaches is the implementation of the transformation from Enterprise 1.0 to Enterprise 2.0. Only at the second attempt changing from techno-centric approaches to a human-centric approach they became successful. The same protagonists (e.g. Bitkom, CEBIT) today proclaiming "Industrie 4.0" as gold standard for strategic orientation, some years ago have favoured "Enterprise 2.0". One of the early definitions of "Enterprise 2.0" shows a strong technological orientation: "Enterprise 2.0 is the use of emergent social platforms within companies, or between companies and their partners or customers" (McAfee 2006, o. S.). With only very few expectations (e.g. Koch and Richter 2009; Back and Heidecke 2009) the scientific debate mirrored assumptions (especially from software developing firms), following the simple equation: Enterprise 2.0 = Introduction of Web 2.0 into enterprises. Preconditions like an appropriate enterprise culture and conditions of organisation were strong underexposed. Anyway, the initially rare success models from "Enterprise 2.0" underline the necessity to focus on social innovations instead of technology. The results of a research project about advanced innovation approaches in high-tech industry that topic (Howaldt et al. 2011) show that the transformation from "Enterprise 1.0" to "Enterprise 2.0" in cases of good practice sometimes starts nearly completely without using Web 2.0-tools like wikis, forums or other social media (Stamer 2008, p. 74). The crucial difference was the scale and alignment of self-organization and participation enabled by management. Ideally "Enterprise 1.0" is based on hierarchic organization and closed innovation while "Enterprise 2.0" is characterized by self-organization, participation and open innovation. This does not mean to neglect hierarchy completely. But it is a functional, flat hierarchy following the principle: as little as possible—as much as is necessary. The responsibility of management here is to enable self-organization in a socio-digital innovation system (Kopp 2011). This means to minimize control and to practice the "art of let loose" (Stamer 2008, p. 61). For that, the management has to channel the flood of internal and external ideas and knowledge by means of collective and intelligent selection and decision processes which enable the reduction of complexity, the condensation of concrete projects and the successful development and marketing of new products. Of course the necessary knowledge management should be supported by new tools of information and communication technology (e.g. Web 2.0). The lesson is to understand that organization development should follow a social demand and not a technological push. This means to activate experience-based knowledge and participation in order to clarify what the needs of employees is, and if and how new technology can satisfy the needs.

4.2 A New Innovation Paradigm

In terms of innovation strategy, "Industrie 4.0" is trusting in the power of engineering, pursues the approach of a technological "push"—a concept that is closely associated with a one-sided technology-focused understanding of innovation. Yet the potentials of the knowledge-based society and economy could be better unlocked through alternative strategies as part of a new innovation paradigm (Bullinger 2006; FORA 2010; Howaldt and Schwarz 2010). Key categories here are the opening of the innovation process towards society, orientation to social challenges, social innovation and the capacity for innovation. Particularly the opening of the innovation process towards society (FORA 2010, p. 15) is a central feature of a changed innovation paradigm. Businesses, universities and research institutes are not the only relevant actors in the innovation process. Citizens and customers no longer serve only as suppliers of information about their needs (as is the case in classical innovation management), as instead they bring information about solutions into the development process for new products. Terms and concepts such as "open innovation" (Chesbrough 2003), customer integration (Jacobsen 2005), and co-evolutionary learning networks (Howaldt et al. 2001) mirror important aspects of this development.

At the same time, social innovations come into focus, in the sense of the reconfiguration of social practices and their establishment in particular sectors of society (Howaldt and Schwarz 2010). Examples range from civil society (environmental movement, new forms of living arrangements) to the area of state action (social insurance), and the economy (learning organisation, new management concepts, new services) (Gillwald 2000, p. 3). A significant milestone in anchoring social innovations in German innovation policy is the German federal government's new high-tech-strategy. The intention is clearly formulated: "We are focusing on a wider understanding of innovation, including not only technological but also social innovations, which involves society as a central actor. We are looking at the whole picture and we consider together that which belongs together" (BMBF 2014, p. 4).

Thus attention is shifting from the market potential of individual technology fields to society's need for sustainable solutions and their realisation. "Now it is a matter of bringing these strands together and considering all key aspects of a comprehensive research and innovation policy in context. This creates an optimal environment for ideas, their implementation in marketable products and services, more value creation and potential for new future-proof jobs" (BMBF 2014, p. 11). Considerations focus on enhancing innovative capacity by stepping up dialogue with a wide variety of stakeholders across organisational boundaries (networking, open innovation)—including a broad spectrum of social actors. However, the development of innovative capacity in this sense is a process that depends on many conditions and creates major challenges for the actors involved—in business, science, politics and society. While the debate surrounding national and regional innovation systems is predominantly concerned with the structural, political and institutional conditions for innovativeness at national and regional level, in the BMBF program "Working–learning–developing skills–innovative capabilities in a modern working world", interest focuses in particular on management and work-related aspects of innovativeness. Terms such as organisation, qualification, technology and health are of central importance here. To enhance innovative capacity, attention at the enterprise level focuses on activities and the creation of conditions conducive to innovation by initiating and supporting learning processes, skills development, and participative forms of organisation (Hartmann 2014).

Against this background, it seems that "Industrie 4.0" is being conceptualized too narrowly and therefore fails to make use of the potentials that a broad concept of innovation could offer. In a present report of a commission of experts for research and innovation (Expertenkomission Forschung und Innovation—EFI 2016), they demand a more extended vision on the topic. There it says: "The intensive focus solely on a small branch of digitalization is not rewarding. This way, "Industrie 4.0" is being reduced to a mere attempt of efficiency enhancement within manufacturing engineering. Furthermore, initiatives concerning manufacturing- or user-specific issues such as Smart Service World or E-Health are being held back on generating positive effects within digital application. There is an urgent need for a comprehensive and convincing strategy" (EFI 2016, p. 15). Within this entire strategy social innovation should rise more attention: "Social innovation is scarcely considered within research and innovation policy that is so far mainly shaped by a technic-centered understanding of innovation. Hence, the EFI-commission of experts ask the Bundesregierung (national government in Germany) to pay more attention to social innovation and to experiment with new models of participation as well as suitable tools of promotion […]. […] The public facilitation of social innovation is intended to provide aid to the development, research and testing of new ideas of changing social practices" (ibid, p. 12).

It seems that the strong focus on "Industrie 4.0" drops behind the intentions of high-tech-strategy that says: "What we particularly need are technical and social innovations that produce new services opening up new markets, and that can contribute to the needs of society. Services follow a slightly different pattern of innovation. Corporate processes, strategies and forms of organization are becoming the center of attention while still maintaining the involvement of the users. Innovations of services combine different services into optimal solutions and appeal to a variety of needs" (BMBF 2016, p. 23). Probably the greatest risk is that the underlying, strongly technology-oriented innovation approach is not capable of appropriately developing the potentials of digital technology. "Further innovations in technology and business are imperative; yet in order to reap their full potential, and at the same time creating social development that is beneficial to cultures as inclusive as diverse, social innovations will make the difference" (Hochgerner and Howaldt 2011, p. 380).

Another plea for a more comprehensive innovation concept can be found in the "Connectedreality 2025" trend study by Z_punkt, which argues that system innovations should help solve social problems. "But [these] cannot be developed and implemented by individual actors. They require partnerships, development alliances and thinking in complex value creation patterns, which a purely technological innovation logic must be subordinate to" (Boeing et al. 2014, p. 55). Greater sensitivity to the need for cooperation between all kinds of stakeholders in the innovation process is characteristic of the new innovation paradigm. In the Digital Agenda for Europe, this concept of open innovation is currently associated with the "quadruple helix model" (Dhondt and Oeij 2014, p. 139; Carayannis and Campbell 2009). Here it states: "Open Innovation is an important component of the foreseen European Innovation System, where all stakeholders need to be involved and create seamless interaction and mash-up for ideas in innovation ecosystem. [...] Open innovation 2.0 (OI2) is a new paradigm based on a Quadruple Helix Model where government, industry, academia and civil participants work together to co-create the future and drive structural changes far beyond the scope of what any one organisation or person could do alone. This model encompasses also user-oriented innovation models to take full advantage of ideas' cross-fertilisation leading to experimentation and prototyping in real world setting" (Digital Agenda for Europe n.d.).

4.3 Changing the Focus: The Growing Importance of Workplace Innovation for Shaping the Future

Partially as a reaction to "Industrie 4.0", another discourse labelled under "Work 4.0" (Arbeiten 4.0) has emerged to tackle the issue of digitalization of work. The question of designing the future of work assumes a central position in this discussion, and is influenced by social partners (Botthof and Hartmann 2015; IG Metall NRW 2013) as well as by social science and labour studies (Hirsch-Kreinsen 2014). These debates are also concerned with quality of work, and with "concepts for job structures that are geared to acceptance, potential for achievement and development, well-being and the health of working people. This is about questions such as how well working environments in "Industrie 4.0" promote learning, the interaction between machines/robots and humans, as well as new opportunities linked to employment policy solutions" (Botthof and Hartmann 2015, p. 6). Work related social scientists, trade unions and policy makers expand the perspective from "Industrie 4.0" to "Work in a digitalized world" (Schlick 2015) or "Work 4.0" (Wetzel 2016), focusing on designing digital work as a project of collaboration between enterprises, trade unions, politic, science and civil society (Schröder and Urban 2016, p. 6). But the potential of the new technologies for the humanisation of work continued to be highly contentious (Mückenberger 2016, p. 247). Urban (2016, p. 33) refers to five future conflicting fields: (a) conflict over social status (protected knowledge worker versus precarious entrepreneur); (b) conflict over time (flexibility or sovereignty); (c) conflict over health (economization or prevention); (d) conflict over qualification (development of personality and competencies or disqualification); (e) conflict over influence ('liquid democracy' or 'democratic work').

At the level of enterprises the socio-technical concept of workplace innovation covers the main elements of a human related perspective for developing work in a digitalized world (Dhondt et al. 2015) that aims for comprehensive utilization of the potentials of human labour as a condition for ensuring innovative ability (Howaldt et al. 2012; Totterdill et al. 2012). Germany—against a background of funding programs such as "humanising work", "work and technology", "innovative workplace design and the future of work"—has built up a wealth of experience which, at the same time, constitutes an important competitive advantage internationally (Georg et al. 2012). These programs were guided from an early stage by the idea of a comprehensive innovation concept. In their analyses of the complex relationships between social and technological innovation processes in enterprises, they provided vital input for a comprehensive understanding of innovation, and fostered the development of new strategies, concepts and instruments which have enabled businesses and intermediary actors to compete successfully in the international arena (see also Ludwig et al. 2007; Streich and Wahl 2007; Gatermann and Fleck 2010; Jostmeier et al. 2014).

When dealing with labour research and the prospective designing of the link between social innovation and new technologies, the new BMBF report "Zukunft der Arbeit—Innovation für die Arbeit von morgen" (BMBF 2016)[3] seems to be particularly relevant. Within this report, one field of action (6.1 "enabling social innovation by means of new work processes") indicates that there is a necessity of inter-play between social and technical innovation (see also BMBF 2016, p. 19). According to that, digital manufacturing cannot be thinkable without innovation facilitating management concepts and organsational structures (see also BMBF 2016, p. 19). The concept of social innovation is clearly grounded in this program: "The challenges not only consist of the new technology itself, but also particularly deal with—accompanied with its introduction—the new configuration of social action and the new interaction between technology and society. The shape of the

[3]Future of work—Innovation for work of tomorrow.

new thing is still uncertain. On an internal company level, there is a necessity for participatory and trust-based designing processes while involving all the different stakeholders. On a level beyond the company, it means to involve clients and civil society in existing and new emerging value chains. Hence, we need to focus on the redesigning of internal actions and on the interdependency of technical and social dimensions. Internally, there is the need to enable and design participatory and trust-based interaction models for all the different stakeholders; externally, the institutional (economic, political, social, cultural) context has to be taken into consideration. On intensive knowledge-based conditions, the effort to create sustainable innovation displays a big challenge for society. With the emergence of Economy 4.0, totally new work and learning environments arise that require considerable research on fields such as technology, staffing, organization and skills acquisition. On an organisational level, more attention has to be paid to social innovations and new insights, to the way how social innovations emerge and how they can be applied within companies. A knowledge-based economy as a condition for the maintenance and development of the German and European economic competitiveness cannot be imaginable without developing the facilitation of in-novation, management concepts and company structures. Social innovation be-gins at the workplace, it demands modern work environments that allow more self-organisation and offer more freedom to individually design one's workplace and work processes […]. In order to make full use of what technological potentials can offer, a comprehensive understanding of innovation is required that sustainably anchors and purposefully implements social and technical innovations within companies" (BMBF 2016, p. 20).

The traditional labour research in Germany, the debate on "Work 4.0" and its perspective on "Industrie 4.0" as well as the programmatic orientation of the BMBF underline how relevant social innovation is at the workplace. "Innovative working environment" also is an important topic in the German federal government's new high-tech-strategy. "New forms of work organisation, stronger service focus, changing skills and job profiles, more interactive value creation processes and increasing digitalisation—all these are driving forces of the far-reaching change that the modern working world is undergoing. Today more than ever, being innovative requires complex processes that need interaction with technological development, but also with human resource, organisational and skills development. 'Good work' is therefore an important basis for business innovations" (BMBF 2014, p. 22). It seems questionable whether national go-it-alone efforts can succeed in developing internationally competitive platforms quickly enough, but that is beyond the scope of this article. The large number of different approaches in European countries with regard to "Industrie 4.0" (and similar approaches with other names) brings to mind the necessity of more coordination. However, the European Workplace Innovation Network (EUWIN) has started to actively address this, and is attempting to develop common standards with its workplace innovation approach. Pot et al. describe the origins of the workplace innovation approach like this: "Workplace innovation, as it developed from the beginning of this century is a member of the Sociotechnical

Systems Design (STSD)-family, going back to the restructuring of Europe after the Second World War, starting more or less the same policies for productivity and industrial democracy in several Western European countries" (Pot et al. 2016, p. 16). Totterdill points out that the requirements for workplace innovation include quality of work, participation and decentralisation, and goes on to state: "Most importantly, workplace innovation is an inherently social process. It seeks to build bridges between the strategic knowledge of the leadership, the professional and tacit knowledge of frontline employees, and the organisational design knowledge of experts. […] Thus in defining workplace innovation it is important to recognize both process and outcomes." (Totterdill 2015, p. 57).[4] The dual practical benefit of corresponding sociotechnical designs—firstly the improvement in motivation, job satisfaction and employee well-being, secondly the improvement in performance—has also been repeatedly confirmed by research (for a current example see Ramstad 2014). Thus there are many good reasons to emphasise the importance of the validated experience that it is not sufficient just to introduce new technologies. "It will require the full utilization of the potential workforce and creation of flexible work organisations" (Pot et al. 2012, p. 261).

4.4 Conclusion and Outlook

From the perspective of social innovation, there have been a number of mentions about problematic aspects and practical consequences that the vision of "Industrie 4.0" exposes. But by no means we can conclude it as a general denial. A technique-based vision has its own value, as long as its goals are seen as an option rather than an obligation. The decision whether this option will be accepted or not, is an issue to be dealt with on a practical level.

Against this background a comprehensive innovation strategy seems to be appropriate. Firstly this strategy aims at supporting those who want to implement an "Industrie 4.0" that services not only economy but also advances good workplaces and social innovations. A look back into the past with regard to technological driven experiences should make us aware that it is only by analysing the complex interplay between social and technological innovations that we arrive at a realistic vision of the future, which can guide us in designing forward-looking production and work systems. "Anyone who wants Industrie 4.0 in Germany should critically examine the 'high-tech-obsession'" and "should regard it primarily as a social

[4]Although WPI is predominantly oriented at enterprise level, social innovation is focusing on societal level. To integrate both levels a theoretical relation between work and organisation issues (like quality of performance and quality of work) and societal issues (like employment, employability and sustainability) should be established, which is beyond the scope of this contribution.

innovation" (Buhr 2015, p. 19; Howaldt et al. 2016). Workplace innovation with its focus on innovation and participatory workplaces covers the main elements (work organisation, reflection and innovation, workplace partnership, structure and systems) to design appropriate work conditions and to the human and economic potential of digitalization. It is to emphasise "that the success of the proclaimed fourth industrial revolution depends crucially on whether it is sustainably anchored in the organisation and implemented in a targeted way. Accordingly, human and technological aspects should be adapted to and aligned with the organisation's structures and processes" (Deuse et al. 2014, p. 44). In Germany the debate around "Work 4.0" (Arbeiten 4.0) contribute to change the focus from "Industrie 4.0" as a technological vision to more human centered approaches. What is important here, is a number of aspects that can be attributed to social innovation (e.g. humane work design, facilitating chances of qualification and its development, participation, protection of pay scales, minimum wage, ecological change, successful integration of immigrated people, facilitation of gender issues). This way, Working 4.0 exceeds what is meant by production work, and establishes connections to the field of services—not only those close to manufactural services but also to individual related services (see also Evans and Hilbert 2016).

Secondly a comprehensive innovation strategy has to pay more attention to services and business models as a topic of digitalization. A wider perspective implies also a stronger emphasis in tackling social challenges. Rather than promoting a "technological push" and its subsequent socially acceptable design, the focus shifts to enhancing innovative capacity by involving social actors in the development of solutions for the future. At the level of enterprises and organization, this is a question of integrated socio-technical management approaches, as are combined for example in international work and management research in the workplace innovation approach. If "Industrie 4.0" (Industry 4.0) should contribute to good and innovative workplaces it needs workplace innovation enabling employees to participate in the ongoing processes of organisational change in a digitalized world of working. New programs launched by the German Federal Ministry of Education and Research (BMBF), the German Federal Ministry for Economic Affairs and Energy (BMWi), and also programs by German states such as the North Rhine-Westphalia lead market competition for the digital working environment and future of work, provide scope for joint activities between academia and practitioners to develop participative management forms under condition of a digitalized working world. It is a big difference to reduce digitalization on "Industrie 4.0" or to follow a comprehensive innovation strategy. And it is a big difference whether workplace innovation is an enabler for "Industrie 4.0" or digitalization is an enabler of workplace innovation. The latter opens up to an innovation strategy where "technology is not an exogenous force over which we have no control. We are not constrained by a binary choice between 'accept and live with it' and 'reject and live without it'. Instead, take dramatic technological change as an invitation to reflect about who we are and how we see the world" (Schwab 2016, p. 4).

References

Back, A., & Heidecke F. (2009). Einleitung. In A. Back, N. Gronau, & K. Tochtermann (Eds.), *Web 2.0 in der Unternehmenspraxis. Grundlagen, Fallstudien und Trends zum Einsatz von Social Software* (2nd ed., pp. 1–8). München.

Bauer, W., Schlundt, S., Marrenbach, D., & Ganschar, O. (2014). Industrie 4.0—Volkswirtschaftliches Potenzial für Deutschland. Studie. Bitcom. http://www.bitkom.org/de/themen/74736_79154.aspx. Accessed March 30, 2015.

Boeing, N., Burmeister, K., Neef, A., Rodenhäuser, B., & Schroll W. (2014). ConnectedReality 2025. Die nächste Welle der digitalen Transformation. http://www.z-punkt.de/connected-reality2025.html. Accessed February 9, 2015.

Botthof, A., & Hartmann, E. A. (2015). *Zukunft der Arbeit in "Industrie 4.0"*. Berlin, Heidelberg: Springer.

Buhr, D. (2015). Soziale Innovationspolitik für die Industrie 4.0. Expertise im Auftrag der Abteilung Wirtschafts-und Sozialpolitik der Friedrich-Ebert-Stiftung. Resource document. Friedrich-Ebert-Stiftung. http://library.fes.de/pdf-files/wiso/11302.pdf. Accessed February 27, 2015.

Bullinger, H.-J. (2006). Verdammt zur Innovation. *RKW-Magazin, 57,* 12–14.

Bundesministerium für Bildung und Forschung (BMBF). (2014). Die neue Hightech-Strategie. Innovationen für Deutschland. Brochure (English version). http://www.bmbf.de/pub_hts/HTS_Broschure_Web.pdf. Accessed February 10, 2015.

Bundesministerium für Bildung und Forschung (BMBF). (2016). Zukunft der Arbeit. Innovation für die Arbeit von morgen. Resource paper. BMBF. https://www.bmbf.de/pub/Zukunft_der_Arbeit.pdf. Accessed July 18, 2016.

Carayannis, E. G., & Campbell, D. F. J. (2009). 'Mode 3' and quadruple helix: Toward a 21st century fractal innovation eco system. *Internationale Zeitschrift für Technologiemanagement, 46*(3–4), 201–234.

Chesbrough, H. W. (2003). *Open innovation: The new imperative for creating and profiting from technology*. Boston: Harvard Business Press.

Deuse, J., Kirsten, W., Hengstebeck, A., & Busch, F. (2014). Gestaltung von Produktionssystemen im Kontext von Industrie. Publication. Bundesministerium für Wirtschaft und Energie. http://www.bmwi.de/BMWi/Redaktion/PDF/Publikationen/zukunft-der-arbeit-in-industrie-4-0,property=pdf,bereich=bmwi2012,sprache=de,rwb=true.pdf. Accessed February 27, 2015.

Dhondt, S., & Oeij, P. (2014). Social innovation related to innovation in management studies. In J. Howaldt, A. Butzin, D. Domanski, & C. Kaletka (Eds.), *Theoretical approaches to social innovation—A critical literature review* (pp. 122–150). Dortmund: Sozialforschungsstelle.

Dhondt, S., Oeij, P., & Preenen, P. (2015). Working in the digitalized world: The meaning of the link between advanced manufacturing and workplace innovation in the EU growth strategy. In C. M. Schlick (Ed.), *Arbeit in der digitalisierten Welt* (pp. 155–170). Frankfurt: Campus Verlag.

Digital Agenda for Europe. (n.d.). Open Innovation. https://ec.europa.eu/digital-agenda/open-innovation-20. Accessed February 10, 2015.

Evans, M., & Hilbert, J. (2016). Personenbezogene Dienstleistungen: Gestaltungsanforderungen einer Zukunftsbranche. In R. Hoffmann & C. Bogedan (Eds.), *Arbeit der Zukunft* (pp. 381–400). Frankfurt: Campus Verlag.

Expertenkommission Forschung und Innovation (EFI). (2016). Gutachten zu Forschung, Innovation und technologischer Leistungsfähigkeit Deutschlands. Resource document. EFI. http://www.e-fi.de/fileadmin/Gutachten_2016/EFI_Gutachten_2016.pdf. Accessed July 18, 2016.

FORA. (2010). New nature of innovation. Report to the OECD. Report. www.tem.fi/files/24835/New_Nature_of_Innovation.pdf. Accessed July 8, 2015.

Gatermann, I., & Fleck, M. (2010). Innovationsfähigkeit sichert Zukunft. Beiträge zum 2. Zukunftsforum des BMBF. Berlin: Duncker & Humblot.

Georg, A., Hasselkuss, M., Howaldt, J., Jacobsen, H., & Jostmeier, M. (2012). Arbeitsforschung für Innovationsprozesse. Ergebnisse und Folgerungen aus der Arbeit des BMBF-Förderschwerpunkts "Innovationsstrategien jenseits traditionellen Managements". Dortmund: Sfs.

Gillwald, K. (2000). Konzepte sozialer Innovation. Research paper. Wissenschaftszentrum Berlin für Sozialforschung. http://bibliothek.wzb.eu/pdf/2000/p00-519.pdf. Accessed February 27, 2015.

Hartmann, E. (2014). Arbeitsgestaltung für Industrie 4.0: Alte Wahrheiten, neue Herausforderungen. Publication. Bundesministerium für Wirtschaft und Energie. http://www.bmwi.de/BMWi/Redaktion/PDF/Publikationen/zukunft-der-arbeit-in-industrie-4-0,property=pdf,bereich=bmwi2012,sprache=de,rwb=true.pdf. Accessed February 27, 2015.

Heßler, M. (2014). Die Halle 54 bei Volkswagen und die Grenzen der Automatisierung. Überlegungen zum Mensch-Maschine-Verhältnis in der industriellen Produktion der 1980er Jahre. *Zeithistorische Forschungen, 1,* 56–76.

Hirsch-Kreinsen, H. (2014). Wandel von Produktionsarbeit—Industrie 4.0. Essay. WSI. http://www.boeckler.de/wsimit_2014_06__hirsch.pdf. Accessed September 12, 2015.

Hirsch-Kreinsen, H. (2016). Industrie 4.0 als Technologieversprechen. Soziologisches Arbeitspapier der TU Dortmund 46/2016. Working paper. TU Dortmund. http://www.neue-industriearbeit.de/fileadmin/templates/publikationen/20160616—Hirsch-Kreinsen_2016_Industrie-4_0-als-Technologieversprechen.pdf. Accessed July 19, 2016.

Howaldt, J., & Kopp, R. (2015). Industrie 4.0 und die Zukunft der Arbeit. *Frankfurter Allgemeine Zeitung, Sonderveröffentlichung NRW*, p. V6, April 10, 2015.

Howaldt, J., Kopp, R., & Beerheide, E. (Eds.). (2011). *Innovationsmanagement 2.0. Handlungsorientierte Einführung und praxisbasierte Impulse*, Wiesbaden.

Howaldt, H., Kopp, R., & Flocken, P. (2001). *Kooperationsverbünde und regionale Modernisierung—Theorie und Praxis der Netzwerkarbeit*. Wiesbaden: Gabler Verlag.

Howaldt, J., Oeij, P. R. A., Dhondt, S., & Fruytier, B. (2016). Workplace innovation and social innovation: An introduction. *World Review of Entrepreneurship, Management and Sustainable Development, 12*(1), 11–12.

Howaldt, J., & Schwarz, M. (2010). Soziale Innovation. Skizze eines gesellschaftstheoretisch inspirierten Forschungskonzepts. Bielefeld: Transcript Verlag.

IG Metall NRW. (2013). Industriearbeit 4.0. die Zukunft mitgestalten. Fachkonferenz zur Industriearbeit 4.0 in Paderborn. Conference paper. http://www.igmetall-nrw.de/uploads/media/2014_04_04_Industrie4.0.pdf. Accessed January 3, 2014.

Ittermann, P., & Niehaus, J. (2015). Industrie 4.0 und Wandel von Industriearbeit. Überblick über Forschungsstand und Trendbestimmungen. In H. Hirsch-Kreinsen, P. Ittermann, & J. Niehaus (Eds.), *Digitalisierung industrieller Arbeit. Die Vision Industrie 4.0 und ihre sozialen Herausforderungen* (pp. 33–51). Baden-Baden: Nomos Verlagsgesellschaft.

Jacobsen, H. (2005). Produktion und Konsumtion von Dienstleistungen: Konsumenten zwischen Innovation und Rationalisierung. In H. Jacobsen & S. Voswinkel (Eds.), *Der Kunde in der Dienstleistungsbeziehung* (pp. 15–36). Wiesbaden: Springer Verlag.

Jostmeier, M., Georg, A., & Jacobsen, H. (2014). Sozialen Wandel gestalten. Zum gesellschaftlichen Innovationspotenzial von Arbeits-und Organisationsforschung, Dortmunder Beiträge zur Sozialforschung. Wiesbaden: Springer.

Kagermann, H., Wahlster, W., & Helbig, J. (2013). Deutschlands Zukunft als Produktionsstandort sichern. Umsetzungsempfehlungen für das Zukunftsprojekt Industrie 4.0. Concluding Report. Platform Industrie 4.0. https://www.bmbf.de/files/Umsetzungsempfehlungen_Industrie4_0.pdf. Accessed March 5, 2015.

Kern, H., & Schumann, M. (1970). *Industriearbeit und Arbeiterbewußtsein*. Frankfurt: Europäische Verlagsanstalt.

Kern, H., & Schumann, M. (1984). *Das Ende der Arbeitsteilung? Rationalisierung in der industriellen Produktion*. Munich: Verlag C.H. Beck.

Koch, M., & Richter, A. (2009). Enterprise 2.0. Erfahrung, Einführung und erfolgreicher Einsatz von Social Software in Unternehmen. München.

Kopp, R. (2011). Enterprise 2.0 als soziodigitales Innovationssystem. In Howaldt, J., Kopp, R., & Beerheide, E. (Eds.), Innovationsmanagement 2.0. Handlungsorientierte Einführung und praxisbasierte Impulse. Wiesbaden, pp. 37–65.

Ludwig, J., Moldaschl, M., Schmauder, M., & Schmierl, K. (2007). *Arbeitsforschung und Innovationsfähigkeit in Deutschland*. Munich: Rainer Hampp Verlag.

Lutz, B. (1987) Das Ende des Technikdeterminismus und die Folgen. Soziologische Technikforschung vor neuen Aufgaben und Problemen. In B. Lutz (Ed.), *Technik und sozialer Wandel. Verhandlungen des 23. Deutschen Soziologentages in Hamburg 1986* (pp. 34–52). Frankfurt: Campus Verlag.

McAfee. (2006). Enterprise 2.0, version 2.0. http://andrewmcafee.org/2006/05/enterprise_20_version_20/. Accessed June 24, 2015.

Minssen, H., Howaldt, J., & Kopp, R. (1991). Gruppenarbeit in der Automobilindustrie—Das Beispiel Opel Bochum. *WSI-Mitteilungen, 7,* 434–441.

MTM aktuell. (2014). Industrie 4.0 ist machbar, der Mensch 4.0 nicht ohne weiteres. MTM-Bundestagung zu den Perspektiven des Industrial Engineering in der Arbeitswelt von morgen. MTM aktuell, 5, pp. 4–6.

Mückenberger, U. (2016). Arbeit vom Menschen her denken: Was wäre heute unter ‚Humanisierung der Arbeit' zu verstehen? In R. Hoffmann, & C. Bogedan (Eds.), *Arbeit der Zukunft. Möglichkeiten nutzen—Grenzen setzen* (pp. 247–259). Frankfurt: Campus Verlag.

Pfeiffer, S. (2010). Technisierung von Arbeit. In F. Böhle, G. Voß, & G. Wachtler (Eds.), *Handbuch Arbeitssoziologie* (pp. 231–262). Wiesbaden: Springer.

Plattform Industrie 4.0 (2015). Industrie 4.0. Whitepaper. Plattform Industrie 4.0. www.zvei.org/Downloads/Automation/Whitepaper-I40-FuE-Themen-2015-04.pdf. Accessed June 24, 2015.

Pot, F., Dhondt, S., & Oeij, P. (2012). Social innovation of work and employment. In H.-W. Franz, J. Hochgerner & J. Howaldt (Eds.), *Challenge social innovation. Potential for business, social entrepreneurship, welfare and civil society* (pp. 261–274). Berlin: Springer.

Pot, F., Totterdill, P., & Dhondt, S. (2016). Workplace innovation: European policy and theoretical foundation. *World Review of Entrepreneurship, Management and Sustainable Development, 12*(1), 13–32.

Ramstad, E. (2014). Can high-involvement innovation practices improve productivity and the quality of working-life simultaneously? Management and employee views on comparison. *Nordic Journal of Working Life Studies, 4*(4), 25–45.

Schlick, C. M. (2015). Arbeit in der digitalisierten Welt. Beiträge der Fachtagung des BMBF 2015. Frankfurt: Campus Verlag.

Schröder, L., & Urban, H.-J. (2016). *Gute Arbeit. Digitale Arbeitswelt—Trends und Anforderungen*. Frankfurt: Bund Verlag.

Schwab, K. (2016). *The fourth industrial revolution*. World Economic Forum: Köln/Genf.

Stamer, S. (2008). Enterprise 2.0—learning by doing. In Buhse, W., & Stamer, S. (Eds.), Enterprise 2.0—Die Kunst, loszulassen. Berlin. pp. 59–87.

Streich, D., & Wahl, D. (2007). Innovationsfähigkeit in einer modernen Arbeitswelt. Personalentwicklung—Organisationsentwicklung—Kompetenzentwicklung. Beiträge der Tagung des BMBF. Frankfurt: Campus Verlag.

Totterdill, P. (2015). Closing the gap: The fifth element and workplace innovation. *European Journal of Workplace Innovation, 1*(1), 55–74.

Totterdill, P., Cressey, P., & Exton, R. (2012). Social innovation at work: Workplace innovation as a social process. In H.-W. Franz, J. Hochgerner, & J. Howaldt (Eds.), *Challenge social innovation. Potential for business, social entrepreneurship, welfare and civil society* (pp. 241–259). Berlin: Springer.

Urban, H.-J. (2016). Die Digitalisierung der Arbeitswelt—ein Blick zurück nach vorn. In L. Schröder, & H.-J. Urban (Eds.), *Gute Arbeit. Digitale Arbeitswelt—Trends und Anforderungen* (pp. 21–45). Frankfurt: Bund Verlag.

Wetzel, D. (2016). Arbeit 4.0. Was Beschäftigte und Unternehmen verändern müssen. Freiburg i. Breisgau: Verlag Harder.

Author Biographies

Prof. Dr. Jürgen Howaldt is a Director of Sozialforschungsstelle Dortmund (sfs), Central Scientific Institute of TU Dortmund. Main research focuses: new management concepts, organisational development and consulting, social sciences-based innovation research.

Dr. Ralf Kopp is a research associate at Sozialforschungsstelle Dortmund (sfs), Central Scientific Institute of TU Dortmund. Main research focuses: network and innovation management, organisational development, social innovations.

Dipl. Inf. Jürgen Schultze is a Schultze, Dipl. Inf. Jürgen, b. 1960, research associate at Sozialforschungsstelle Dortmund (sfs), Central Scientific Institute of TU Dortmund. Main research focuses: organisational development, comprehensive innovation dialogues, importance of social innovations for sustainability.

Part II
Theory Matters & Evidence

Chapter 5
Theoretical Approaches Supporting Workplace Innovation

Peter R.A. Oeij and Steven Dhondt

5.1 Introduction

The publication of this book underlines that Workplace Innovation (WPI) has become popular in research and policy making and is finding its way to practice (see the policy debate in Part I of this volume). Simultaneously, organisational practices that can be identified as WPI-like are in dire need of appropriate conceptualisation (see the EUWIN website and their Knowledge bank of cases[1]). Whereas policy makers stress the importance of empirical evidence of WPI and the positive effects it may have (Chap. 6), a unitary theory of WPI is missing. This chapter addresses the topic of theory. In general, a theory is an idea, a coherent set of ideas or a general principle that is intended to explain facts or events. Because of the lack of theory on WPI, the evidence is scattered, and practitioners are missing hands-on advice regarding WPI implementation. This chapter undertakes a review of different theoretical approaches that inform us about WPI and that could be useful when thinking about WPI.

The purpose of this chapter is to show how well-known approaches support workplace innovation practices and its claims for results and discuss what we can learn from each approach for WPI. Indeed, each approach can contribute in its own

[1]EUWIN's knowledge bank is hosted by UKWON at http://portal.ukwon.eu/euwin-knowledge-bank-menu-new.

P.R.A. Oeij (✉) · S. Dhondt
TNO, The Netherlands Organisation for Applied Scientific Research,
Leiden, The Netherlands
e-mail: peter.oeij@tno.nl

S. Dhondt
e-mail: steven.dhondt@tno.nl

S. Dhondt
KU Leuven, Leuven, Belgium

P.R.A. Oeij et al. (eds.), *Workplace Innovation*, Aligning Perspectives on Health, Safety and Well-Being, DOI 10.1007/978-3-319-56333-6_5

unique way to WPI. We seek to discuss three issues: (1) provide a working definition of WPI; (2) discuss a number of different theoretical approaches that can be informative to the study and practice of WPI; (3) highlight some ways in which these approaches can inform a theory of WPI.

Approaches in the literature differ substantially. Some approaches to WPI, for example, are practice-oriented, like the Fifth Element model (Totterdill and Exton 2014). Other approaches focus on definitions of WPI and try to capture the mechanisms that explain how WPI can lead to better performance and quality of working life, as in the Eurofound report on 51 case studies (Oeij et al. 2015). Yet others are focused on applying concepts to measure the theoretical constructs that together constitute WPI and the possible effects of WPI (e.g., Mohr and Van Amelsvoort 2016; Oeij and Vaas 2016). Most of the theoretical endeavours, including our own, are normative and descriptive, only some are explanatory and predictive (e.g., Van Hootegem 2016). Hence, the question arises what commonalities these approaches share in supporting WPI.

In our work (Eeckelaert et al. 2012; Oeij et al. 2012, 2015; Oeij and Vaas 2016; Pot 2011)[2] we have consistently emphasized that WPI is not a goal in itself but

---

[2]In the past years, we ourselves have proposed four, highly overlapping, definitions, which evolved over time.

Definition 1: In 2011 Pot wrote: "Workplace innovation is defined as the implementation of new and combined interventions in the fields of work organisation, human resource management and supportive technologies. Workplace innovation is considered to be complementary to technological innovation. (…) by introducing workplace innovation, improvement of quality of working life (QWL) and organisational performance can be achieved simultaneously" (Pot 2011, pp. 404–405). The definition asserts that combined interventions targeting work organisation, HRM and supportive technology could result in both improved quality of performance and quality of working life. Pot's article was however less concerned with theory than with providing practical, empirical examples of WPI and its effects, and to convince policy makers of the promise of WPI.

Definition 2: Oeij et al. (2012) wrote: "Pot stresses new and combined interventions, by which "new" is understood as "innovation" and "combined" as a bundle of measures referring to work organisation, Human Resource Management and supportive technologies. In this respect, workplace concerns several elements of the organisation. This viewpoint is also shared by Totterdill (2010), who calls workplace innovations "collaboratively adopted changes in a company's work, organisational and human resource management practices that lead to improved operative/human performance and that also support other types of innovation". One can see that Totterdill underscores the participative role of people with the word "collaboratively". Totterdill sees WPI particularly as a process leading to the desired outcomes. It makes sense to say that workplace innovations have to do with organisation and people. The term innovation is taken up by Pot as "renewal" and by Totterdill as a "change" leading to improvements.

Pot's and Oeij et al's definitions of WPI take the point of view that a number of interventions together constitute workplace innovation, which aligns with the argumentation behind combining several HR-measures into 'HR-bundles' as in the theorising on 'high-performance work systems' (Boxall and Macky 2009); and 'dynamic capabilities' that provide unique competitive advantage for organisations (Helfat et al. 2007). These two streams can be traced back to the 'resource based view of the firm' (RBV).

Definition 3: Oeij and Vaas (2016, pp. 106–107) contend that "DC [Dynamic Capabilities] is a theory about economic strategic management, while HPWS (High Performance Work Systems—

rather that it is instrumental to achieve two objectives simultaneously, namely, improved organisational performance and improved quality of working life. As a consequence, the innovative capability of the organisation as a whole should be positively affected. Thus, WPI is a means to achieve these objectives. In turn, WPI is influenced by the strategic choices of the organisation's leadership, by their preferences for certain management 'philosophies' and organisational designs (e.g., centralisation vs. decentralisation), and the manner in which people are being deployed and involved in decision-making at various levels. Therefore, an integral approach to WPI is superior to interventions targeting separate problems, such as singling out people issues with HR-measures or technical issues with technological innovations. WPI thus refers, on the one hand, to a process of implementing interventions; and on the other hand, it refers to achieving results such as better performance and working life quality. In conclusion, we observe an evolution of the WPI definition within the field in that (1) WPI should benefit both organisations and people (as a goal); and (2) WPI is simultaneously seen as a process of innovative change and adaptation (as an activity), and (3) WPI as a theoretical conceptualisation of what is going on in organisations that apply WPI-like measures and interventions (as a way of framing and understanding).

We, as authors, have been stressing different elements in pointing out the mechanisms or interventions that explain why WPI would lead to such outcomes (see Footnote 2). Sometimes we were generic instead of specific when we spoke of 'interventions in the field of work organisation, human resource management and supportive technologies' or a 'bundle of measures' referring to those same categories. Sometimes we mentioned that improved labour relations and optimized use

---

(Footnote 2 continued)

see further on in the main text) is a theory about organisational processes acknowledging people as a strategic factor. Both variants of the RBV can be linked with the socio-technical systems theory, which states that changes in the technical system must be aligned with changes in the social system, to not only improve organisational performance, but to also simultaneously guarantee an acceptable quality of working life and better labour relations (De Sitter et al. 1997). Therefore, the roots of workplace innovation can be traced back to the socio-technical systems theory, as it underscores the urgency in aligning technological and workplace innovation". Moreover, technology is just as little a specific area of attention in the RBV as in WPI (in this definition); and that is why the RBV and workplace innovation seem to match well, say Oeij and Vaas. Workplace innovation is defined as a strategic renewal in organising and organisational behaviour; it is an organisational capability.

Definition 4: In our fourth and most recent definition, workplace innovation is: a developed and implemented practice or combination of practices that structurally (division of labour) and/or culturally (empowerment) enable employees to participate in organisational change and renewal, in order to improve quality of working life and the organisational performance (Oeij et al. 2015, p. 6; Howaldt et al. 2016, p. 2). Structural aspects refer to the production system and the design of organisational departments, teams and jobs, while cultural aspects point to behavioural phenomena like cooperation and communication and enabling certain behaviours, attitudes and motivations. This definition basically consists of the same ingredients as the earlier ones, but shows the relevance of both structural and cultural elements by implicitly including a root-cause approach that goes beyond combatting symptoms of organisational underperformance and lesser job quality (Oeij et al. 2015, pp. 12–14, 18–19, 61).

of human talent should be part of those measures. At other times, we stressed the importance of 'process' instead of solely paying attention to 'content', and suggested that employees should always play a participatory role (i.e., engagement, involvement) when it comes to designing and implementing interventions. And, last but not least, we advised practitioners to integrate structural change, like the design of an organisation and of jobs, with cultural change, such as leadership behaviour and honest, transparent communication. Whereas our thinking has evolved over time, three core ideas regarding WPI have clearly emerged:

- WPI has the combined objective of improved organisational performance and quality of working life;
- it stresses a participatory role for employees in the process;
- it underlines the need for an integral approach to WPI to achieve the objectives.

Therefore, the 'working definition' of WPI that we propose based on our previous work is:

> Workplace Innovation is an integral set of participative mechanisms for interventions relating structural (e.g., organisational design) and cultural aspects (e.g., leadership, coordination and organisational behaviour) of the organisation and its people with the objective to simultaneously improve the conditions for the performance (i.e., productivity, innovation, quality) and quality of working life (i.e., wellbeing at work, competence development, employee engagement).

In this working definition, 'participative mechanisms for interventions' are synonymous to employee engagement in decision-making processes and represent a precondition for WPI (Totterdill and Exton 2014)[3].

5.2 Other Theoretical Approaches

Several theoretical approaches, definitions and concepts can be related to the above working definition. In fact, there are many approaches that strive for similar objectives as WPI, but they use different terminologies, other variables and vary in their point of departure.

[3]Unless work processes are completely automated or robotised, workplaces are manned by people. The present economy is knowledge-based and more organised in networks and forms of interactive cooperation. The place of people is less one of a functionalist link in a standardised process (as in the times of mass production), but all the more a relational one within processes that are less standardised and static but often evolving and unique. As a consequence, workplaces require employees to be more proactive, responsive and interactive. Employees are requested to deploy their brains instead of just their hands (Drucker 2003). In such changed employment relationships, given that relations at the same time become more electronical and virtual due to IT (Sennett 1998), employee engagement is crucial for success, which means that producers of goods and services—in both modern and traditional industries—should take human relationships seriously (Gittell 2016; Herriot 2001).

Figure 5.1 is built up around the notion that agents, namely researchers, policy makers and practitioners, have their own unique ideas regarding WPI. The figure tries to express that each agent has a preferred point of departure when starting an intervention to change the primary or supportive process of an organisation. Some start with people, others with systems, yet others with the process itself. These starting points are (sometimes) implicit for most agents, and reflect their choices regarding strategy, management regime and affiliated political regime. Together these preferred starting points and strategic choices have implications for how WPI is applied as a means to an end (the arrow points to the intended end in this case). That end is not always the simultaneous goal of better organisational performance and better quality of working life. For some, WPI is a manner in which to design organisations (Chap. 17), but for others it can be an organisational change approach (Chap. 18), or a combination of the two (Chap. 20).

We should acknowledge that this figure is a simplified model and that the different elements can influence each other. Our goal is to show that the discussed theoretical approaches are rooted in those preferences (1, 2, 3 4) and how each starting point can support WPI. Note that, although, we obviously favour a genuine role for employees to play a part in that process, we do not intend to prescribe how WPI should be developed and implemented.

When agents strive for the innovation of workplaces some prefer to concentrate on people (1, human relations and communication), others on systems (2, work organisation, technology, rules) and, again others on change processes and interventions (3). The content of the three points of departure depends on strategic business choices, as well as chosen management regimes (the degree of centralisation as in 'command & control' or 'participation & trust'), for which the room to manoeuvre is dependent upon the current system of a country's political economy (e.g., 'free market economy' or 'social market economy') (4). Most authors represent a point of interest as their perspective, and by doing so, they express their preferred choices and solutions. While there are undoubtedly more good options than just 'one way of organising', some options might be simply suboptimal as compared to an integrative viewpoint. Our point of view is, that, while each perspective has its merits, integrating them provides more value.

We will discuss each of these points of departure with brief examples from the literature. The purpose is to discuss how these approaches support WPI, which fields they integrate and what we can learn from each approach for WPI.

We will first discuss these theoretical approaches. In the subsequent section we will relate them to WPI and how they can support WPI.

1. ***Human relations and communication***

A. *Relational Coordination*

Relational Coordination (RC) proposes that highly interdependent work is most effectively coordinated through relationships that are characterized by shared goals, shared knowledge and mutual respect, and that are supported by frequent, timely, accurate and problem-solving communication (Gittell 2016). Research shows an

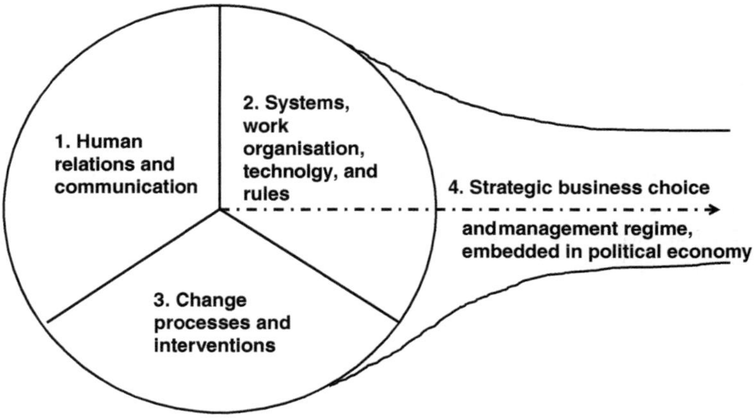

Fig. 5.1 Categorizing approaches related to WPI

association between RC and a wide range of positive performance outcomes for organisations and employees. The design of work systems can support or undermine RC. RC is first of all communicating and relating for the purpose of task integration—a powerful driver of performance when work is interdependent, uncertain and time-constrained. RCs key concept is 'team work' (Gittell 2016). RC relates to WPI through employee engagement.

B. *Job Demands-Resource model*

The Job Demands-Resources (JD-R) model (Bakker and Demerouti 2014; Demerouti et al. 2001) is used to predict employee burnout and engagement, and consequently, organisational performance. The JD-R model assumes that employee well-being is explained by job demands and job resources. Research has provided evidence for the existence of two simultaneous processes: the health process and the motivational process. High job demands exhaust employees' mental and physical resources and therefore lead to the depletion of energy and to health problems (i.e., the health process). In contrast, job resources foster employee engagement and extra-role performance (i.e., the motivational process). Several studies have shown that job resources may buffer the impact of job demands on stress reactions. In addition, research has confirmed that job resources have motivational potential particularly when job demands are high. While the JD-R model and the Job Demands- Control-Support model of Karasek (1979; see below) are both concerned with individual well-being, the latter plays a larger role in the design of jobs and organisations (mainly at the team level as in modern sociotechnology), whereas the first plays a larger role in the management of burn-out, stress and engagement (mainly at the individual level). The JD-R model includes more subjective or personal job resources than Karasek's model, whose focus is on objective job characteristics as job resources. More recently the JD-R model has been connected to job crafting, i.e., changes employees may make regarding their job demands and job resources, which, in fact, combines individual and job characteristics in changing one's work (Bakker and Demerouti 2014; but there is also attention for collective job crafting which is relevant for teams, see Oldham and Fried 2016). This underlines the relationship of the JD-R model with employee engagement, a central aspect of WPI.

2. ***Systems, work organisation, technology, rules***

A. *Modern Socio-Technical Systems Design*

The aim of modern socio-technical systems design (M-STSD or modern sociotechnology[4]) is to simultaneously achieve improved organisational performance (e.g., quality, flexibility, effectiveness, and efficiency), quality of working life, and better labour relations (see also Chap. 17). One core idea of M-STSD, related to job level, is the balance between 'control requirements' (quantitative and qualitative job demands) and 'control capacity' (job control or autonomy). Control capacity in jobs is improved by decentralizing decision latitude as much as possible, which results in rich and complex jobs. Therefore, De Sitter's motto was: "from complex organisations with simple jobs to simple organisations with complex jobs" (De Sitter et al. 1997). To improve the quality of jobs, M-STSD has integrated the 'job demands-control-model' (Karasek 1979) into its theory, which was helpful to design 'active jobs' that ensure a balance of challenging job demands and required job autonomy (see below). The link with WPI is that M-STSD aims at both better jobs and performance.

B. *Job Demands-Control-Support model*

The job demands-control (JDC) model (Karasek 1979; Karasek and Theorell 1990; Theorell and Karasek 1996) predicts that high job demands combined with high job control foster motivation and learning and result in 'active jobs'. High job demands combined with low job control engender a risk of work-related stress and result in 'high-strain jobs'. The JDC-model was extended with the social support dimension (S: support of colleagues and supervisor) and with innovative and productive work behaviour (Karasek and Theorell 1990). There is empirical evidence supporting the JDCS-model (Häusser et al. 2010). Especially the presence of job control has been associated with positive outcomes, such as learning, job engagement, well-being, and organisational commitment (see Chap. 8). Based on the theory underlying the model, design rules can be derived concerning 'active jobs'. The promises of WPI regarding competence development, well-being at work and prevention of work-related stress are partly based on the JDCS model.

---

[4]The M in M-STSD points to modern; M-STSD is 'modern' because it is partly based on the longstanding history of socio-technical systems design theory dating back to the 1950s (De Sitter et al. 1997; Van Eijnatten 1993; Van Eijnatten and Van der Zwaan 1998).

C. *Lean Management*

Lean production (also Lean Management—see also Chap. 13 for a Lean approach applied to improving the psycho-social work environment) is a rather similar design theory to modern sociotechnology (M-STSD). The original idea was that reduced 'waste' was to be reinvested in the job quality and competences of employees, making lean production a win-win for the organisation, the employees and the customers (Womack and Jones 2005). Few research projects indicate that this win-win was actually achieved (e.g., Landsbergis et al. 1999). This seems to only be the case if employees participate in the interventions from the beginning; if not, working conditions (notably standardization, high workloads and limited autonomy) are deteriorating (De Menezes et al. 2010; Koukoulaki 2014). Lean Management, thus, can only contribute to WPI if quality of working life is one of the objectives and employee participation is part of the process.

3. ***Change processes and interventions***

A. *Fifth Element Model*

The Fifth Element Model (Totterdill and Exton 2014) refers to the chemistry of integrating four elements: 'work organisation' (first element), 'structures and systems' (second element), 'learning and reflection' (third element) and 'workplace partnership' (fourth element), that should result in increased customer focus, employee engagement, an enabling culture, resilience, positive employment relations, and enterprising behaviour. When these four elements are integrated, the approach, which is more practice-based than most of the others, will culminate in high performance, good work and sustainable organisations (Totterdill and Exton 2014). This approach to *"workplace innovation describes the participatory and inclusive nature of innovations that embed workplace practices grounded in continuing reflection, learning and improvements in the way in which organisations manage their employees, organise work and deploy technologies"* (Pot et al. 2016, p. 15).

B. *Employee-Driven Innovation*

Employee-Driven Innovation (EDI) (Høyrup 2012, see also Chap. 19) is based on the fundamental belief that all employees have the potential to contribute to innovation and growth in a company. Unlike innovation that is determined and driven from the top of an organisation, EDI is a bottom-up process and an experience/knowledge based practice tied to employees' daily challenges. The philosophy of employee-driven innovation is typically based on individual (direct) participation and the assumption of a community of interests among the stakeholders in the company. The Norwegian Confederation of Trade Unions goes a step further and also sees EDI as a way to democratize innovation processes. They argue that EDI should be based on both direct and indirect participation (Kallevig 2012). WPI advocates employee participation as well.

C. *Democratic Dialogue*

Democratic Dialogue is firmly related to worker participation. Dialogue starts from the point of view of communication. *"Because a given reality can be seen and interpreted in different ways, with no supreme court to decide which way is the right one, there is a need for a process that can make people, without the force of an unequivocal reality, adjust their concepts in relation to each other to a degree sufficient to make joint action possible"* (Gustavsen 2016, p. 193). A set of dialogue criteria should ensure that all participants have equal opportunities of influencing the dialogue. Based on egalitarian, innovative and trust-promoting patterns of communication, additional criteria are the degree of trust between the actors involved, their willingness and ability to cooperate and ensure participation in developing new forms of organisation (Gustavsen 2015). Democratic dialogue contributes to WPI in particular by supporting a smoother process for workplace interventions.

D. *Labour Process Approach*

The Labour Process Approach (Knights and Willmott 1990), tracing its origins back to Marxism, views capitalism as a system of unbalanced power relations and explains how employees are controlled by management through deskilling, polarization of job skills, low wages and a minimum of social security. Proponents of this view strove to improve employee working conditions, remuneration and quality of jobs by collective actions such as strikes. Although we consider this approach as too radical because many employers are aware of the need of investing in employees, the Labour Process Approach informs WPI by pointing out that power relations are ever present and co-creation is only possible in social systems and organisations in which there is a tradition of institutionalized negotiation and a certain level of trust.

4. ***Strategic business choices, management regimes***

A. *Resource Based View/Dynamic Capabilities/High Performance Work Systems/Knowledge-based Capital*

These approaches start from a systemic perspective and they all focus not only on the competitiveness of products and services but also on internal resources for competitive advantage, such as management skills, work organisation, knowledge and competences. Competitive advantage can be achieved when these resources improve efficiency and efficacy and when they are rare or difficult to copy. The Resource Based View and Dynamic Capabilities approach (Eisenhardt and Martin 2000) take necessary adaptations to changes in the environment into account, namely "the capacity of an organization to purposefully create, extend, or modify its resource base" (Helfat et al. 2007). In turn, the OECD looks at all the assets needed to support firm growth. They acknowledge that company-accounting does undervalue what they call 'knowledge-based capital' (KBC) (OECD 2012). Investing into KBC, such as, for instance, in 'managing human resources, reinforces the innovation capabilities of a firm. Other resource-based approaches underline the importance of competences and are usually framed under 'High Performance Work Systems (HPWS)' (Appelbaum et al. 2011; Boxall and Macky 2009). Many HPWS approaches aim especially at improving economic performance, and to a much lesser extent at the quality of jobs. They simply assume that quality of work associates with high productivity, but they do not always study job autonomy. Within the family of HPWS, however, one member, 'High Involvement Work Systems', stresses the importance of job autonomy and employee engagement. According to recent research by Boxall and Macky (2014, p. 1) "*Higher involvement is a key factor predicting higher job satisfaction and better work–life balance*", which supports workplace innovation. These strategic and management regime approaches can align with WPI only when they take quality of working life into account, that is, when economic goals are not their exclusive motive to innovate.

B. *Managerial Technology*

The managerial technology approach (Bloom and Van Reenen 2010) states that different combinations of 'managerial capabilities'—or management practices and tools—should be aligned to improve economic performance. Management makes choices about these measures and organisation is not the result of 'pure market forces'. Bloom and Van Reenen go to great lengths to show how performance, also innovation performance, requires a clear strategy to combine measures and resources. The focus is on the shift from tangible, hard technological innovations to intangible, managerial practices that can influence productivity differences. In other words, management can actively choose to take WPI interventions as a point of departure for innovation, if they believe that employee engagement is crucial for performance and productivity improvement. Inspired by this perspective, organisation researchers (Dhondt et al. 2013) have developed the 'capability maturity model of workplace innovation' that identifies 37 capabilities to manage control, human resources, the production process, and communication and information. These capabilities can be seen as components of WPI for which organisations can develop interventions and measures and thus improve performance and quality of working life.

C. *High Reliability Organisations*

High Reliability Organizations (HROs) are operating "*under very trying conditions all the time and yet manage to have fewer than their fair share of accidents*" (Weick and Sutcliffe 2007, pp. 17–18; see also Chap. 8). They include power grid dispatching centres, air traffic control systems, nuclear aircraft carriers, hospital emergency departments, and accident investigation teams. Although the core of the process of HROs is safety, their ideas concerning highly concerted team work and linking organisational goals to shop floor operation starts seeping through to non-HROs (Oeij 2017). Whereas HROs have a high degree of standardization and formalization to prevent mistakes and disasters, at the same time, they do not fully rely on those rules because they know that real-life situations are unique and never the same. Hence, they are continuously critically examining and revising those same standards and rules. Therefore, HROs have designed operational jobs without a rigid division of managing and executing tasks. "*In HROs, this separation between thinking and doing, conception and execution, is broken down. The individuals who execute the routines are also involved in the critical examination, adjustment and improvement of them*" (Christis 2010, p. 44). HROs therefore have 'active jobs', and such jobs are a hallmark of WPI.

5.3 Integrating the Perspectives

The working definition of WPI implies taking an integral approach to change as organisations consist of parts that influence one another. WPI is not a goal in itself, but a means to contribute to the desired goals of improved organisational performance and quality of working life. Based on this integral approach viewpoint we discuss the approaches and perspectives above (Table 5.1; Fig. 5.1) in light of how they could strengthen the goals of WPI.

The approaches mentioned under 'Human relations and communication' (dimension 1) stress the role of humane and personal relations to improve performance and job quality. They acknowledge that business success is impossible without dignified human interactions in the world of work. This strengthens the element of employee engagement and involvement (Job Demands-Resource model) of WPI and implies that those human features (Human/Social Capability approach) and relations (Relational Coordination theory) should be part of interventions developed from the WPI integral perspective.

The approaches Modern Sociotechnology and Lean Management mentioned under 'systems work organisation, technology, rules' (dimension 2) are systemic approaches that acknowledge that the whole is more than the sum of its parts. Both

Table 5.1 Approaches that support WPI from different fields

Field	Approaches
1. Human relations and communication	A. Relational Coordination B. Job Demands-Resource model
2. Systems, work organisation, technology, rules	A. Modern Sociotechnical Systems Design B. Job Demands-Control-Support model C. Lean Management
3. Change processes and interventions	A. Fifth Element B. Employee-Driven Innovation C. Democratic Dialogue D. Labour Process Approach
4. Strategic business choices, management regimes	A. Resource Based View/Dynamic Capabilities/High Performance Work Systems/Knowledge based capital B. Managerial Technology C. High Reliability Organisations

M-STSD and LM underline the imperative role of team work for the quality of performance. For M-STSD, meaningful work for employees is an implicit assumption, built into how core work processes are designed (maximising job complexity and job control), which fits with the WPI body of thought. In Lean Management this is somewhat ambiguous, as its Japanese origin of the Toyota-system valued the collective team performance higher than individual job satisfaction. The Japanese valued collective team input for continuous improvement in so-called quality circles for the sake of organisational performance. Therefore, the Job Demands-Control-Support model is crucial for WPI as it provides hands-on advice on how to design 'active' jobs with learning opportunities that enable employee engagement and involvement.

The approaches Fifth Element model, Employee-Driven Innovation and Democratic Dialogue, mentioned under 'change processes and interventions' (dimension 3) focus on employees as active participants with voice in the process of change and renewal. Participation ensures commitment and is beneficial for organisational performance. Hence it is no wonder that these approaches show affinity with dimension 1, human relations and communication, which is more theory-based than dimension 3. For many authors within this dimension 3, who are more practice-based, WPI hinges on the active, even political role that employees play in the work process.

The approaches Resource Based View, Dynamic Capabilities, High Performance Work Systems and Managerial Technology mentioned under 'strategic business choices and management regimes' (dimension 4) all propose that there is choice regarding the strategy, management models ('philosophies') and political economic motivation. These streams are aligned with WPI if these choices not only benefit the business, but also the interests of employees and customers. To best serve customers, employees should have excellent skills and organisational facilities, which, in turn, improve the quality of working life. High Reliability Organisations are a special branch because their strategic choice represents a focus on safety and reliability. While being largely rule-based, hence showing affinity with dimension 2 on systems and rules, they explicitly choose to design teams with broad tasks and create organisational slack, so as to promote learning organisations that are equipped to deal with unexpected events. In acknowledging the dependence of organisational success on human efforts, motivation and competencies all approaches have a clear relationship with WPI goals. In these approaches the internal resources are being optimised to improve organisational operational excellence and innovative capability. But we only regard them as being relevant to WPI when employee interests are taken seriously.

The field of WPI and the community of researchers, practitioners and policy makers might need an integral perspective to change, competitiveness and innovation that includes all four dimensions, for which relating the mentioned approaches can be helpful. Figure 5.1 is informative for a needs analysis that would provide a starting point for WPI interventions. It is further useful to look for learning opportunities from other approaches. For each of the four dimensions we have tried to assess how it strengthens WPI as a means to improve organisational performance and quality of working life.

We, however, have no intention to be conclusive at this point, because for every situation each agent has unique possibilities in terms of how to go about it. We propose that future agents make an integral needs analysis of all four dimensions in Fig. 5.1 to assess a situation. But we would not suggest that all the different approaches mentioned are necessary at the same time. Within and across the four dimensions some approaches can even be interchangeable. If one, for example uses M-STSD, one might not need Lean Management; or when agents apply job crafting by using the JDC(s)-model they may not need to bother too much about Democratic Dialogue as that has been inherently taken care of. The key check is whether employees are engaged and involved in the developed and implemented changes (as concluded in Oeij et al. 2015).

One of the initiatives that has been making a start in expanding the WPI-body of thought is Total Workplace Innovation (TWIN) (discussed in Chap. 17). This approach takes M-STSD as a point of departure and discusses perspectives from business administration, operations management, HR, and ICT that could become part of the concept of Total Workplace Innovation. While this TWIN-model requires further elaboration, the promising point of departure is its strong conviction that employees need to be involved for better organisational performance and better quality jobs to emerge.

5.4 Conclusion and Discussion

This chapter has proposed a working definition for WPI and has presented approaches that support WPI. WPI-approaches in research, practice and policy acknowledge the indispensability of the people factor for organisational performance and innovation capability. The presented approaches are in line with WPI-goals, provided that employee interests are included, and that efficiency-driven interventions are not dominant. Purely Tayloristic and bureaucratic views, full top-down management strategies and an extensive division of labour and flexibilisation of contracts and payment systems, for example, are serious threats to the WPI-body of thought and its endeavours (Dhondt and Van Hootegem 2015). At the same time, WPI is no straitjacket. The presented approaches can all be helpful as long as employee employability and empowerment is serious and not rhetorical (Herriot 2001). Even though we have not presented a complete overview of approaches that we regard as being related to WPI, we believe that our overview sufficiently underpins our argument.

Part of the mission of this book is to contribute to how to better integrate these approaches into a WPI-theory and provide a set of hands-on tools (see also Part IV in this volume). Future research could support this by including WPI-elements in empirical studies, for example in the European Working Conditions Survey and the European Company Survey; both of Eurofound (Chap. 16). However, learning from interventions implemented in practice is welcomed as well, as these can be evaluated on their WPI-characteristics. Last but not least it remains imperative to try to assess the direct and indirect effects of WPI on three types of outcomes: organisational performance, quality of working life, and innovative capabilities of firms and innovative abilities of people.

Acknowledgements We would like to thank Frank D. Pot for our fruitful discussions and for commenting on earlier drafts. However, the authors are responsible for the final text.

References

Appelbaum, E., Hoffer Gittell, J., & Leana, C. (2011, April). *High-performance work practices and sustainable economic growth.* Washington, DC: The Center for Economic and Policy Research (CEPR).

Bakker, A. B., & Demerouti, E. (2014). Job demands—Resources theory. In P. Y. Chen & C. L. Cooper (Eds.), *Work and wellbeing: A complete reference guide* (Vol. III, pp. 1–28). Chichester: Wiley.

Bloom, N., & van Reenen, J. (2010). Why do management practices differ across firms and countries? *Journal of Economic Perspectives, 24*(1), 203–224.

Boxall, P., & Macky, K. (2009). Research and theory on high-performance work systems: Progressing the high-involvement stream. *Human Resource Management Journal, 19,* 3–23. doi:10.1111/j.1748-8583.2008.00082.x

Boxall, P., & Macky, K. (2014). High-involvement work processes, work intensification and employee well-being. Work Employment Society. doi:10.1177/0950017013512714 published online June 12, 2014.

Christis, J. H. P. (2010). Organization and job design: What is smart organizing? In H. A. M. van Lieshout, L. Polstra, J. H. P. Christis, & B. J. M. Emans (Eds.), *Management of labour. Societal and managerial perspectives* (pp. 39–71). Groningen: Hanzehogeschool Groningen University of Applied Sciences.

De Menezes, L., Wood, S., & Gelade, G. (2010). The integration of human resource and operation management practices and its link with performance: A longitudinal latent class study. *Journal of Operations Management, 28*(6), 455–471.

De Sitter, L. U., den Hertog, J. F., & Dankbaar, B. (1997). From complex organizations with simple jobs to simple organizations with complex jobs. *Human Relations, 50*(6), 497–534.

Demerouti, E., Bakker, A. B., Nachreiner, F., & Schaufeli, W. B. (2001). The job demands resources model of burnout. *Journal of Applied Psychology, 86*(3), 499–512.

Dhondt, S., Oeij, P. R. A., van der Meulen, F. A., Preenen, T. Y. P., Vergeer, R., van der Kleij, R., et al. (2013). *Platform workplace innovation: Workplace innovation in a capability maturity framework.* Leiden: TNO.

Dhondt, S., & Van Hootegem, G. (2015). Reshaping workplaces: Workplace innovation as designed by scientists and practitioners. *European Journal of Workplace Innovation, 1*(1), 17–24.

Drucker, P. F. (2003). The productivity of the knowledge worker. In P. F. Drucker (Ed.), *A functioning society. Selections form sixty-five years of writing on community, society, and polity* (pp. 169–178). New Brunswick, London: Transaction Publishers.

Eeckelaert, L., Dhondt, S., Oeij, P., Pot, F., Nicolescu, G. I., Webster, J., et al. (2012). *Review of workplace innovation and its relation with occupational safety and health: Literature review.* Bilbao: European Agency for Safety and Health at Work.

Eisenhardt, K. M., & Martin, J. A. (2000). Dynamic capabilities: What are they? *Strategic Management Journal, 21*(10–11), 1105–1121.

Gittell, J. H. (2016). *Transforming relationships for high performance. The power of relational coordination*. Stanford, California: Stanford University Press.

Gustavsen, B. (2015). Practical discourse and the notion of democratic dialogue. *European Journal of Workplace Innovation, 1*(1), 25–35.

Gustavsen, B. (2016). Democratic dialogue. In B. J. Mohr & P. Van Amelsvoort (Eds.), *Co-creating humane and innovative organizations. Evolutions in the practice of socio-technical system design* (pp. 186–200). Portland ME: Global STS-D Network.

Häusser, J. A., Mojzisch, A., Niesel, M., & Schulz-Hardt, S. (2010). Ten years on: A review of recent research on the Job Demand-Control (-Support) model and psychological well-being. *Work & Stress, 24*(1), 1–35.

Helfat, C., Finkelstein, S., Mitchell, W., Peteraf, M., Singh, H., Teece, D., et al. (2007). *Dynamic capabilities: Understanding strategic change in organizations*. Malden, MA: Blackwell.

Herriot, P. (2001). *The employment relationship: A psychological perspective*. Hove: Routledge.

Howaldt, J., Oeij, P. R. A., Dhondt, S., & Fruytier, B. (2016). Workplace innovation and social innovation: An introduction. *World Review of Entrepreneurship, Management and Sustainable Development, 12*(1), 1–12.

Høyrup, S. (2012). Employee-driven innovation: A new phenomenon, concept and mode of innovation. In S. Høyrup, C. Hasse, M. Bonnafous-Boucher, K. Møller, & M. Lotz (Eds.), *Employee-driven innovation: A new approach* (pp. 3–33). Basingstoke and New York: Palgrave Macmillan.

Kallevig, A. (2012). *Medarbeiderdrevet innovasjon*. http://www.arbeidslivet.no/Arbeid1/Naringspolitikk/Medarbeiderdrevet-innovasjon/. Accessed October 15, 2012 (in Norwegian).

Karasek, R. A. (1979). Job demands, job decision latitude, and mental strain: Implications for job redesign. *Administrative science quarterly, 24*, 285–308.

Karasek, R. A., & Theorell, T. (1990). *Healthy work; Stress, productivity and the reconstruction of working life*. New York: Basic Books.

Knights, D., & Willmott, H. (1990). *Labour Process Theory*. London: Macmillan.

Koukoulaki, T. (2014). The impact of lean production on musculoskeletal and psychosocial risks: An examination of sociotechnical trends over 20 years. *Applied Ergonomics, 45*(2), 198–212.

Landsbergis, P. A., Cahill, J., & Schnall, P. (1999). The impact of lean production and related new systems of work organization on worker health. *Journal of Occupational Health Psychology, 4*(2), 108–130.

OECD. (2012). *New sources of growth, knowledge-based capital driving investment and productivity in the 21st century*. Paris: OECD.

Oeij, P. R. A. (2017). *The resilient innovation team. A study of teams coping with critical incidents during innovation projects*. Ph.D. dissertation. Open University of The Netherlands.

Oeij, P. R. A., Dhondt, S., Kraan, K. O., Vergeer, R., & Pot, F. D. (2012). Workplace innovation and its relations with organisational performance and employee commitment. *LLinE, Lifelong Learning in Europe, 4* (no page numbers). http://www.elmmagazine.eu/articles/workplace-innovation-and-its-relations-with-organisational-performance-and-employee-commitment/

Oeij, P. R. A., & Vaas, F. (2016). Effect of workplace innovation on organisational performance and sickness absence. *World Review of Entrepreneurship, Management and Sustainable Development, 12*(1), 101–129.

Oeij, P., Žiauberytė-Jakštienė, R., Dhondt, S., Corral, A., Totterdill, P., & Preenen, P. (2015). *Workplace innovation in European companies*. Study commissioned by Eurofound. Luxemburg: Office for Official Publications of the European Communities.

Oldham, G. R., & Fried, Y. (2016). Job design research and theory: Past, present and future. *Organizational Behavior and Human Decision Processes, 136*, 20–35.

Pot, F. D. (2011). Workplace innovation for better jobs and performance. *International Journal of Productivity and Performance Management, 60*(4), 404–415.

Pot, F., Totterdill, P., & Dhondt, S. (2016). Workplace innovation: European policy and theoretical foundation. *World Review of Entrepreneurship, Management and Sustainable Development, 12*(1), 13–32.

Sennett, R. (1998). *The corrosion of character. The personal consequences of work in the New Capitalism*. New York, London: Norton.

Theorell, T., & Karasek, R. A. (1996). Current issues relating to psychosocial job strain and cardiovascular disease research. *Journal of Occupational Health Psychology, 1*(1), 9–26.

Totterdill, P. (2010). *Workplace innovation: Europe 2020's missing dimension*. Report of a workshop hosted by DG Employment, Social Affairs and Equal Opportunities, June 23, 2010, UKWON, Nottingham.

Totterdill, P., & Exton, R. (2014). Defining workplace innovation: The fifth element. *Strategic Direction, 30*(9), 12–16.

Van Eijnatten, F. M. (1993). *The paradigm that changed the work place*. Assen: Van Gorcum.

Van Eijnatten, F. M., & van Der Zwaan, A. H. (1998). The Dutch IOR approach to organizational design: An alternative to business process re-engineering? *Human Relations, 51*(3), 289–318.

Van Hootegem, G. (2016). Changing the nature of work: Toward Total Workplace Innovation. In B. J. Mohr & P. Van Amelsvoort (Eds.), *Co-creating humane and innovative organizations. Evolutions in the practice of socio-technical system design* (pp. 326–343). Portland ME: Global STS-D Network.

Weick, K. E. & Sutcliffe, K. M. (2007). *Managing the unexpected: Resilient performance in an age of uncertainty*, (2nd Edn.; 1st Edn. 2001), San Francisco: Jossey-Bass.

Womack, J. P., & Jones, D. T. (2005). *Lean solutions: How companies and customers can create value and wealth together*. London: Simon & Schuster.

Author Biographies

Peter R.A. Oeij is a senior research scientist & consultant affiliated with TNO, The Netherlands Organisation for Applied Scientific Research (Leiden, The Netherlands). His field of work is innovation management, workplace innovation, social innovation and team dynamics.

Steven Dhondt is a visiting professor, Chair Social Innovation, at KU Leuven and senior research scientist at TNO, The Netherlands Organisation for Applied Scientific Research (Leiden, The Netherlands). Current research: workplace innovation, digitalisation and work.

Chapter 6
Evidence of Workplace Innovation from Organisational and Economic Studies

Steven Dhondt, Lander Vermeerbergen and Geert Van Hootegem

6.1 Introduction

Workplace Innovation (WPI) researchers start from the idea that "people are our most valuable assets" (Alagaraja 2013, p. 136). Work processes and HR-practices in firms should be organised in such a way that employees profit from a higher quality of working life and organisations have a superior performance. In this sense, WPI can be categorised as a 'best practice approach' (Guest 1997, p. 273): choosing for WPI is the best thing to do for management (Mohr and Van Amelsvoort 2016). Several authors claim that there is empirical support for the link between WPI and performance at the organisational and employee level (Appelbaum et al. 2011; Van Hootegem 2016; Totterdill et al. 2016). The sources that these authors cite do not provide definitive scientific answers for the hypothesis. This is exemplified by, for instance, Della et al. (2013, p. 2584) when discussing High Performance Work Systems as an example of WPI: "(…) it should be noted that while the positive effects for enterprises of adopting HPWS have been clearly demonstrated (…), evidence of the beneficial effects for workers is more controversial (…). [There is] a tendency on the part of the advocates of HPWS [i.e. WPI] to emphasise their positive results and to gloss over the negative ones (…)."

S. Dhondt (✉)
TNO, The Netherlands Organisation for Applied Scientific Research, Leiden, The Netherlands
e-mail: steven.dhondt@tno.nl

S. Dhondt
KU Leuven, Leuven, Belgium

L. Vermeerbergen · G. Van Hootegem
Centre for Sociological Research, KU Leuven, Leuven, Belgium
e-mail: lander.vermeerbergen@kuleuven.be

G. Van Hootegem
e-mail: geert.vanhootegem@kuleuven.be

P.R.A. Oeij et al. (eds.), *Workplace Innovation*, Aligning Perspectives on Health, Safety and Well-Being, DOI 10.1007/978-3-319-56333-6_6

WPI is a broad concept which includes different organisational measures that are, together, leading to superior organisational performance and higher quality of working life. These measures or organisational practices deliver synergetic results (Subramony 2009). In econometrics, this is called 'complementarity'. Theory on this topic predicts that specific practices cluster. An organisation with a set of the practices will also have similar other measures (Brynjolfsson and Milgrom 2013, p. 11). The question then is which measures should be selected, and how can they be brought together, to deliver superior performance?

The WPI-literature is still far away from giving practitioners precise indications on which measures increase performance and quality of working life. Moreover, the WPI-literature provides very convincing statements to invest in WPI, but the scientific basis for these statements is unclear (Pot 2011). For example, WPI receives much attention in The Netherlands because the Erasmus Competition and Innovation Monitor (ECIM) 2005–2010 stated that technological innovation only explained 25% of innovation successes, whereas workplace innovation explained about 75% (Volberda et al. 2006, 2011). It is, however, difficult to assess how these authors measured these percentages. Such results require sound and open research on WPI. For further diffusion of WPI, studies on WPI need to rely on transparent methods. This chapter investigates which approaches could help to underpin the stated hypotheses.

Next to this first question, we examine how policy making can and should support WPI. In our discussion with policy makers, mainly from the economic domain, we are confronted with the reasoning that if the tenets of WPI are confirmed in research, this should then be sufficient motivation for companies to choose for WPI. Organisations should, rationally, organise their work in a WPI way (Van Reenen 2011). We need to clarify why companies do not always do what is good for them, and what the role is of policy makers. In this chapter, we outline that this rational argument may be too simplistic. This is also exemplified by, for instance, Jongkind et al. (2004): "more than economic variables alone determine producer's actions".

In this chapter, we show the state of art on research on WPI as an organisational innovation. We also outline why we think that circumstantial evidence is not an issue for this topic. We elaborate the following issues. First, we discuss the definition of WPI. This is also dealt with in Chap. 5. Second, we give an overview of current research on WPI. The aim is to identify which research approaches are the most promising for WPI. Third, we question why not all companies go for the best organisational model. If an organisational form such as WPI delivers such results, why can we not find more organisations that implement WPI?

6.2 Conceptualising WPI

In this section we approach the concept of WPI from three major scientific disciplines. First, organisational sociology or psychology tries to explain behaviour of employees in organisations. The most progress in recent years stems from the fact that bundles of (HR-)practices are more important than single (HR-)measures (Subramony 2009; Vermeerbergen et al. 2016). But which combination of HR-practices should be bundled? Volberda et al. (2011) mention, for instance, five capabilities that should be aligned for superior performance: (1) dynamic management, (2) flexible organisation, (3) smart work, (4) external networks and knowledge alliances, and (5) external stakeholders and strategic regulation.

Second, a relatively new discipline is insider econometrics (Gibbons and Robert 2013). Gibbons and Robert bring economic thinking and organisational econometrics to the micro-level of the firm. This discipline focuses on three questions (Ichniowski and Shaw 2013, p. 305):

1. Does a new management practice raise productivity?
2. Why does the new management practice raise productivity?
3. Why is the new management practice adopted?

Insider econometrics is 'neutral' about which measures managers should implement. Most of the scholars only test hypotheses from organisational sociologists or psychologists (Baron and Kreps 2013). But hidden under this 'neutral position' are authors' beliefs about the relationship between organisational measures and organisational contexts. These assumptions are discussed by Van Reenen (2011), who juxtaposes two organisational economics theories that prove that organisational models can deliver superior economic performance. The first theory is what he calls 'design theory': capabilities of organisations are directly impacted by the economic environment. The environment directly selects the 'optimal organisational form'. Differences between organisations can be explained by different environments across organisations. The second theory is 'managerial technology theory': some aspects of management are considered as a technology or "best practice". Adopting organisational best practices would improve productivity in a typical firm. But, of course, the environment is helpful in selecting successful organisational practices. Bloom and Van Reenen (2010a) identify eighteen basic management practices in three areas, which they believed to be associated to better performance: Monitoring—Targets—Incentives. Bloom and his team have even developed an experiment to show what happens to the performance of organisations if these capabilities are improved. Their conclusion is that improving these capabilities gives firms a 20% productivity rise within 40 weeks (Bloom et al. 2013). However, the room for manoeuvring between measures is still extremely wide, and, there seems to be within this 'managerial technology theory' a dominant insistence on (top-down) management. It is unclear how thinking on teams and workers' autonomy fit with this theory.

A third discipline is engineering or operations research. Modern Sociotechnical thinking (MST) (Mohr and Van Amelsvoort 2016) identifies on the one hand structural dimensions of organisations (production organisation; control structure). The way how work is divided between organisational units is in this dimension at stake. MST identifies on the other hand HR-dimensions, which needs to be aligned with the first dimension. Many organisational sociologists, psychologists and econometrists lack this first dimension (see also Chaps. 5 and 20 in this book).

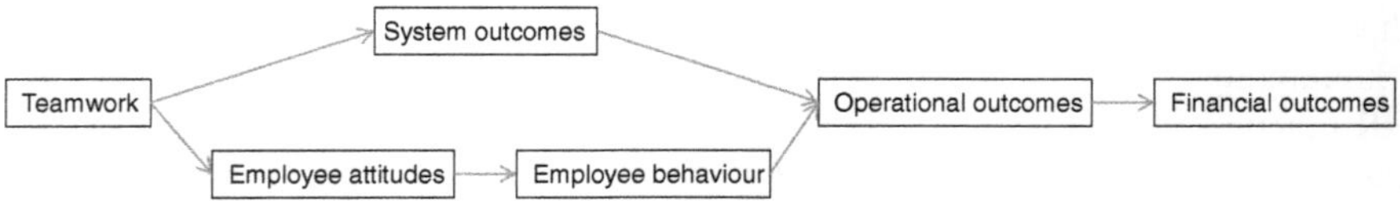

Fig. 6.1 Teamwork and financial performance. *Source* Delarue et al. (2008)

The importance of aligning both dimensions is shown by Delarue et al. (2008), who show the impact of teamwork on financial performance of firms through operational outcomes (e.g. productivity, quality or innovation). Operational outcomes are influenced by employees' attitudes (e.g. satisfaction or involvement) and employee behaviour (e.g. turnover or absenteeism) but also by system outcomes (e.g. reduction of losses or throughput time). Both paths may increase or decrease the operational outcomes, and are thus as important as each other for a higher financial performance. It is therefore also possible that both pathways influence each other. In other words, a change in system outcomes can, for instance, interact with employees' attitudes and then lead to lower operational performances. Figure 6.1 summarises the study findings graphically, the impact of both pathways on each other is not shown in this figure.

From this section, we conclude that the concept of WPI is a broad set of measures, not only pertaining to HR-practices, but also to developing the organisation structure. Although managers need to develop both elements (structures, HR-practices) to create superior organisations, WPI-thinkers insist that also employees need to develop solutions (Høyrup et al. 2012). This section showed that the 'independent' side of the organisation is far from clear. Therefore, a clear set of WPI measures to improve both working life quality and performance is difficult to generate. In fact, we can only repeat the statement of Guest (1997, p. 266) on the state of the art on WPI: "(…) we still lack a coherent theoretical basis for classifying HRM policy and practice a problem that becomes more apparent when we start to look at the empirical research."

6.3 Do Current Research Approaches Provide Convincing Evidence?

In this section we question which methodology would be most supportive and conclusive for WPI. We first look at the current methods used (i.e. case studies, surveys, experiments, and literature overviews). Then we move to promising methodological approaches. We are not so much interested in the direction or magnitude of the effects and/or the variables (Pot 2014), but more in the applied methodologies.

6.3.1 Case Studies

Single case studies provide deeper detail about the hypothesized relationships of WPI with outcome conditions. As Ichniowski and Shaw (2013, p. 305) show, insider knowledge is essential to understand what we are looking at: "(...) it is precisely this kind of detailed information (...) that allows the researcher to test new theories about what makes workers productive and firms productive and profitable". However, the problem with single case studies is that they are highly selective. Databases such as the *workplaceInnovation.org* or *rcrc.brandeis.edu* show numerous companies that improved their performance by using WPI. The databases are however underrepresenting the less successful companies. There is a clear bias in what is presented.

There are also examples of how multiple case studies provide insights. We give two examples. First, the Hi-Res study (Totterdill et al. 2002), a meta-analysis of 120 case studies across ten European countries, shows that WPI takes diverse forms, but is always characterised by a clear focus on work environment factors that determine the extent to which employees can develop and use their competencies and creative capabilities for enhancing (1) the company's capacity for innovation and competitiveness, and (2) quality of working life. These work environment factors are: empowering job design, implementing autonomous teamworking, structuring opportunities for reflection, learning and improvement, implementing high involvement innovation practices, encouraging entrepreneurial behaviour at all levels of the organisation, and realising employee representation in strategic decision-making. The authors state that these workplace factors enhance the ability of employers to secure a full return on their investments in training and technology because of better performance and a higher quality of working life. Second, in a systematic analysis of 51 WPI-cases in the European Company Survey (Eurofound 2015), Oeij et al. (2015) show how company management, employee representatives, and employees think about WPI. Their findings show that each company may apply different WPI-practices, there was also a high variety of applied practices across the cases; companies may combine different bundles of WPI-practices, which all lead to higher performances and a higher quality of working life.

Sometimes, major data collection and data analysis of case studies was subsidized by government programmes. Ramstad (2009) conducted an evaluation of 470 Finnish WPI projects (409 organisations) between 1996 and 2005. Management, staff, and experts of these 409 organisations made a self-assessment. Economic performance was measured by labour productivity, quality of goods and services, quality of operations, flexible customer service, and smoothness of operations. Quality of working life was measured by assessing working methods, cooperation between management and staff, social relationships, development of vocational skills, and mental well-being. The study findings show that quality of working life improvements are strongly associated with increases in economic performance. By using a cluster analysis, three groups were distinguished: the best achievers (achieving better performance and better quality of working life) consisted of

152 projects, the worst achievers (poor WPI for both performance and quality of working life) consisted of 31 projects, and the medium achievers (medium scores for performance and quality of working life) consisted of the remaining 287 projects. For the best achievers, employment growth in the project was significantly higher than for the worst achievers. The most striking difference between the best group and the worst group was that: the staff in the best group initiated more often the WPI project, employee participation was stronger and internal collaboration was better (Ramstad 2009).

6.3.2 Surveys and Statistical Approaches

Most studies on WPI use large-scale surveys at the company level. WPI companies are identified and surveys are then send out, the assumption is that in the WPI companies, quality of working life is higher than in other companies. We give five examples. First, one of the most cited studies in the field is the EPOC-study of 1997. The Employee Participation and Organisational Change (EPOC) survey questioned 6000 employees in Europe, and confirmed that direct employee participation has strong positive impacts on productivity, innovation and quality. 68% of the firms which implemented semi-autonomous groups had reductions in costs, 87% reported reduced throughput times, 98% improved products and services, and 85% had increased sales (Eurofound 1997).

Second, sometimes the results of major surveys are at more aggregated and thus abstract level. Polder et al. (2010), for instance, use the Community Innovation Survey to show that organisational innovation improves Total Factor Productivity (TFP)-levels. Product or process innovations only lead to higher TFP when combined with organisational innovations.

Third, the *Erasmus Competition and Innovation Monitor* (ECIM) of the Erasmus University Rotterdam uses a yearly (approximately 10%) sample of companies from a population of 9000 companies to study WPI. As clarified above, ECIM sees dynamic management, flexible organisation, working smarter and external cooperation as the core elements of WPI. The results show that, compared to non-WPI companies, WPI companies perform better for turnover, profits, market share, innovation, productivity, attracting new clients, and are employing more satisfied employees (Volberda et al. 2010). The ECIM repeatedly confirms these results over the years. It remains, however, unclear how selective the company responses are, as the limited sample of 10% may be biased.

Fourth, Schumacher et al. (2015) used a similar survey set-up as the ECIM. They surveyed in The Netherlands 141 organisations in the industry, commercial services, and non-commercial services. Their Monitor Workplace Innovation Region Limburg shows that organisations investing in WPI (which was measured by strategic orientation on workplace innovation; speed of internal change; self-organising capacity; talent development; investment in knowledge base; employability) perform better than other organisations. The organisations show higher turnover growth, stronger development of new products and services, a more comprehensive sustainability, and lower sickness absenteeism.

Fifth, between 2002 and 2004 surveys were send to companies in Flanders (Gryp 2011; Delarue et al. 2004). Companies were randomly selected based on a database of companies which was provided by the Belgian Federal Public Social Security Service. Organisations from different sizes and sectors were included in the sample. In each wave different questions were asked. The first wave, for example, was focused on organisational design. The results are conflicting (Delarue et al. 2004). On the one hand, the results show that teamwork does not always increase organisational performance such as labour productivity. On the other hand, sociotechnical teamwork, for instance, decreases absenteeism.

In general, there are five major difficulties with using survey-data for the evaluation of WPI projects. First, most surveys are cross-sectional. Therefore, a performance difference between companies can be due to different realizations of measures or to sectoral differences. Second, performance and WPI are mostly self-reported, and measured with a limited number of questions, which are not always capable of grasping the precise performance. Although most survey questionnaires have been thoroughly validated, the results reliability is not always high. For example, Oeij et al. (2015) used the European Company Survey to create a WPI-index. When contacting the cases for follow-up interviews on the why and how of WPI, the researchers had great difficulties in selecting suitable cases (i.e. making sure that the cases were indeed WPI-cases) (Dhondt et al. 2014a, b). Third, survey findings are not always replicable because the questionnaires and datasets are not shared within the academic field (e.g. ECIM) or because there was no consistent and systematic data collection process. Fourth, the considerable number of different conceptualisations of WPI (above we already discuss six different surveys) make comparisons between surveys difficult. These four difficulties make verification of study findings very complicated. Moreover, Guest (1997, p. 263) proclaims that many studies emphasise statistical (and econometric) sophistication at the expense of theoretical rigour. “As a result, studies are non-additive, except in a very general way”. This non-additivity makes it difficult to make the theoretically well founded argument that WPI increases performance and quality of working life.

What would then constitute convincing proof with a survey set-up? First, from a statistical perspective, however, it is quite surprising to find so many data sources that point into the same direction. Augusto et al. (2014) report eighteen studies that confirm the hypothesis that WPI improves productivity. But the field deserves more standardised and comparable survey approaches. Second, surveys should be used in a panel or longitudinal set-up. Third, a more additive approach could come from a more systematic analysis of the literature, and the surveys used. An example of such a project was the MEADOW-project, an EU-wide inventory of surveys and methodologies on organisational innovation (CEE/GREDEG 2010). The MEADOW-guidelines are now being used by the statistical offices of Denmark, Finland, Sweden, France, and (to some degree) The Netherlands. Results of some of these national surveys are currently being analysed, therefore the results on the relationship between organisational innovation and company performance are not yet available (communication Danish project team, 2016). We will probably see the results from this data collection process in the next years.

6.3.3 Experiments

Experiments are very rare in the field of WPI. Randomised Controlled Trials are, however, seen by many scientists as the golden standard to settle the argument. Despite the advantages of experiments, the range of organisational hypotheses studied in experiments is small (Camerer and Weber 2013). Camerer and Weber state that experiments are necessarily much simpler and usually constrain the number of agents and the length of time that WPI can be studied in a way that is both inexpensive and well controlled. Organisations have a too complicated reality, which is difficult to realise in an experiment setting. "This mismatch in scope naturally leads to heightened concerns about generalizability" (Camerer and Weber 2013, p. 214).

Even with simple designs it is still possible to test a whole range of elements related to WPI, performance and quality of working life. Camerer and Weber (2013) discuss the impact of coordination mechanisms, incentives, communication patterns, and even organisational culture on performance of teams and groups. Organisational culture is for example needed to reduce coordination efforts, but such results also show the difficulty to help integrate different groups into one organisation and keep up performance. Communication is important to improve the focus of all participants in the games that have been developed in these experiments. Leadership helps to keep performance high, but at the same time, the impact of leadership on performance seems to be an illusion. Large groups do not keep up their performance in repeated games, even under same leadership. Camerer and Weber (2013) conclude their study overview with the observation that many interesting topics, included in the discourse on workplace innovation, are not looked at or completely unstudied with experimental research.

Compared to laboratorial experiments, field experiments are of course a great help, if well designed. The experiment of Bloom et al. (2013) in textile firms in the region of Mumbai (India) shows how a broad set of managerial and organisational improvements leads to a significant rise in performance of organisations. Figure 6.2 summarises the results.

Field experiments also show major problems. Ichniowki and Shaw (2013) report some of the issues. First, the treatment realisation is mostly not randomly selected. There is possibly a selection bias as well as endogeneity in the choices of workers and managers. Second, there is omitted variable bias in the production function, we cannot know what would have happened if non-adopting firms had adopted the new management practice or if adopting firms had not adopted this practice. This is illustrated by Bloom and Van Reenen (2010b, p. 21), who showed an incentive effect: "The firm who introduced the policy presumably believed there would be

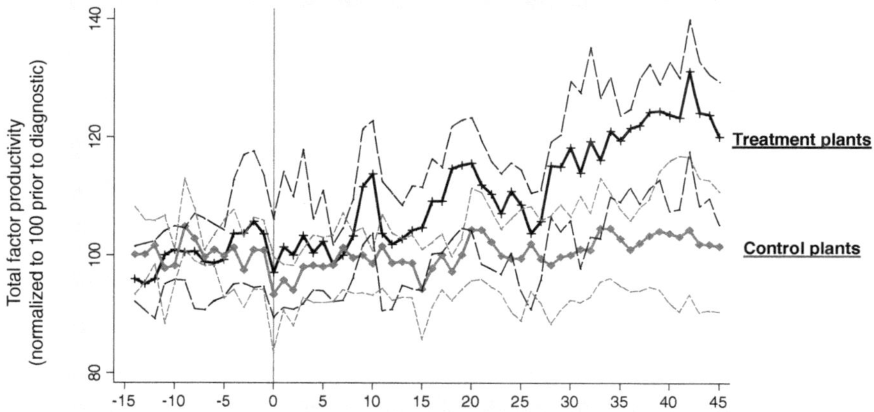

Fig. 6.2 Impact of organisational measures on total factor productivity in Indian textile firms. *Source* Bloom et al. (2013). *Note* Weekly average total factor productivity for the 14 treatment plants which adopted modern management practices for quality, inventory and production efficiency and the 6 control plants. All plants make cotton fabric near Mumbai, India, with between 100 and 1000 employees. Values normalized so both series have an average of 100 prior to the start of the intervention. Confidence intervals bootstrapped over firms

some benefits from doing so, thus it is hard to rule out the idea there may have been some other contemporaneous change that affects worker productivity."

Thus, even with field experiments, we remain puzzled to understand what is going, i.e. we face the reality of company life. Companies rarely share their organisational secrets. The study informants (in most cases HR-managers) are sometimes poorly informed about the organisational performance and are seriously biased about their own actions (Bloom et al. 2014, p. 63). For researchers, it remains therefore difficult to get the right information (see for example Oeij et al. 2015). Even if researchers get inside an organisation, this is rarely for the longer term, but sometimes the evaluation needs to be studied on the long run.

6.3.4 Literature Reviews

Only very few studies did a systematic review of the literature on WPI. A rare but thorough review of the international literature was undertaken by Croucher et al. (2013) for the ILO. They studied the relationship between practices relating to worker voice, working conditions, safety and health, skills development and to positive outcomes in small and medium-sized enterprises (SMEs). The study findings show that this relationship is influenced by individual business resources and capabilities, owner/managers' motivations, and other external factors (i.e. the diverse market and regulatory/institutional constraints that may affect competitive advantage and enterprise survival).

Moreover, the review reveals an increasing focus in research on bundles of practices rather than on individual practices, to effectively translate and link practices to better business outcomes. Many of the bundles (e.g. identified as being effective by the researchers include independent, non-company-driven voice mechanisms such as trade unions, which, although sometimes present) are less common in SMEs than in larger enterprises, even in developed economies. The review did not reveal a body of research that clearly identifies the complex nature of the relationship between working conditions, safety and health, skills development and firm outcomes. However, part of the reviewed research has established some evidence of links between good practices in all the areas examined and a range of positive enterprise-level outcomes. Positive outcomes include reduced employee turnover, higher discretionary contributions by workers to enterprise capacities, improved productivity and profitability and higher levels of customer satisfaction, all leading to sales growth. The review also highlighted a clear need for more longitudinal and multi-level research work, which allows further insights into identifying specific factors that enable the adoption of desirable measures. To complement this, detailed examination is required into differences in relationships created by institutional and cultural environments, the development context and the impact these differences have on the relation between WPI, performance and quality of working life. Finally, another useful suggestion in the review was the call for more detailed case studies with a process approach. According to Croucher et al. (2013) such "thickly described" case studies could show how social dynamics create a link between improving working conditions and positive performance in companies (Dhondt 2014).

The field of workplace innovation lacks a structure for conducting meta-analyses, such as the Cochrane collaboration in medicine. Cochrane is an independent, non-profit, non-governmental organisation that consists of thousands of volunteers in medicine research who systematically evaluate health interventions (see: www.cochrane.org). These evaluations are published in The Cochrane Library. We argue that there is need for such a structure for research on WPI. This would, however, require a research infrastructure in social sciences to support this effort. Such an infrastructure does not yet exist in organisational sciences. One could see the InGRID-project (Integrating expertise in inclusive growth) as such a European (Framework Programme 7) research infrastructure, built on a part of the MEADOW-network and able to integrate transnational data access, organising mutual knowledge exchange activities and improving methods and tools for comparative research. This effort is not yet focused on building a Cochrane library.

6.3.5 A Multipronged Approach: Towards a New Methodological Agenda

Having dealt with the several methodological approaches to underpin the hypothesized relation between workplace innovation and organisational performance, we can only see that a broad methodological approach will lead us forward. This is in line with Croucher et al. (2013) who advise researchers in the WPI-field to develop a multi-pronged approach. Ichniowski and Shaw (2013) prescribe a practical roadmap for insider econometric studies and what to do to develop the right data sources. Dhondt (2014) formulated requirements for studies on the relation between occupational safety, health, and company performance. Some of these recommendations are also relevant for WPI:

- More studies are needed which include the multilevel nature of WPI, and also detail examinations of differences in relationships created by institutional and cultural environments;
- More case studies are needed on how social dynamics influence the relation between quality of working life and company outcomes;
- Longitudinal studies are certainly needed. There are currently very few longitudinal studies, this makes it difficult to evaluate the impacts of better working conditions, especially as most impacts can only be seen on the longer run. Few studies take such long term impacts into account and therefore fail to distinguish between short-term and long-term effects;
- Most studies suffer from selection bias. Research is often limited to the better workplace innovation performers, or do not take into account the heterogeneity of the business world. Ideally research ought to take account of different resources and capabilities of these companies and of owner/manager motivations, as well as the varied nature of markets and associated regulatory/institutional constraints for companies. Studies should also take into account less favourably outcomes and report also these outcomes;
- Few studies account for intermediate variables; direct effects are not always represented. Researchers will need to develop more complex designs to take into account the different effects that are possible;
- Most studies limit themselves to surveys among employers or HR-managers. It is unclear how biased their opinions are. Researchers will need to get access to employees to confirm answers from employers and HR-managers. The approach developed by Oeij et al. (2015) is helpful in this account. They had interviews with these three stakeholder groups in a large set of companies;
- Studies rarely account for the role of 'influence' or 'voice' in implementing workplace innovation. Future studies need to take the 'influence' (voice) factor into account when studying how working time and type of training influences productivity in companies;

- Experimental designs are certainly needed. For this to happen, researchers will need to break down the complexity of workplace innovation as a theory (see Chap. 5) and submit the building stones to such game-theoretic approaches;
- For surveys, more learning is needed from the different studies and surveys. Reference to the MEADOW-standard would be helpful.

6.4 Why Do Not All Companies Opt for Workplace Innovation?

This question is quite relevant at the end of this chapter because it also helps us understand if and how policy makers should play a role in this field. Most policy makers (certainly in the economic domain) assume that markets are efficient and, if workplace innovation delivers superior results for firms, then such solutions should spread quite fast and immediately. From several surveys (for example ECIM), it is clear that workplace innovation is not automatically selected by organisations. Bloom and Van Reenen (2010a) confirmed for instance that bad organisational models may survive a long time.

As Ichniowski and Shaw (2013, p. 293) observe, “economists believe that firms make optimal decisions [such as WPI] about compensation plans, teamwork and related HRM practices, the scope and size of research projects, and whether to outsource certain work”. They continue: “Shouldn’t these firms have always had these optimal practices? Fortunately, there are some clear reasons firms adopt new management policy treatments rather than having one set of optimal management practices from their inceptions.”

Optimal management practices are only adopted because of several time-series shocks. The authors discuss four such shocks. Price of capital inputs may change rapidly. New information technologies have clearly affected management practices within industries. Best managerial practices may change drastically over time. For example, the introduction of lean production was one practice that was introduced in US and European car producers. Changes in the supply of certain human capital (internal and external labour markets) may also change the requirements of an organisational practice. External market pressures can lead to changes: new companies entering an existing industry with another model can lead to cross-sectional within industry practices. The example of Uber to the taxi sector is such an example (Ichniowski and Shaw 2013, p. 294).

In the end, models are not copied easily and always require a choice. That is why Van Reenen (2011) talks about ‘managerial technology’. Managers may have good reasons not to adopt the right practices. Ichniowski and Shaw (2013) summarise three reasons from Bloom et al. (2013) on why companies may not adopt efficient practices:

Chapter 7
Workplace Innovation and Wellbeing at Work

Frank D. Pot

7.1 Introduction

Workplace innovation promises to improve organisational performance, quality of working life and consequently wellbeing at work simultaneously. However wellbeing at work is not an easy concept, not for research and not for policy. Working conditions cannot make everybody happy for the simple reason that people have individual and different wishes and needs that might differ over time as well. The question is which working conditions provide the best opportunities for wellbeing at work for as many employees as possible. This chapter discusses the concepts of workplace innovation and wellbeing, the theories of how the two are related, and the policies that have been derived from that relation, including dilemmas and obstacles.

7.2 Description of Concepts of Workplace Innovation

The European Workplace Innovation Network (EUWIN) started in 2013 and described 'workplace innovation' as follows: "Workplace innovations designate new and combined interventions in work organisation, human resource management, labour relations and supportive technologies. It is important to recognise both process and outcomes. The term workplace innovation describes the participatory and inclusive nature of innovations that embed workplace practices grounded in continuing reflection, learning and improvements in the way in which organisations manage their employees, organise work and deploy technologies. It champions workplace cultures and processes in which productive reflection is a part of

F.D. Pot (✉)
Radboud University, Nijmegen, The Netherlands
e-mail: frank.pot@ardan.demon.nl

P.R.A. Oeij et al. (eds.), *Workplace Innovation*, Aligning Perspectives on Health, Safety and Well-Being, DOI 10.1007/978-3-319-56333-6_7

everyday working life. It builds bridges between the strategic knowledge of the leadership, the professional and tacit knowledge of frontline employees and the organisational design knowledge of experts. It seeks to engage all stakeholders in dialogue in which the force of the better argument prevails. It works towards "win-win" outcomes in which a creative convergence (rather than a trade-off) is forged between enhanced organisational performance and enhanced quality of working life (Dhondt 2012, p. 2)."

The *organisational performance* which is related to workplace innovation is primarily labour productivity and innovation capability. In research, other performance measurements are being used as well, such as new products and services, profit, turnover, sickness absenteeism. However, the relation with such results is a more indirect one (Oeij et al. 2012).

There is a wide variety of definitions of wellbeing at work (Schulte and Vainio 2010) and some of them cover all working conditions including health and safety (e.g. ILO 2009). Differences between countries have been investigated by the European Agency for Safety and Health at Work (EU-OSHA 2013b). Maybe the only thing that those definitions have in common is reflected in the title of the EU-OSHA report: 'Well-being at work—creating a positive work environment'. Or, as American colleagues describe it: "Generally, it (wellbeing, FP) is a synonym for health and a summative term to describe a flourishing worker who benefits from a safe, supportive workplace, engages in satisfying work, and enjoys a fulfilling work life" (Schulte et al. 2015, p. 31). However, regarding determinants of well-being at work, in this chapter we prefer not to take into account all possible elements of working life, but to pay attention to those elements that in particular create this positive work environment: Such as participation and involvement, psychosocial risks, learning opportunities and competence development, ergonomic comfort, and work-life balance. Good wellbeing conditions may of course influence individual work engagement and subjective wellbeing, but our focus is on the objective conditions or the 'conditional approach', to be explained later in the chapter.

The *overlap* between workplace innovation and occupational safety and health can be shown in Fig. 7.1.

Participation is the key issue to achieve organisational performance and quality of working life simultaneously (De Sitter et al. 1997; Karasek and Theorell 1990; Mikkelsen et al. 2000; Ramstad 2009; Totterdill et al. 2002). However, participation strategies "do not always lead to 'win-win' outcomes and tend to be less sustainable as participation is often perceived as a technical solution to problems of engagement and productivity, not as a fundamental approach to relations between management and labor" (Cressey et al. 2013, p. 219). Akerlof contends from an economic perspective that participation needs to take the form of gift-exchange or reciprocity to be effective (Akerlof 1982). But participation goes beyond organisational performance and wellbeing at work. Some of the founding fathers and current opinion leaders emphasise the importance of developing a high quality democratic society in which people can be creative and can participate in decision making (Emery and Thorsrud 1976; Gustavsen 1992). "Participative work practices

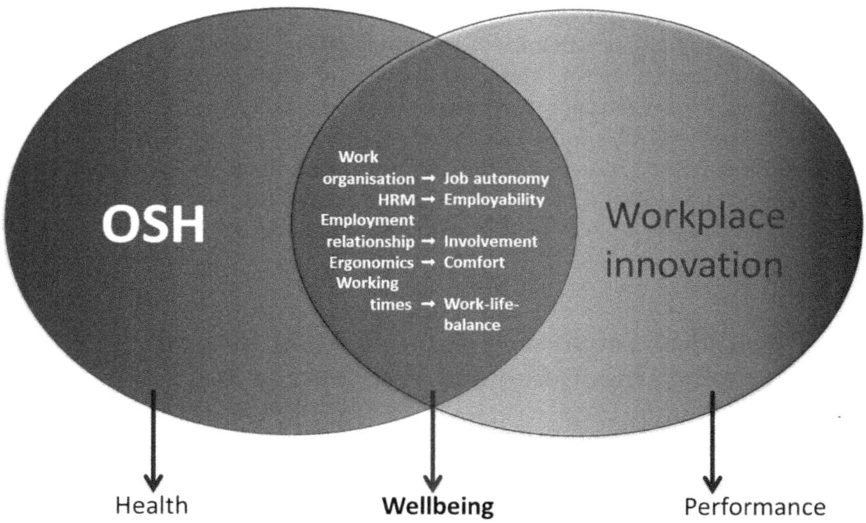

Fig. 7.1 Overlap between OSH and workplace innovation (EU OSHA 2013b, p. 9)

and the creation of an informed and democratically active workforce should be placed at the heart of business ethics and sustainability (Cressey et al. 2013, p. 220)."

Two side notes may be important here. The first one is that this chapter is about 'workplace innovation related wellbeing'. Workplace innovation approaches assumes ideally that issues such as minimum wage and occupational safety and health legislation have been settled. This is—considering worldwide working conditions—a rather luxurious and not realistic situation. In many countries a majority of the employees work under poor conditions which do not meet the ILO-criteria of 'decent work' such as some employment security, a living wage, social protection, rights at work and basic occupational safety and health (ILO 2012). Such conditions cannot be expected to be related to wellbeing at work. First priority in those countries is to implement the Decent Work Agenda. However, in some of those countries some attention is being paid, most of the time by researchers, to the relation between workplace innovation and wellbeing at work, for instance in Ethiopia (Jilcha et al. 2016).

The second side note is the common misunderstanding that the *happy employee* is a productive and innovative employee as well. However, 'employee satisfaction' measures, most of the time, the level of adjustment to existing circumstances and not the work environment itself. Being happy with your work may have to do with having nice colleagues and an average workload. Individual and group performance is not directly the result of employee satisfaction or motivation, but is achieved through the involvement and commitment of workers' representation, HRM practices and work organisation (Judge et al. 2001; Taris et al. 2008). For instance, organisational commitment can be brought about by an organisational design that provides job autonomy, possibilities of consulting others, learning opportunities etc. (Karasek and Theorell 1990). These are exactly the same measures that are recommended to reduce psychological stress risks as a way of 'prevention at the source' (Karasek and Theorell 1990; Pot et al. 1994; Cox et al. 2000).

To summarise, the concept of wellbeing that is elaborated in this chapter refers not to individual wellbeing or to basic working conditions, but to wellbeing conditions at organisational level such as participation and learning opportunities.

7.3 Theoretical Foundation of the Relation Between Workplace Innovation and Wellbeing at Work

The theoretical foundation of workplace innovation goes back more than 35 years to the job demands-control-model of Karasek (1979) and the integration of that model in the modern sociotechnical design theory of De Sitter (1981). Additional contributions to the theoretical foundation are those of Hacker (1978, 2003) and Volpert et al. (1989) (action regulation theory), Gustavsen (1992, 2016) (democratic dialogue) and Argyris and Schön (1978) (single loop and double loop learning). These theories show that organisational performance and wellbeing at work are two sides of the same coin if workplace innovation is properly implemented.

7.3.1 Organisational Design Theories

The hard core of design theories that support workplace innovation is De Sitter's sociotechnical systems design theory. The central idea is the balance between 'control requirements' (demands) and 'control capacity' (job control). "It's not the problems and disturbances in the work that cause stress, but the hindrances to solve them" (De Sitter 1981, p. 155)". In order to maintain this balance, control capacity is required on individual job level (called internal control capacity or job autonomy) as well as regarding the work organisation, in particular the reduction of organisational complexity on production group and plant level (called external control capacity or shop floor consultancy/participation). This has been summarised by De Sitter et al. (1997) in the title of their paper "from complex organisations with simple jobs to simple organisations with complex jobs".

In 1981 De Sitter integrated the 'job demands-control-model' (Karasek 1979) into his theory of modern sociotechnical design. The Job Demands-Control (JDC) model holds two predictions. High job demand as well as low job control represent risk factors such as stress and low skill discretion that are detrimental to health outcomes such as coronary heart disease, mental ill health etc. The model also predicts that high job demand, as well as high job control foster motivation and learning. Central features of the JDC-model are also the strain and learning hypotheses, referring to two interaction hypotheses on the balance between job demands and job control. Jobs with high demands and low control can be called 'high strain jobs' which are a risk for work-related stress. Moreover, stress inhibits learning. Jobs with high demands and high control are called 'active jobs' which offer opportunities for learning and coping with stressors (1979; Karasek and Theorell 1990). Later, this model was extended with the social support dimension and with innovative and productive work behaviour (Karasek and Theorell 1990). The foundation of this JDCS-model (where S stands for Support) goes back to job (re)design and research theories from work organisation sociology (e.g. Blauner 1964).

So far, the Job Content Questionnaire which is based on the JDCS-model (JCQ; Karasek and Theorell 1990) does not make a distinction between different dimensions or levels of job control and refers to task-level (also referred to as job level, micro-level or individual level). "This most commonly used definition of job decision latitude describes features of jobs, primarily the ability of the worker to use his or her skills on the job and to have authority to make decisions regarding how the work is done and to set the schedule for completing work activities. This level of decision latitude focuses on the worker's abilities to control his or her own activities and skill use within the job. These concepts are operationalized by the 'skill discretion' and 'decision authority' subscales of the JCQ (Karasek and Theorell 1990, p. 60 in Landsbergis 2005)." This task-orientation of the concept of control also holds for other approaches such as the theory of control related to stress, proposed by Frese, about decisions regarding sequence, timeframe and content, related to tasks, plans and feedback (Frese 1989), and the control-model (timing control, method control), related to wellbeing and 'production responsibility' of Jackson et al. (1993). These authors discuss the importance of new technology and work organisation on system's level and how this influences job content and job control. However, their measurements of job control are confined to task level. One exception is the collective control approach and measurement of Norwegian researchers, who refer to social relations at the collective level, in particular group-level norms of employees and management (Saksvik et al. 2013). There are only two research projects that the author of this chapter is aware of that make a clear distinction between levels of job control. The first is that of Gallie (2013), who distinguishes 'individual task discretion (autonomy)', 'semi-autonomous teamwork' and 'consultative participation': "Consultation through wider workplace meetings, more localized briefing groups, problem-solving groups or quality circles would empower employees by allowing them to influence organisational issues through direct communication with management" (Gallie 2013, p. 456). The second one is the research of Dhondt et al. (2014) which distinguishes three levels: job autonomy, functional support and organisational level decision latitude. The JDCS-model contains 'social support' as well, which covers instrumental social support as well as socioemotional social support. Where socioemotional support is different from job control, instrumental social support is not. When employees can call in the help of colleagues and supervisors this provides them with a better control of their work. That is why some researchers consider instrumental support to be a third dimension of job control, a level between job autonomy and organisational level decision latitude. Dhondt et al. (2014) call it 'functional support'.

There is quite some empirical evidence for the JDCS-model. Reviews of longitudinal studies lend some support to these strain and learning interaction hypotheses (De Lange et al. 2003). The main effects of job demand and job control on health and wellbeing are more often found than demand control interaction effects (Häusser et al. 2010). However, empirical findings with the model also suggest that especially the presence of high job demands, more than a lack of job control, results in work stress and work-related health problems. Conversely, especially the presence of job control is associated with positive outcomes, such as learning, job engagement, wellbeing and organisational commitment (Demerouti et al. 2001; Gallie 2013; Lyness et al. 2012; Stansfeld et al. 2013). In one research project, mentioned above, the relative influence of three levels of job control has been analysed. Functional support and organisational level decision latitude showed stronger relations with the outcome variables subjective wellbeing and organisational commitment than job autonomy (Dhondt et al. 2014).

The notion of 'complex jobs' can also be found in two other theories: the action regulation theory—framed as 'complete jobs'—which was developed by Hacker (1978, 2003); and the double loop learning theory by Argyris and Schön (1978). Hacker distinguishes three stages of action regulation: action preparation, task execution and evaluation. Complete jobs cover all these stages. "Activities can be considered to be sequentially complete when they do not merely allow people to execute the task, but also allow them to do the required preparatory cognitive operations (in particular goal setting and deciding on the measures to be taken). These cognitive operations are particularly necessary when people participate in organizing the work, and checking the results of one's work. Moreover a task is considered to be hierarchically complete, when the mental regulation is not limited to automated processes, but requires controlled, i.e., knowledge-based and, moreover, intellectual control processes as well. Sequentially and hierarchically complete activities offer the crucial option of learning, as opposed to deterioration skills and abilities in simple and limited routine activities" (Hacker 2003, p. 112). For the execution of tasks (implementation), internal control capacity is needed. Organising the work and checking the results of one's work presupposes external control capacity. However, Hacker does not make a distinction between these different dimensions of job control. "Decision latitude (or autonomy) is the most important feature of complete activities. Complete activities offer the decision latitude that is necessary for setting one's goals. These are prerequisites of comprehensive cognitive requirements of a task, and determine the intrinsic task motivation, i.e., being motivated by a challenging job. These aspects serve as a well-known buffer against negative consequences of a high workload" (Hacker 2003, p. 112).

In the organisational learning theory of Argyris and Schön (1978), two levels of control can be recognized. "Ordinary repetitive acting corresponds with the 'given order with prescribed procedures' method. Innovative acting includes the characteristics of ordinary repetitive acting, but is also aiming for improvement of procedures, working conditions, and results in order to enhance effectiveness or efficiency" (Argyris and Schön 1978, p. 117). In other words: job autonomy (internal control capacity) relates to 'single loop learning' (doing things better) and complex or complete jobs with external control capacity facilitate 'double loop learning' (doing the right things).

The conclusion of this paragraph is that organisational design theories support the relation between workplace innovation and wellbeing at work.

7.3.2 Change Process Theories

Keeping in mind that workplace innovation covers outcomes as well as processes the theories described so far are mainly about outcomes. They are organisational design theories, describing how work organisation should be designed to achieve the intended purposes for organisation and employees. The role of experts who understand these theories is important. In the Scandinavian countries however, the emphasis has always been on the process part of organisational (re)design, on involving all stakeholders, including employees, in the process of change. The rationale of this approach is not only to make people learn how to create participative work roles, building 'innovative communication' (Burns and Stalker 1961), but also the notion that all general theories have to be applied locally. In Europe this was called 'democratic dialogue' (Gustavsen 1992). In a recent publication Gustavsen says: "The point is that all initiatives emerged within specific, and different, contexts and gained much of their visibility through their ability to concretize issues of critical importance within each of these contexts." (2016, p. 188). This point is of major importance in understanding the agreement between social partners in Norway in 1983. "This agreement did not focus on design principles but on the processes needed to work out and implement alternatives to Taylorism. In line with such notions as participative design and user-driven change, what was seen as the key issue was to create a broad interest in exploring alternative forms of work organisation and help the interested parties develop fruitful processes during which they, themselves, could decide on design issues" (Gustavsen 2016, p. 189).

How was this achieved? "The notion of democratic dialogue emanated with the labour market parties and initially took the form of a negation of the characteristics of negotiations. Negotiations are generally performed through representatives, in an adversarial way, and concern quantifiable issues, mainly time and money. What the parties did was to reverse these characteristics and replace them with the notion of direct participation in a constructivist setting, in which all topics of interest to the participants can be brought up" (Gustavsen 2016, p. 189). For this democratic dialogue a set of specific criteria had been developed (Gustavsen 2016, p. 190).

Another way of conceptualising the process part of workplace innovation is the use of the concept of 'productive reflection', covering jointly "the role that organisational structures have in articulating employee voice together with the active use of employee's formal and tacit skills and competences in the process of improvement, innovation and change (Cressey et al. 2013, p. 221)".

A first side note is that this kind of democratic dialogue has only a chance in countries and sectors where constructive industrial relations exist. A second side note is that this dialogue approach may collide with the organisational design approach which is mainly an expert approach. Predominant design experts may frustrate the democratic dialogue on the one hand and on the other hand giving priority to democratic processes may lead to suboptimal organisational solutions. The definition of workplace innovation as mentioned in the beginning of this chapter covers both approaches, but in real life this is not an easy thing to do.

The conclusion of this section is that these change process theories support the relation between workplace innovation and wellbeing. In practice it works only in situations of constructive industrial relations and it is not easy to find a balance between 'dialogue' and 'design knowledge'.

7.3.3 A Conditional Approach to Wellbeing at Work

Making the bridge from theories to policies, decisions have to be made about whether to focus on wellbeing conditions or subjective wellbeing outcomes, and on the individual or organisational level.

To ensure healthy working, it is eminent that work should be adjusted to the capacities of people in general, and technology should be human-centered in a way that reflects the latest advances. Such kind of texts therefore are or can be included in legislation, sometimes together with policies like job autonomy, ergonomics, prevention of psychosocial workload, reduction of machine-paced work and short cycled work. Applying design rules that can be derived from these policies is primary prevention (reduction of wellbeing risks), and objective measurements of the implementation of such design rules can support enforcement.

Sometimes in legislation, the text also mentions that the work should be adjusted to the individual worker and take into account differences between workers. While this is a sympathetic general philosophy, it is difficult to measure and impossible to enforce. To implement such an approach Person-Environment Fit (P-E- Fit) models are used (French et al. 1974). Two types of fit are usually distinguished: fit defined in terms of abilities-environmental demands and fit in terms of needs-environmental supplies. Apart from problems of commensurate measurements (Caplan 1987, p. 263), in particular needs (preferences, wishes, etc.) may change over day time, life time, situations, moods etc. Another complication is that employees come and go and adjusting person and environment has to be done again and again. In terms of prevention policy this is secondary prevention. The P-E-Fit approach does not allow risk assessment at organisational level because it focusses on individual fits. Nor can this approach be used by labour inspectors because the inspectors are not able to check individual fits and individual wellbeing at work.

Hence, for general standards and primary prevention, it would be better to stick to the theories mentioned before and leave P-E-Fit models for the fine tuning at individual level. Wellbeing and motivation are "environmentally facilitated" (Karasek and Theorell 1990, p. 170), in particular through job control and complete/complex jobs or skill discretion. The focus of organisational design should be on structural parameters for work organisation (not on individual needs and appraisals). Hacker relates 'intrinsic task motivation' to complete jobs. De Sitter (1981) draws our attention to the danger of 'alienation' (more or less the opposite of 'wellbeing') where people have jobs with low control and low job demands. In his definition, control covers job autonomy as well as shop floor consultancy/participation. These are conditions that may influence the levels of stress and wellbeing.

Looking at conditions that are well known risk factors, instead of looking at subjective wellbeing, makes risk assessment possible at the organisational level as well as facilitates the enforcement of legislation.

At the same time it can be observed in some national policies, such as in the Netherlands and Belgium, that, although conditional approaches still officially exist, mainstream communications emphasise subjective wellbeing, individual responsibility for employability and improving coping with stress instead of reducing psychosocial risks. The 'European work stress—weeks' of November 2013 and November 2014 also showed that clearly. The Health and Safety Laboratory observes the same trend in the UK: "To date, wellbeing at work programmes have often been concerned with health promotion, (…) but they don't tackle the underlying influences on wellbeing at work, such as leadership styles, attitudes to health, and management systems." (HSL 2016) However, this is not a new phenomenon as Caplan noticed thirty years ago: "Management personnel preferred that their employees deal with stress by changing P [P is person in the Person-Environment Fit model, FP] (biofeedback, meditation, retraining, and so forth). Union members preferred that management give up some of its control and power to the employees, rather than have the employees develop better psychic stiff upper lips" (Caplan 1987, pp. 258–259). The background is probably not only theoretical bias or insufficient knowledge, but power relations as well.

Recently ideas such as 'purpose economy' (Hurst 2014), 'mindfulness at work' and 'happy at work' have become popular concepts. All of this has to do with wellbeing at work. It seems however so far, that this is mainly about beliefs and individual motivation and responsibility and not about theories on organisational structures and power relations.

7.4 Obstacles and Dilemmas

Although there are enough reasons to develop workplaces from the point of view of quality of working life and performance, it is not an easy job to do. Obstacles are an inevitable part of change—and are perhaps integral to the process of organisational learning. There are many obstacles that must be overcome such as resistance to change and previous failures (Totterdill et al. 2002, p. IV).

These obstacles are interwoven with dilemmas that employees and their representatives as well as employers/managers are facing with respect to their involvement in and commitment to workplace development. For employees these include the long- and short-term effects (e.g. employment) and getting more responsibility, being more accountable but not getting more authority and power to fulfill that responsibility. It also includes situations in which organisational commitment leads to working harder instead of working smarter as happens in those varieties of 'lean' which do not stand in the participatory tradition (Koukoulaki 2014). The employer/management side faces dilemmas such as: the benefits of workplace development appear later than the results of short-term budget cuts; bonuses stimulate short-term thinking; social innovation is more complex than technological innovation; and sharing knowledge and power is not easy. However, the argument of many executives, who claim to be imprisoned by iron economic laws dictating them to match employment practices offered by their lowest-cost competitors, is contradicted by research findings (O'Toole 2008). And finally focussed on wellbeing at work: improving individual coping behaviour is much easier than creating a better work environment.

In spite of the obstacles and dilemmas, the evidence is growing from surveys and case studies that workplace innovation contributes to better jobs and performance (see especially Chap. 6 in this volume).

7.5 Examples of Policies on Wellbeing at Work

To cope with the obstacles and dilemmas, a good starting point in a number of countries (such as Finland, Germany and the Netherlands) is that unions and employers' associations have a tradition of cooperation and mutual consultation. It is clear that in countries where the government supports workplace innovation politically (campaigns) and financially (e.g. by using ESF funding) the attention for and dissemination of workplace innovation increases. Finland, Germany and Flanders (Belgium) have recently decided to continue and refresh their programmes (see Chap. 3). The Netherlands, Belgium and Finland are the best examples of countries which have wellbeing policies which are related to policies regarding work organisation or workplace innovation.

The Netherlands

In 1989 the 'wellbeing at work' article in the working conditions act was enforced in the Netherlands. There was nonetheless quite some confusion about what was meant by 'wellbeing at work' and how this could be enforced. After substantial research and dialogue, a conditional approach seemed to be the most promising and prevention oriented. Job content and ergonomics were considered to be the most important conditions for wellbeing at work. Assessment and design tools (see WEBA-tool hereafter) were introduced as a practical operationalisation of the act.

Based on the organisational design theories (JDCS-model, sociotechnical systems design theory, action regulation theory), a practical expert tool has been developed to assess the quality of jobs and to design high quality jobs. The Dutch government funded the development of the instrument which was—among other aims—supposed to help the Labour Inspectorate to enforce wellbeing at work. The instrument is called WEBA, a Dutch abbreviation of wellbeing at work (Pot et al. 1994; Dhondt and Vaas 2001). The WEBA distinguishes seven dimensions: Completeness of the job, Repetitive work, Cognitive complexity, Job autonomy, Contact opportunities (social contacts and opportunities for assistance or functional support), Organisational level decision latitude and Information. Job control is covered by the three dimensions autonomy (internal control capacity), contact opportunities and organisational level decision latitude (the last two dimensions covering external control capacity). The instrument is mainly being used in the Netherlands and Flanders (Belgium).

In 1999 new labour conditions legislation was accepted by Parliament in which the concept of 'wellbeing at work' was no longer included. The concept caused too much confusion as wellbeing is much broader than being strictly work related. Yet, to design workplaces and work organisation according to the state of the art of science, the prevention of psychosocial workload, reduction of repetitive work etc., remained in the legislation. Therefore, the WEBA tool continued to be used by labour inspectors, consultants, occupational health services and companies.

Wellbeing at work was given new attention in the Netherlands by the policy of 'social innovation of work and employment'—nowadays called 'workplace innovation'—since the beginning of this century. The initiative in the Netherlands came from employers' associations and trade unions, supported by universities and TNO. The government supported this initiative and provided funding for some projects (Pot et al. 2012). Workplace innovation spread all over Western Europe and was adopted in the EU policy (since 2012, see Chap. 2 in this volume).

Belgium

In 1996, new Belgian Legislation came in force "on the policy of wellbeing of workers at work". This covered (1) work safety; (2) health protection of the worker; (3) psychosocial stress caused by work including violence, bullying, and sexual harassment; (4) ergonomics; (5) occupational hygiene; (6) care for the natural environment (Belgian Legislation 1996). The concept of wellbeing is used here as a kind of umbrella for occupational safety and health at large. The WEBA-tool and similar tools were developed and are still recommended for the reduction of stress risks (see for the PRIMA-EF tool Chap. 8 in this volume).

Finland

The best example of recent policies on wellbeing at work is the Finnish 'National Working Life Development Strategy to 2020', published in 2012, because well-being at work, productivity and innovation are connected explicitly. Wellbeing and health are used as complementary concepts. Wellbeing refers to work organisation, meaningful work, competence development and implementing change together. In the National Working Life Development Strategy it is said: "Health and well-being at work are invested in within well-functioning workplace communities: such investments pay themselves back. Good working conditions and well-being at work are a key part of future working life and inspiring workplace communities. They increase the productivity and attractiveness of working life". The strategy is also ambitious: "The strategy's vision is to make working life in Finland the best in Europe by 2020." (Ministry of Labour 2012, pp. 3–4).

Parallel to the working life policy, 'workplace development programmes' have been initiated. Again productivity, innovation and wellbeing at work are connected. These programmes did not start from the occupational safety and health perspective but from the workplace innovation perspective, using change process theories (Alasoini 2016, see Chap. 3 in this volume). The two types of programmes (OSH and workplace innovation) reinforced each other. Evaluations of the workplace innovation programmes have shown positive results on wellbeing at work (Ramstad 2009, 2014; Alasoini 2016).

European policy

The connection with organisational performance was mentioned in the European OSH-policies in the first decade of this century as 'healthy and productive jobs'. However, 'productive' was referring to productivity loss as an effect of sickness absenteeism, not to enhancing productivity. Gradually the realisation grew that OSH contributes to enhancing productivity and innovation.

In 2012 DG Enterprise & Industry (nowadays GROWT) adopted workplace innovation in its policy on advanced manufacturing. In 2015 DG EMPL published '*Employment and social developments in Europe 2014*'. Chapter 3 of this report is about "the future of work in Europe: job quality and work organisation for a smart, sustainable and inclusive growth." One paragraph in that chapter describes that technological innovation should be complemented by workplace innovation (DG EMPL 2015, p. 164).

Furthermore workplace innovation was connected to 'wellbeing at work' in the policy to extend occupational safety and health to 'wellbeing at work' (EU-OSHA 2013b).

A more coherent policy perspective at EU-level is emerging (see Chap. 2 in this volume).

7.6 Adjustment of Research Designs

The attention for wellbeing at work and its conditions regarding the work organisation stimulated researchers to adjust research designs and questionnaires. Eurofound adjusted and extended the questionnaires for the European Working Conditions Survey (EWCS) (Eurofound 2015b) and the European Company Survey (ECS) (Eurofound 2015a) by including proper measurements of workplace innovation. Hence, these surveys can now be used as monitors for wellbeing at work conditions, such as the balance between job demands and job control (job autonomy), skill variety and stimulating work, repetitive tasks and monotonous work, time pressure (work intensity) and direct participation and involvement. The EWCS also covers a scale for subjective wellbeing (the WHO-five Well-being Index 1998 version).

EU-OSHA commissioned a literature review on the relation between workplace innovation and occupational safety and health (OSH) (Eeckelaert et al. 2012). Consequently workplace innovation was connected to 'wellbeing at work' in the research priorities of OSH (EU-OSHA 2013a).

7.7 Epilogue

As shown in the preceding paragraphs the relation between workplace innovation and wellbeing at work is well-founded, empirically as well as theoretically. Focussing in both research and policy on the determinants of wellbeing in the work environment, the 'conditional approach' ('primary prevention') is considered to be a more successful strategy than improving individual coping behaviour ('secondary prevention').

As described above a number of countries and EU-institutions have emphasised the need to extend occupational safety and health policies to wellbeing at work, to create a positive work environment for the benefit of employees and organisations as well. Workplace innovation seems to be a concept and an approach to design such an environment in a participatory way. To deal with obstacles and dilemmas constructive industrial relations are crucial. Hard regulation (legislation) is not sufficient and perhaps unsuitable to create a positive work environment. Framework agreements between social partners and preferably government as well are probably more successful. An example is the European 'Framework Agreement on Work-Related Stress' (ETUC et al. 2004), followed by 'Promoting mental health in the workplace. Guidance to implementing a comprehensive approach' (Wynne et al. 2014). The titles of the two agreements reflect the development of approaches from reduction of psychosocial risks to creating a positive work environment. Connecting these agreements to workplace innovation initiatives on European, country and sectorial levels would make these approaches even more comprehensive.

References

Akerlof, G. A. (1982). Labor contracts as partial gift exchange. *Quarterly Journal of Economics, 97*(4), 543–569.

Alasoini, T. (2016). *Workplace development programmes as institutional entrepreneurs. Why they produce change and why they do not*. Aalto University publication series, doctoral dissertations 12/2016.

Argyris, C., & Schön, D. (1978). *Organizational learning*. Reading, Massachusetts: Addison-Wesley.

Belgian Legislation. (1996). *Act of 4, August 1996 on well-being of workers in the performance of their work.*

Blauner, R. (1964). *Alienation and freedom: The factory worker and his industry*. Chicago: University of Chicago Press.

Burns, T., & Stalker, G. M. (1961). *Management of innovations*. London: Tavistock Publications.

Caplan, R. D. (1987). Person-environment fit theory and organizations: Commensurate dimensions, time perspectives, and mechanisms. *Journal of Vocational Behavior, 31*(3), 248–267.

Cox, T., Griffiths, A., & Rial-Gonzalez, E. (2000). *Research on work-related stress*. Luxemburg: European Agency on Safety and Health at Work.

Cressey, P., Totterdill, P., & Exton, R. (2013). Workplace social dialogue as a form of 'productive reflection'. *International Journal of Action Research, 9*(2), 209–245.

De Lange, A. H., Taris, T. W., Kompier, M. A. J., Houtman, I. L. D., & Bongers, P. M. (2003). The very best of the millennium: Longitudinal research and the demand-control-(support) model. *Journal of Occupational Health Psychology, 8*(4), 282–305.

De Sitter, L. U. (1981). *Heading towards new factories and offices* (in Dutch). Deventer: Kluwer.

De Sitter, L. U., den Hertog, J. F., & Dankbaar, B. (1997). From complex organizations with simple jobs to simple organizations with complex jobs. *Human Relations, 50*(5), 497–534.

Demerouti, E., Bakker, A. B., Nachreiner, F., & Schaufeli, W. B. (2001). The job demands-resources model of burnout. *Journal of Applied Psychology, 86*(3), 499–512.

DG EMPL. (2015). *Employment and social developments in Europe 2014*. Brussels: European Commission.

Dhondt, S. (Ed.) (2012). *European learning network for workplace innovation*, Section 4 technical proposal, call for tender no 212/pp/ent/cip/12/c/n02c04. Hoofddorp: TNO.

Dhondt, S., Pot, F. D., & Kraan, K. O. (2014). The importance of organizational level decision latitude for wellbeing and organizational commitment. *Team Performance Management: An international journal, 20*(7/8), 307–327.

Dhondt, S., & Vaas, F. (2001). *WEBA analysis manual*. Hoofddorp: TNO Work and Employment.

Eeckelaert, L., Dhondt, S., Oeij, P., Pot, F., Nicolescu, G. I., Webster, J., et al. (2012). *Review of workplace innovation and its relation with occupational safety and health*. Bilbao: European Agency for Safety and Health at Work.

Emery, F., & Thorsrud, E. (1976). *Democracy at work: The report of the Norwegian Industrial Democracy Program*. Leiden: Martinus Nijhoff.

ETUC, UNICE, UEAPME, & CEEP (2004). *Framework agreement on work-related stress*, Brussels.

EU OSHA (European Agency for Safety and Health at Work). (2013a). *Priorities for occupational safety and health research in Europe: 2013–2020*. Bilbao: EU OSHA.

EU OSHA (European Agency for Safety and Health at Work). (2013b). Well-being at work—creating a positive work environment. Bilbao: EU OSHA.

Eurofound. (2015a). *Third European Company Survey. Workplace practices: Patterns, performance and well-being*. Luxembourg: Publications Office of the European Union.

Eurofound. (2015b). First findings 6th EWCS. http://www.eurofound.europa.eu/sites/default/files/ef_event/field_ef_documents/6ewcs_first_findings_-_eurofound.pdf. Accessed May 5, 2016.

French, J. R. P., Jr., Rodgers, W., & Cobb, S. (1974). Adjustment as person-environment fit. In G. V. Coelho, D. A. Hamburg, & J. E. Adams (Eds.), *Coping and adaptation* (pp. 316–333). New York: Basic Books.

Frese, M. (1989). Theoretical models of control and health. In S. L. Sauter, J. J. Hurrell, & C. L. Cooper (Eds.), *Job control and worker health* (pp. 107–128). Chichester: Wiley.

Gallie, D. (2013). Direct participation and the quality of work. *Human Relations, 66*(4), 453–473.

Gustavsen, B. (1992). *Dialogue and development: Theory of communication, action research and the restructuring of working life*. Assen: Van Gorcum.

Gustavsen, B. (2016). Democratic dialogue. In B. J. Mohr & P. Van Amelsvoort (Eds.), *Co-creating humane and innovative organizations. Evolutions in the practice of socio-technical system design* (pp. 186–200). Global STS-D Network Press.

Hacker, W. (1978). *Allgemeine Arbeits- und Ingenieurpsychologie*. Bern: Huber.

Hacker, W. (2003). Action regulation theory: A practical tool for the design of modern work. *European Journal of Work and Organizational Psychology, 12*(2), 105–130.

Häusser, J. A., Mojzisch, A., Niesel, M., & Schulz-Hardt, S. (2010). Ten years on: A review of recent research on the Job Demand-Control (-Support) model and psychological well-being. *Work & Stress, 24*(1), 1–35.

HSL. (2016). http://www.hsl.gov.uk/resources/health-safety-insights/wellbeing-tree. Accessed October 1, 2016.

Hurst, A. (2014). *The purpose economy. How your desire for impact, personal growth and community is changing the world*. eBook by Elevate.

ILO (International Labour Organization). (2009). *Workplace well-being*. Geneva: ILO.

ILO. (2012). *Decent work indicators: Concepts and definitions: ILO manual*. Geneva: ILO.

Jackson, P. R., Wall, T. D., Martin, R., & Davids, K. (1993). New measures of job control, cognitive demand, and production responsibility. *Journal of Applied Psychology, 78*(5), 753–762.

Jilcha, K., Besha, B., & Kitaw, D. (2016). Workplace innovation influence on occupational safety and health. *African Journal of Science, Technology, Innovation and Development, 8*(1), 33–42.

Judge, T. A., Thoresen, C. J., Bono, J. E., & Patton, G. K. (2001). The job satisfaction—Job performance relationship: A qualitative and quantitative review. *Psychological Bulletin, 127* (3), 376–407.

Karasek, R. A. (1979). Job demands, job decision latitude, and mental strain: Implications for job redesign. *Administrative Science Quarterly, 24*(2), 285–308.

Karasek, R. A., & Theorell, T. (1990). *Healthy work; Stress, productivity and the reconstruction of working life*. New York: Basic Books.

Koukoulaki, T. (2014). The impact of lean production on musculoskeletal and psychosocial risks: An examination of sociotechnical trends over 20 years. *Applied Ergonomics, 45*(2), 198–212.

Landsbergis, P. (2005). *A possible design example for JCQ 2.0 job decision latitude scales* (discussion note).

Lyness, K. S., Gornick, J. C., Stone, P., & Grotto, A. R. (2012). It's all about control: Worker control over schedule and hours in cross-national context. *American Sociological Review, 77* (6), 1023–1049.

Mikkelsen, A., Saksvik, P. Ø., & Landsbergis, P. (2000). The impact of a participatory organizational intervention on job stress in community health care institutions. *Work & Stress, 14*(2), 156–170.

Ministry of Labour. (2012). *National working life development strategy to 2020*. Helsinki, Finland: Ministry of Labour.

O'Toole, J. (2008). Free to choose—How managers can create globally competitive and healthy workplaces (an American perspective). In V. Weber (Ed.), *Achieving business excellence—Health, well-being and performance* (pp. 24–39). Essen Germany: Bertelsmann Stiftung/BKK.

Oeij, P., Dhondt, S., Kraan, K., Vergeer, R., Pot, F. (2012). Workplace innovation and its relations with organisational performance and employee commitment. *LLINE Lifelong Learning in Europe*, Issue 4/2012, Theme 'Workplace Learning and Innovation' (article 10) http://www.elmmagazine.eu/articles/workplace-innovation-and-its-relations-with-organisational-performance-and-employee-commitment/

Pot, F., Dhondt, S., de Korte, E., Oeij, P., & Vaas, F. (2012). Workplace innovation in the Netherlands. In I. Houtman (Ed.), *Work life in the Netherlands* (pp. 173–190). Hoofddorp: TNO.

Pot, F. D., Peeters, M. H. H., Vaas, F., & Dhondt, S. (1994). Assessment of stress risks and learning opportunities in the work organisation. *European Work and Organizational Psychologist, 4*(1), 21–37.

Ramstad, E. (2009). Promoting performance and the quality of working life simultaneously. *Internal Journal of Productivity and Performance Management, 58*(5), 423–436.

Ramstad, E. (2014). Can high-involvement innovation practices improve productivity and the quality of working-life simultaneously? Management and employee views on comparison. *Nordic Journal of Working Life Studies, 4*(4), 25–45.

Saksvik, P. Ø., Hammer, T. H., & Nytrø, K. (2013). Social relations at the collective level: The meaning and measurement of collective control in research on the psychosocial work environment. *Nordic Journal of Working Life Studies, 3*(3), 31–46.

Schulte, P. A., Guerin, R. J., Schill, A. L., Bhattacharya, A., Cunningham, T. R., Pandalai, S. P., et al. (2015). Considerations for incorporating "Well-Being" in public policy for workers and workplaces. *American Journal of Public Health, 105*(8), e31–e44. doi:10.2105/AJPH.2015.302616.

Schulte, P., & Vainio, H. (2010). Well-being at work: Overview and perspective. *Scandinavian Journal of Work, Environment & Health, 36*(5), 422–429.

Stansfeld, S. A., Shipley, M. J., Head, J., Fuhrer, R., & Kivimäki, M. (2013). Work characteristics and personal social support as determinants of subjective well-being. *PLoS ONE, 8*(11), e81115, 1–8. doi:10.1371/journal.pone.0081115

Taris, T. W., Schreurs, P. J. G., Eikmans, K. J. L., & Van Riet, P. (2008). Werkkenmerken, welzijn en organisatieprestatie: een toets van de happy-productive worker hypothese op organisatieniveau. (Work characteristics, well-being and organisational performance: A test of the happy-productive worker hypothesis on organisation level). *Gedrag & Organisatie*, *21*(1), 3–18.

Totterdill, P., Dhondt, S., & Milsome, S. (2002). *Partners at work? A report to Europe's policymakers and social partners*. Nottingham: The Work Institute.

Volpert, W., Kötter, W., Gohde, H.-E., & Weber, W. G. (1989). Psychological evaluation and design of work tasks: Two examples. *Ergonomics, 32*(7), 881–890.

Wynne, R., De Broeck, V., Vandenbroek, K., Leka, S., Jain, A., Houtman, I., & McDaid, D. (2014). *Promoting mental health in the workplace. Guidance to implementing a comprehensive approach.* European Commission.

Author Biography

Frank D. Pot is an Emeritus professor of Social Innovation of Work and Employment, Radboud University Nijmegen (NL) and former director of TNO Work and Employment.

Chapter 8
Workplace Innovation and Wellbeing at Work: A Review of Evidence and Future Research Agenda

Aditya Jain, Vlad Dediu, Gerard Zwetsloot and Stavroula Leka

8.1 Wellbeing at Work

Scholars and practitioners agree on the importance of wellbeing (WB) at work, and the extensive research that has been conducted in this area converges on the idea that it is indispensable for both employees and organizations (Costanza et al. 2014; Judge and Kammeyer-Muller 2011). Work organization aspects or psychosocial factors have been shown to be key determinants of WB at work both in terms of theoretical knowledge as well as in terms of applied practice.

The changing nature of work and its related consequences have been under investigation over the past decades. Sparks et al. (2001) identified factors such as restructurings, downsizing and new forms of employment emerging, which influence the workplace, characterized by increasing autonomy and decentralization, with practices such as short term contracts, teleworking and outsourcing being more widespread. At the same time, the understanding that these aspects will have an impact on employee WB was emerging. Taking into account the recent economic events, and the severe effects they had on the workplace, it can be argued that now more than ever it is important for academics, practitioners and other key actors

A. Jain (✉)
Nottingham University Business School, Nottingham, United Kingdom
e-mail: aditya.jain@nottingham.ac.uk

V. Dediu · G. Zwetsloot · S. Leka
Centre for Organizational Health & Development, School of Medicine, University of Nottingham, Nottingham, United Kingdom
e-mail: vlad.dediu@nottingham.ac.uk

G. Zwetsloot
e-mail: gz@gerardzwetsloot.nl

S. Leka
e-mail: Stavroula.Leka@nottingham.ac.uk

P.R.A. Oeij et al. (eds.), *Workplace Innovation*, Aligning Perspectives on Health, Safety and Well-Being, DOI 10.1007/978-3-319-56333-6_8

dealing with policy development to investigate and understand the antecedents of WB (EU-OSHA 2013).

As it was discussed in Chap. 7, WB has been theorized in a number of ways (Schulte and Vainio 2010). Generally, it refers to the individual's subjective appraisal of their lives, and more specifically raises the concept of "optimal psychological functioning and experience" (Ryan and Deci 2001, p. 142). From a philosophical standpoint, much of the research has focused on the hedonic perspective, which argues that WB can be conceptualized in terms of the subjective evaluation of "feeling good", with high levels of positive affect, and low levels of negative affect being the desired state (Diener 2000; Ryff 1995; Waterman 1993). Based on this, in the occupational health psychology area, studies have been concentrated on approaches derived from models such as the circumplex model, looking at affective and psychosomatic aspects of WB (Fisher 2010; Nixon et al. 2011). This general theoretical framework proposes that affect can be operationalized in terms of levels of pleasure and activation (Russel 1980). Furthermore, recently more and more attention has been given to positive aspects of affect, leading to the proposition that WB is higher when an individual experiences appropriate levels of activation and high levels of pleasure (Bakker and Oerlemans 2011). Based on this perspective, WB at work has been conceptualized by some scholars as work engagement, happiness at work and job satisfaction on the positive side, with workaholism and burnout as their negative counterparts (Bakker and Oerlemans 2011; Sonnentag 2015). While, as mentioned in Chap. 7, American scholars refer to WB as a synonym for health and a summative term to describe a flourishing worker who benefits from a safe, supportive workplace, engages in satisfying work, and enjoys a fulfilling work life (Schulte et al. 2015).

8.2 Determinants of Wellbeing at Work

As previously stated, the need to promote and improve the levels of WB in organizations is paramount, here is where workplace innovation comes in as a principal mean in achieving this goal. Europe's commitment to smart growth has focused renewed attention on work organization practices, which are now being discussed under the concept of workplace innovation (WI). Consequently, research examining the potential of workplace innovation to improve health and well-being at work has been identified as a research priority by the European Agency for Safety and Health at Work (EU-OSHA 2013).

In Chap. 7, WI has been defined as a comprehensive set of practices that are reflective, participatory and inclusive, and that are concerned with managing employees, promoting learning and continuous development, and work organization. When discussing how WB can be promoted, several enablers (or determinants) need to be considered; at the individual level: personal ability and social skills; at the task level: autonomy and control over one's work, demands and support; and at the organizational level: management practices, organizational culture and openness to innovation. These enablers are intrinsically linked with WI, as WI practices promote measures that will support and foster the development of the said determinants.

Several of the factors highlighted as aspects of WI (e.g. autonomy and control over one's work, demands and support; management practices, organizational culture) have also been conceptualized in the literature as work organization or psychosocial factors. For example, in a recent publication, Leka et al. (in press) have argued that psychosocial factors should not be conceptualized in terms of negative dimensions alone (as hazards in the workplace). Instead, the psychosocial work environment includes several dimensions that can result in positive or negative outcomes depending on the effectiveness of their management (see Table 8.1).

Work organization and psychosocial factors have been the focus of several years of research and theoretical development and have consistently been shown to be related to numerous outcomes, including WB at work. For example, theoretical models like the Job Demands Control (JDC) model (Karasek 1979) and the Job Demands Resources model (Bakker and Demerouti 2007), have showcased evidence on the relationship between several WI aspects and WB at work. While most of this research has focused on negative facets of WB, e.g. health concerns, mental ill health, and burnout (e.g. Alarcon 2011; Bonde 2008; Eller et al. 2009; Kivimäki et al. 2006), most recently studies have focused on positive facets such as work engagement and job satisfaction (e.g. Hakanen et al. 2008; Taris and Schreurs 2009).

The next section sets out to review evidence on the relationship between the aforementioned factors (in particular autonomy and control over one's work, demands, and support that represent key elements of WI) and WB. Bearing in mind the vast majority of past research conducted (e.g. Bakker and Oerlemans 2011; Sonnentag 2015), we consider job engagement, job satisfaction and burnout as key facets of WB, and where appropriate we also examine health outcomes (e.g. cardiovascular health, musculoskeletal disorders).

8.3 The Evidence Base on Key Determinants of Wellbeing at Work

Throughout the literature several models have emerged that offer good frameworks to test how specific job characteristics pertinent to WI influence WB. In this section we focus on three key theoretical models: the Job Demands Resources (JDR) (Bakker and Demerouti 2007), Job Demands Control (JDC) (Karasek 1979), and the Job Demands Control-Support (JDCS) (Johnson and Hall 1988) models. These models focus on how characteristics such as demands (workload and time constraints), autonomy (or control) and social support (broadly referred to as resources) influence several WB facets. These characteristics are important

Table 8.1 Potential negative and positive psychosocial work environment dimensions

Dimensions	Negative psychosocial work environment	Positive psychosocial work environment
Organizational culture & function	Poor psychosocial safety climate, Poor communication, low levels of support for problem solving and personal development, lack of definition of, or agreement on, organizational objectives	Good psychosocial safety climate, clear organizational objectives, appropriate support for problem solving and personal development, good communication processes
Job content	Lack of variety or short work cycles, fragmented or meaningless work, under use of skills, high uncertainty, continuous exposure to people through work	Meaningful work, appropriate use of skills, work retaining employee interest and engagement, appropriate support
Workload & work pace	Work overload or under load, machine pacing, high levels of time pressure, continually subject to deadlines	Appropriate level of workload, appropriate work pace, sensible and achievable deadlines
Work schedule	Shift working (especially irregular), night shifts, inflexible work schedules, unpredictable hours, long or unsociable hours	Sensible shifts and reasonable working hours to maintain work-life balance, flexible working practices
Control	Low participation in decision making, lack of control over workload, pacing, shift working	Participation in decision making, control at work
Environment & equipment	Inadequate equipment availability, suitability or maintenance; poor environmental conditions such as lack of space, poor lighting, excessive noise	Good physical working conditions according to good practice guidance
Interpersonal relationships at work	Social or physical isolation, poor relationships with superiors, interpersonal conflict, lack of social support, harassment, violence	Good relationships at work, teamwork, social support, appropriate policies and procedures to deal with conflicts
Role in organization	Role ambiguity, role conflict, responsibility for people	Clear roles and responsibilities, appropriate support to meet objectives
Career development	Career stagnation and uncertainty, under promotion or over promotion, poor pay, job insecurity, low social value to work	Appropriate career prospects & development matching skills & performance, effort reward balance, valuable/meaningful work, job security
Home-work interface	Conflicting demands of work and home, low support at home, dual career problems	Work-life balance, supportive organizational policies and practices to achieve 'life balance'

Source Adapted from Leka et al. (in press)

elements of WI. Scientific evidence of these theoretical models would support the development of WI policies.

8.3.1 Negative Facets of WB

Table 8.2 summarizes research regarding the link between several psychosocial factors that are also considered key WI elements, and the negative spectrum of well-being at work. The levels of job demands, control and support, as well as factors like skill discretion, have been found in the literature to be consistently correlated with the incidence of burnout, depression, anxiety, coronary and cardiovascular disease, as well as MSDs.

WI is concerned with workplace practices that encourage participatory and inclusive working methods, and at the same time broaden job design and create high quality work environments. On the basis of the evidence presented, it is clear why promoting WI would be beneficial in not only addressing the aforementioned negative outcomes, but also preventing them from appearing to begin with. Taking into account the similarities between psychosocial factors theories and the idea of WI, the conclusion can be drawn that the promotion of WI might also contribute to the reduction of health and wellbeing risks in the work environment.

8.3.2 Positive Facets of WB

Two main concepts are prevalent in the literature when discussing the positive aspects of WB, namely job engagement and job satisfaction. Work engagement is defined as a positive state, in which individuals feel a sense of fulfilment, vigour, dedication and absorption with their work (Bakker et al. 2014). Job satisfaction was defined by Locke, who described it as a "pleasurable emotional state resulting from the appraisal of one's job" (Locke 1969, p. 317). This section presents evidence on the associations between job characteristics as defined in the JDR and JDC/JDCS models that are key to WI and these two constructs (see Table 8.3).

Overall, it is evident from the reviewed studies that several psychosocial factors such as demands, control, autonomy and support do have a significant effect on both job engagement and job satisfaction as indicators of employee WB. Again, as previously mentioned, WI is concerned with practices aimed at broadening job design, encouraging participatory and empowering working practices and offering organizational support with the aim to improve autonomy, social interactions and reduce unnecessary hindrances, while promoting challenging work. Clearly therefore, WI has a major role to play in not only reducing and combating WB in the negative spectrum, but also promoting positive facets of WB at work.

Table 8.2 Evidence on the link between WI and negative facets of WB

Authors	Methodology	Predictors	Outcome	Main findings
Alarcon (2011)	Meta-analysis of studies	Demands, resources	Burnout (exhaustion, cynicism, reduced personal accomplishment)	All the demands (role ambiguity, role conflict, workload) were positively related to burnout, while for resources (control, autonomy, decision latitude) the opposite was true
Hansen et al. (2009)	Cross-sectional study in a population of nurses; based on the JDR model	Demands, resources	Burnout	Demands (workload, role conflict) positively predicted burnout (emotional exhaustion, cynicism), while a negative relationship was observed for resources (autonomy, goal clarity, job challenge)
Hakanen et al. (2008)	Longitudinal study (over 3 years) in a population of Finnish dentists; based on the JDR model	Demands (home and job), resources (home and job)	Burnout engagement	Both job and home demands significantly and positively predicted burnout at T2, while job resources had a negative effect. The opposite relationships were observed for engagement
Häusser et al. (2010)	Review of studies on the JDC-S model and psychological well-being	Demands, control, social support	A range of general and job-specific wellbeing indicators (psychological distress, mental health, emotional exhaustion, depression, job satisfaction, task satisfaction)	The additive effects of the predictors were observed to hold up in both cross-sectional and longitudinal studies when looking at general WB indicators, but somewhat weaker relationships were observed for job specific WB in longitudinal studies
Rau et al. (2010)	Employees and job analysis experts evaluated the level of demands and resources from the job	Demands, resources	Depression	Objective as well as subjective ratings of job demands were linked with the incidence of major depression. However, individuals that were depressed reported lower levels of perceived job control than both individuals that were not depressed and the objective ratings

(continued)

Table 8.2 (continued)

Authors	Methodology	Predictors	Outcome	Main findings
Backe et al. (2012)	Review of studies on the link between JDC/JDCS	Demands, control, support	Cardiovascular disease (CVD), coronary heart disease, myocardial infarction, heart failure, stroke	High strain (high demands and low control), the risk to develop CVD was increased. Two studies included social support as an additional variable (as in the JDCS model), and showed that when "isostrain", or high strain combined with low social support, was reported, the risk of cardiovascular disease was higher
Kivimäki et al. (2006)	Meta-analysis	Demands, control	Coronary heart disease	After controlling for age, sex and socio-economic status and other lifestyle factors (smoking, alcohol consumption and physical activity), consistent results were found supporting the idea that indeed job strain is positively related to coronary heart disease
Netterstrøm et al. (2010)	Longitudinal study based on the JDCS model	Demands, control, support	Ischemic heart disease	The results did not find support for the idea that either control or demands by themselves directly influence IHD. However, in high strain work environments (high demands combined with low control), an increased prevalence of IHD was observed. For women only, social support from colleagues was also borderline significantly associated with IHD
Haukkal et al. (2011)	Longitudinal study	Skill discretion, control, support	Musculoskeletal disorders (MSDs)	Low skill discretion, low job control and poor support (supervisory support, co-worker relationships) were found to influence the prevalence of musculoskeletal pain among female kitchen workers, even after adjusting at baseline for age, body mass index, smoking, physical exercise, perceived physical workload and existing musculoskeletal pain

(continued)

Table 8.2 (continued)

Authors	Methodology	Predictors	Outcome	Main findings
Hauke et al. (2011)	Review and meta-analysis	Social support, demands, control, decision making latitude, skill discretion	MSDs	Low social support, high job demands, low job control, low decision making latitude, low skill discretion and high job strain had a significant effect, with the risk of developing MSDs being higher in populations reporting these conditions
Eatough et al. (2012)	Cross-sectional study in a sample of employees from multiple sectors; investigating the link between psychosocial factors and MSDs	Demands, control and social characteristics (safety-specific leadership)	MSDs	Findings indicate that more strain responses are reported when role conflict is high and control and safety specific leadership is low. In turn, higher strain was correlated with increased levels of MSD-related complaints in the shoulder, wrist/hand and lower back regions

Table 8.3 Evidence on the link between WI and positive facets of WB

Authors	Methodology	Predictors	Outcome	Main findings
Mauno et al. (2007)	Longitudinal study	Demands, resources	Job engagement	Job demands (job insecurity and time demands) and resources (control, organization-based self-esteem and management quality) influenced engagement at baseline, while over time control was found to be the best predictor of engagement
Christian et al. (2011)	Review of studies	Autonomy, task variety, task significance, feedback, social support	Job engagement	All the predictors emerged as some of the most important precursors of engagement and, furthermore, for autonomy and social support a lagged (or over time) effect was observed as well
Crawford et al. (2010)	Meta-analysis	Job resources, demands (differentiated as challenge demands and hindrance demands)	Job engagement	Challenges had a positive influence on engagement, while hindrances had a negative effect. Furthermore, in accordance with previous research, job resources were again positively related to engagement
Häusser et al. (2010)	Review of studies	Demands, control, social support	Job engagement	Support is found for the additive effects of social support, control and demands on employee wellbeing if sample size is sufficient, but at the same time stronger effects are observed in cross-sectional rather than longitudinal studies
Birtch et al. (2016)	Cross-sectional study based on the JDR model	Demands, autonomy, support	Job satisfaction	Control and support had a positive effect on the outcome measures, while demands had a negative influence
Tims et al. (2013)	Longitudinal study based on the JDR model	Demands (challenges and hindrances), job resources (structural – autonomy, task variety, opportunities for development; and social – social support, feedback and coaching)	Job satisfaction	Following a SEM analysis, all of the resources emerged as significant predictors of both job satisfaction and engagement, but none of the demands had a significant effect on either of the two outcome variables
De Jonge et al. (2010)	Longitudinal study based on the JDC model	Demands (mental, emotional, physical), control	Job satisfaction	Results indicate that none of the three types of demands measured had a direct significant effect on job satisfaction, while for control a strong a positive relationship was observed. However, looking at the interaction between the demands and control, all three interactions (one for each type of demand) were statistically significant. In all cases, as demands increased and control decreased a strong and negative effect was observed on job satisfaction

8.4 Examples of Good Practice to Promote Workplace Innovation and Wellbeing at Work

Having looked at theoretical developments and the evidence base in relation to determinants of several facets of WB, it is also important to consider some examples of good practice that are relevant for policy development and organizational practice. These examples show that practices to reduce psychosocial risks and to enhance wellbeing at work can go hand in hand with improvement of organizational performance such as safety, productivity, quality, innovation. Both of these examples are based on the evidence base discussed so far as well as a review of good practice methods and tools implemented in several countries.

8.4.1 *Vision Zero*

In the industry throughout the world there is a growing attention to 'Vision Zero' (VZ), the ambition and commitment to create and ensure safe and healthy work and prevent all (serious) accidents and occupational diseases in order to achieve excellence in safety, health and wellbeing at work, usually expressed as zero accidents (safety) or zero harm (safety and health). VZ has recently developed also into a major issue in national and international policies for promoting occupational safety and health (OSH) (DGUV 2008; G7 2015; ILO 2015; ISSA 2016).

Simply doing the same things better than before, is not a successful strategy to achieve 'zero'. Both technical and social innovations are needed as well as out-of-the-box thinking for solving existing problems (Zwetsloot et al. 2013). VZ and the associated ambitions should be understood as a journey (process), rather than a target.

Singh (2012) analyses ten new 'Mega Trends' that are likely to have major impacts on the developments in business and society in the coming decade. 'Innovating to zero' is one of the ten Mega trends. According to Singh, 'innovating to zero' implies the desire for perfection in our society: a 'zero concept' world with a vision on zero carbon emissions, zero crime rates, zero accidents, carbon-neutral cities, etc. Innovating to zero is a way of running and innovating businesses while capitalizing on opportunities. Success requires an innovation agenda that talks of breakthroughs in the face of radical goals that intend to create a better world, a zero concept world, which is free of unhelpful externalities and defects. According to Singh, "the most remarkable feature of VZ is that the ultimate opportunity lies not in attaining the actual goal itself, but in capitalizing on the opportunities that would lead to it. Success in innovating to zero requires an innovation agenda that bravely talks of breakthroughs in the face of radical goals—goals that intend to create a better world, a zero concept world, which is free of unhelpful externalities and defects. It also needs a strong culture from people within that ecosystem" (Singh 2012, p. 59).

Focusing on the narrower Zero Accident Vision, Zwetsloot et al. (2017) suggest six innovative perspectives of VZ, which are mutually compatible, and probably synergetic: (1) VZ as the basis for a commitment strategy for safety and health; (2) Excellent OSH as a way of doing business; (3) VZ as the basis for a prevention culture; (4) VZ as a trigger for innovation; (5) VZ as a strong link with business ethics and CSR; and (6) VZ implies networking and co-creation. The processes involved in VZ cannot be realised with existing good practices only: innovative practices are needed. Research into V might elaborate on the theories of socio-technical (re)design, on theories of resilience engineering (Hollnagel et al. 2006) and high reliability organizations (Roberts 1990; Weick and Sutcliffe 2007).

Several of the psychosocial work environment dimensions mentioned in Table 8.1 have a negative or positive impact on safety, e.g. through influencing human error/accuracy, and the capacity to respond adequately to deviations. A meta study by Nahrgang et al. (2011) showed that job demands and resources relate to safety outcomes. In safety literature there is not so much attention to job characteristics but more to organizational characteristics like (safety) culture; justice and trust (Reason 1997) are needed for a safety culture, are also important for wellbeing at work, and probably for the willingness of employees to share ideas not only on the improvement of safety, but also on workplace innovation.

8.4.2 PRIMA-EF

Another good practice example is PRIMA-EF where focus is placed on achieving optimal conditions that can enhance a range of outcomes at organizational level such as the reduction of psychosocial risks and the enhancement of productivity, quality and innovation (Leka and Cox 2008). The Psychosocial Risk Management —European Framework (PRIMA-EF) is intended to be a framework for harmonizing practice and current methods in the area of psychosocial risk management. It can also be used as a guidance tool for the development of further methods both in Europe and internationally as it can provide a benchmark for validation of new methods (Leka et al. 2011). The model has been specified both for use at the organizational and policy levels.

As the model specifies, the organization and management of work processes through the appropriate implementation of work and production methods that incorporate a risk management process can lead to various outcomes that are relevant to both the organizational and societal levels (e.g. workers' health, safety and wellbeing, quality, innovation) (see Fig. 8.1).

Essentially, the risk management process seeks to monitor and manage aspects of the psychosocial work environment (including factors such as demands, resources, and support) to achieve optimal outcomes. This model includes key elements of WI in its consideration of management and organization of work processes that it seeks to manage in the workplace in order to achieve desirable outcomes. Like the implementation of processes to promote WI, psychosocial risk

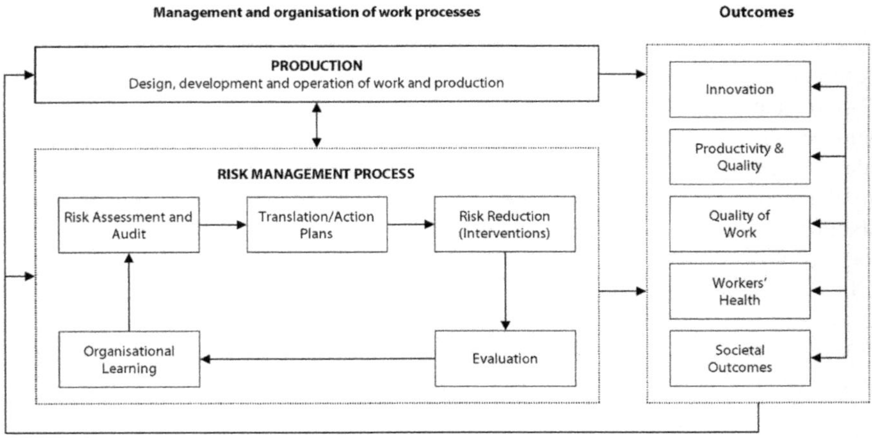

Fig. 8.1 The PRIMA-EF model for the management of psychosocial risks—enterprise level. *Source* Adapted from Leka et al. (2008)

management aims to promote quality of work and innovation and enhance economic performance and competitiveness of enterprises. It can contribute to the creation of positive work environments where commitment, motivation, learning and development play an important role and sustain organizational development (Leka et al. 2008).

PRIMA-EF has been used both by policy makers and organizations. It has been disseminated by the World Health Organization (Leka and Cox 2008) and has been incorporated in their Healthy Workplace model (WHO 2010) that has been actively promoted across WHO regions. The model has also been adapted and is being used in Japan and Brazil. Additionally, it has been incorporated in the first guidance standard on the management of psychosocial risks in the workplace, PAS1010, published in 2011 by the British Standards Institution (BSI 2011).

Furthermore, PRIMA-EF has been implemented in several organizations. For example, it has informed good practice in the oil and gas industry by leading to the development of a comprehensive system including relevant indicators, interventions and auditing practices (e.g. Bergh et al. 2014, 2015, 2016), that has the aim to promote organizational learning and development, innovation, and quality of work.

8.5 Conclusions and Recommendations for Future Research and Practice

This chapter has discussed some key determinants of WB at work and reviewed available evidence in relation to negative and positive facets of WB. As has been shown in the chapter, there is compelling evidence based on longitudinal studies, on the relationship between factors such as autonomy and control over one's work, demands and support with WB at work. These factors, that represent aspects of WI, have also been conceptualized in the literature as work organization or psychosocial factors. This chapter has also highlighted two good practice, evidence-based models that have been used in policy and practice to promote WB and organizational performance. Given the complementarities and convergence found in the literature on the important role of these factors, this last section argues that WI and WB are not only linked but also dependent on one another. Therefore, it is important for interventions at policy, and at organizational level to promote these key common determinants in order to achieve optimal desired outcomes.

In considering the theoretical models discussed in this chapter on WB at work, it becomes obvious that they focus to a different extent on key determinants such as demands, resources and support. JDR has conceptualized resources in terms of both organizational and personal resources, with the latter focusing on individual characteristics (Xanthopoulou et al. 2007). On the other hand, JDC has been extended theoretically to encapsulate macro dimensions, also relevant to the societal level (Karasek 2008). Since WB is a multi-faceted concept, it is logical that theoretical models attempting to capture it will become more comprehensive over time. This is a step in the right direction although there are implications in terms of parsimony and measurement that scholars have to consider going forward.

For example, the WB model developed by Schulte and Vainio (2010) is an ambitious overarching model that takes into consideration various avenues of impact across macro, organizational and individual factors. Further research is needed to test such holistic models and especially to consider inter-relationships and additive effects between the variables and dimensions presented. This is also relevant to the issue that WB encapsulates various positive and negative facets, and so further research is desirable to explore their inter-relationship and any additive effects. Since WI factors are discussed in the existing literature on the basis of various theoretical models, it is important that there is clarity both in terms of concepts used and in terms of good practice in implementing processes to promote desired outcomes at various levels.

Consequently, a third important research avenue concerns intervention studies to promote WI as an avenue to improve WB. It has been acknowledged that intervention studies are a priority due to a need to achieve better WB, but also due to their scarcity, poor design and challenges in their implementation (e.g. EU-OSHA 2010). Intervention studies can take place both at the organizational and macro policy levels, although the latter are less prevalent. It is important for such intervention studies to be based on integrated holistic models and focus on key determinants on the basis of the evidence base. As presented in this chapter, these include control and autonomy, skill discretion, management demands in terms of time and work processes, as well as provision of resources in terms of manager and colleague support. It is also important to conduct evaluation studies of such interventions, considering both their process and their outcomes (Biron and Karanika-Murray 2014). Considering the macro level, it should be noted that interventions should be underpinned by existing legislation where available (e.g. European Union) and promote workplace factors that can have a positive impact on WB, such as the WI factors clarified here. Furthermore, with the policy agenda shifting and the move towards deregulation and practices such as flexicurity (Leka et al. 2015), it is important for more policy relevant research to be conducted focusing on evaluation of the impact of promoting WI policies on desired outcomes.

The final point to be made concerns the existing overlap across concepts and measurement. This is relevant to theoretical models (e.g. on work organization and the psychosocial work environment or on WI) as well as, for example, understanding the difference between WI and innovative work behaviour. Clarity of terminology and conceptualization are essential to move forward and this book represents a step in the right direction. It is important for scholars to consider existing evidence from various disciplines more holistically and use it in their work in order to promote synergies and a common language in theory, research, policy and practice. It is hoped that this chapter has contributed to this end by highlighting overlaps and commonalities in theories of WB at work, and WI in the workplace.

References

Alarcon, G. (2011). A meta-analysis of burnout with job demands, resources, and attitudes. *Journal of Vocational Behavior, 79*(2), 549–562.

Bakker, A., & Demerouti, E. (2007). The job demands-resources model: State of the art. *Journal of Managerial Psychology, 22*(3), 309–328.

Bakker, A. B., Demerouti, E., & Sanz-Vergel, A. I. (2014). Burnout and work engagement: The JD–R approach. *Annual Review of Organizational Psychology & Organizational Behavior, 1*(1), 389–411.

Bakker, A. B., & Oerlemans, W. (2011). Subjective wellbeing in organizations. *The Oxford Handbook of Positive Organizational Scholarship*, 178–189.

Backé, E. M., Seidler, A., Latza, U., Rossnagel, K., & Schumann, B. (2012). The role of psychosocial stress at work for the development of cardiovascular diseases: A systematic review. *International Archives of Occupational and Environmental Health, 85*(1), 67–79.

Bergh, L. I. V., Hinna, S., & Leka, S. (2014). Sustainable business practice: Integrating psychosocial risk management into a company management system. In S. Leka & R. Sinclair (Eds.), *Contemporary occupational health psychology: Global perspectives on research and practice* (Vol. 3). Chichester, England: Wiley-Blackwell.

Bergh, L. I. V., Hinna, S., Leka, S., & Jain, A. (2015). Developing a performance indicator for psychosocial risk in the oil and gas industry. *Safety Science, 62,* 98–106.

Bergh, L. I. V., Hinna, S., Leka, S., & Zwetsloot, G. I. J. M. (2016). Developing and testing an internal audit tool of the psychosocial work environment in the oil and gas industry. *Safety Science, 88,* 232–241.

Biron, C., & Karanika-Murray, M. (2014). Process evaluation for organizational stress and well-being interventions: Implications for theory, method, and practice. *International Journal of Stress Management, 21*(1), 85–111.

Birtch, T. A., Chiang, F. F., & Van Esch, E. (2016). A social exchange theory framework for understanding the job characteristics–job outcomes relationship: The mediating role of psychological contract fulfillment. *The International Journal of Human Resource Management, 27*(11), 1217–1236.

Bonde, J. P. E. (2008). Psychosocial factors at work and risk of depression: A systematic review of the epidemiological evidence. *Occupational and Environmental Medicine, 65,* 438–445.

British Standards Institution (BSI). (2011). *Publicly Available Specification 1010: Guidance on the management of psychosocial risks in the workplace*. London: BSI.

Christian, M. S., Garza, A. S., & Slaughter, J. E. (2011). Work engagement: A quantitative review and test of its relations with task and contextual performance. *Personnel Psychology, 64*(1), 89–136.

Costanza, R., Kubiszewski, I., Giovannini, E., Lovins, H., McGlade, J., Pickett, K., …. Wilkinson, R. (2014). Development: Time to leave GDP behind. *Nature, 505*(7483), 283–285.

Crawford, E. R., LePine, J. A., & Rich, B. L. (2010). Linking job demands and resources to employee engagement and burnout: A theoretical extension and meta-analytic test. *Journal of Applied Psychology, 95*(5), 834.

de Jonge, J., van Vegchel, N., Shimazu, A., Schaufeli, W., & Dormann, C. (2010). A longitudinal test of the demand–control model using specific job demands and specific job control. *International Journal of Behavioral Medicine, 17*(2), 125–133.

DGUV. (2008). Prävention lohnt sich: Die Position der Selbstverwaltung der gesetzlichen Unfallversicherung zur Prävention - Leitlinien und Umsetzung - Mitgliederversammlung 2/08 der DGUV, Fulda.

Diener, E. (2000). Subjective wellbeing: The science of happiness and a proposal for a national index. *American Psychologist, 55*(1), 34–43.

Eatough, E., Way, J., & Chang, C. (2012). Understanding the link between psychosocial work stressors and work-related musculoskeletal complaints. *Applied Ergonomics, 43*(3), 554–563.

Eller, N. H., Netterstrøm, B., Gyntelberg, F., Kristensen, T. S., Nielsen, F., Steptoe, A., et al. (2009). Work-related psychosocial factors and the development of ischemic heart disease: A systematic review. *Cardiology in Review, 17*(2), 83–97.

EU-OSHA—European Agency for Safety and Health at Work. (2010). European Survey of Enterprises on New and Emerging Risks: Managing safety and health at work. European Risk Observatory Report. Luxembourg: Office for Official Publications of the European Communities.

EU-OSHA—European Agency for Safety and Health at Work. (2013). *Priorities for OSH research in Europe: 2013–2020*. Luxembourg: Office for Official Publications of the European Communities.

Fisher, C. (2010). Happiness at work. *International Journal of Management Reviews, 12*(4), 384–412.

G7. (2015). *Action for fair production, ministerial declaration*. http://www.bmz.de/g7/includes/Downloadarchiv/G7_Ministerial_Declaration_Action_for_Fair_Production.pdf. Accessed January 28, 2016.

Hakanen, J., Schaufeli, W. B., & Ahola, K. (2008). The Job Demands-Resources model: A three-year cross-lagged study of burnout, depression, commitment, and work engagement. *Work & Stress, 22*(3), 224–241.

Hansen, N., Sverke, M., & Näswall, K. (2009). Predicting nurse burnout from demands and resources in three acute care hospitals under different forms of ownership: A cross-sectional questionnaire survey. *International Journal of Nursing Studies, 46*(1), 96–107.

Hauke, A., Flintrop, J., Brun, E., & Rugulies, R. (2011). The impact of work-related psychosocial stressors on the onset of musculoskeletal disorders in specific body regions: A review and meta-analysis of 54 longitudinal studies. *Work & Stress, 25*(3), 243–256.

Haukkal, E., Leino-Arjasl, P., Ojajärvil, A., Takalal, E. P., Viikari-Juntural, E., & Riihimäkil, H. (2011). Mental stress and psychosocial factors at work in relation to multiple-site musculoskeletal pain: A longitudinal study of kitchen workers. *European Journal of Pain, 15*(4), 432–438.

Häusser, J. A., Mojzisch, A., Niesel, M., & Schulz-Hardt, S. (2010). Ten years on: A review of recent research on the Job Demand-Control (-Support) model and psychological wellbeing. *Work & Stress, 24*(1), 1–35.

Hollnagel, E., Woods, D. D., & Leveson, N. (2006). *Resilience engineering: Precepts and concepts*. Aldershot: Ashgate.

ILO. (2015). *Our goal: Vision Zero*. Berlin: Federal Ministry of Labour and Social Affairs, G7 Project Group, October 2015. http://www.ilo.org/wcmsp5/groups/public/—europe/—ro-geneva/—ilo-berlin/documents/genericdocument/wcms_414171.pdf. Accessed January 28, 2016.

ISSA. (2016). Safety 7.0—7 Golden rules to vision zero. https://www.issa.int/-/safety-7-0-7-golden-rules-to-vision-zero. Accessed February 12, 2016.

Johnson, J., & Hall, E. (1988). Job strain, work place social support, and cardiovascular disease: A cross-sectional study of a random sample of the Swedish working population. *American Journal of Public Health, 78*(10), 1336–1342.

Judge, T., & Kammeyer-Mueller, J. (2011). Happiness as a societal value. *Academy of Management Perspectives, 25*(1), 30–41.

Karasek, R. (1979). Job demands, job decision latitude, and mental strain: Implications for job redesign. *Administrative Science Quarterly, 24*(2), 285.

Karasek, R. (2008). Low social control and physiological deregulation: The stress–disequilibrium theory, towards a new demand–control model. *Scandinavian Journal of Work, Environment and Health Supplements, 6,* 117–135.

Kivimäki, M., Virtanen, M., Elovainio, M., Kouvonen, A., Väänänen, A., & Vahtera, J. (2006). Work stress in the etiology of coronary heart disease—A meta-analysis. *Scandinavian Journal of Work, Environment & Health, 32,* 431–442.

Leka, S., & Cox, T. (2008). *PRIMA-EF guidance on the European Framework for Psychosocial Risk Management: A resource for employers and worker representatives* (WHO Protecting Workers' Health series, number 9). WHO: Geneva.

Leka, S., Cox, T., & Zwetsloot, G. (2008). The European Framework for Psychosocial Risk Management (PRIMA-EF). In S. Leka & T. Cox (Eds.), *The European Framework for Psychosocial Risk Management: PRIMA-EF* (pp. 1–16). Nottingham, UK: I-WHO Publications.

Leka, S., Jain, A., Cox, T., & Kortum, E. (2011). The development of the European framework for psychosocial risk management: PRIMA-EF. *Journal of Occupational Health, 53,* 137–143.

Leka, S., Jain, A., Iavicoli, S., & Di Tecco, C. (2015—open access). An evaluation of the policy context on psychosocial risks and mental health in the workplace in the European Union: Achievements, challenges and the future. *Biomed Research International, Special issue on Psychosocial Factors and Workers' Health & Safety.* http://dx.doi.org/10.1155/2015/213089

Leka, S., Jain, A., & Lerouge, L. (in press). Work-related psychosocial risks: Key definitions and an overview of the policy context in Europe. In L. Lerouge (Ed.), *Psychosocial risks in labour and social security law.* The Netherlands: Springer.

Locke, E. A. (1969). What is job satisfaction? *Organizational Behavior and Human Performance, 4*(4), 309–336.

Mauno, S., Kinnunen, U., & Ruokolainen, M. (2007). Job demands and resources as antecedents of work engagement: A longitudinal study. *Journal of Vocational Behavior, 70*(1), 149–171.

Nahrgang, J. D., Morgeson, F. P., & Hofmann, D. A. (2011). Safety at work: A meta-analytic investigation of the link between job demands, job resources, burnout, engagement and safety outcomes. *Journal of Applied Psychology, 96*(1), 71–94.

Netterstrøm, B., Kristensen, T., Jensen, G., & Schnor, P. (2010). Is the demand-control model still a useful tool to assess work-related psychosocial risk for ischemic heart disease? Results from 14 year follow up in the Copenhagen City Heart study. *International Journal of Occupational Medicine and Environmental Health, 23*(3), 217–224.

Nixon, A., Mazzola, J., Bauer, J., Krueger, J., & Spector, P. (2011). Can work make you sick? A meta-analysis of the relationships between job stressors and physical symptoms. *Work & Stress, 25*(1), 1–22.

Rau, R., Morling, K., & Rösler, U. (2010). Is there a relationship between major depression and both objectively assessed and perceived demands and control? *Work & Stress, 24*(1), 88–106.

Reason, J. T. (1997). *Managing the risks of organisational accidents*. Aldershot: Ashgate.

Roberts, K. H. (1990). Some characteristics of high-reliability organizations. *Organization Science, 1,* 160–177.
Russell, J. (1980). A circumplex model of affect. *Journal of Personality and Social Psychology, 39* (6), 1161–1178.
Ryan, R., & Deci, E. (2001). On happiness and human potentials: A review of research on hedonic and eudaimonic wellbeing. *Annual Review of Psychology, 52*(1), 141–166.
Ryff, C. (1995). Psychological wellbeing in adult life. *Current Directions in Psychological Science, 4*(4), 99–104.
Schulte, P., & Vainio, H. (2010). Well-being at work: Overview and perspective. *Scandinavian Journal of Work, Environment & Health, 36*(5), 422–429.
Schulte, P. A., Guerin, R. J., Schill, A. L., Bhattacharya, A., Cunningham, T. R., Pandalai, S. P., et al. (2015). Considerations for incorporating "Wellbeing" in public policy for workers and workplaces. *American Journal of Public Health, 105*(8), e31–e44.
Singh, S. (2012). *New mega trends—Implications for our future lives*. Hamshire (UK), New York (US): Palgrave Macmillan.
Sonnentag, S. (2015). Dynamics of wellbeing. *Annual Review of Organizational Psychology and Organizational Behavior, 2*(1), 261–293.
Sparks, K., Faragher, B., & Cooper, C. (2001). Wellbeing and occupational health in the 21st century workplace. *Journal of Occupational and Organizational Psychology, 74*(4), 489–509.
Taris, T. W., & Schreurs, P. J. (2009). Wellbeing and organizational performance: An organizational-level test of the happy-productive worker hypothesis. *Work & Stress, 23*(2), 120–136.
Tims, M., Bakker, A. B., & Derks, D. (2013). The impact of job crafting on job demands, job resources, and wellbeing. *Journal of Occupational Health Psychology, 18*(2), 230.
Waterman, A. (1993). Two conceptions of happiness: Contrasts of personal expressiveness (Eudaimonia) and hedonic enjoyment. *Journal of Personality and Social Psychology, 64*(4), 678–691.
Weick, K. E., & Sutcliffe, K. (2007). *Managing the unexpected*. San Francisco: Jossey-Bass.
World Health Organization (WHO). (2010). *Healthy workplaces: A model for action: For employers, workers, policy-makers and practitioners*. Geneva: WHO.
Xanthopoulou, D., Bakker, A. B., Demerouti, E., & Schaufeli, W. B. (2007). The role of personal resources in the Job Demands-Resources model. *International Journal of Stress Management, 14*(2), 14–121.
Zwetsloot, G. I. J. M., Aaltonen, M., Wybo, J. L., Saari, J., Kines, P., & Op De Beeck, R. (2013). The case for research into the zero accident vision. *Safety Science, 58,* 41–48.
Zwetsloot, G. I. J. M., Kines, P., Wybo, J. L., Ruotsala, R., Drupsteen, L., & Bezemer, R. A. (2017). Zero Accident Vision based strategies in organisations: Innovative perspectives. *Safety Science, 91,* 260–268.

Author Biographies

Aditya Jain Assistant Professor of Human Resource Management, Nottingham University Business School (UK), Associate Editor 'Safety Science' and Executive Officer of the European Academy of Occupational Health Psychology.

Vlad Dediu M.Sc. (Hons). Work and Organizational Psychology and doctoral researcher at the University of Nottingham, editorial officer of the European Academy of Occupational Health Psychology.

Gerard Zwetsloot is director of Gerard Zwetsloot Research and Consultancy (NL, see: www.gerardzwetsloot.nl), and honorary professor in Occupational Health and Safety Management at the University of Nottingham (UK).

Stavroula Leka is a Professor of Work, Health & Policy, and Director of World Health Organization Centre for Organizational Health & Development, University of Nottingham.

Part III
Research

Chapter 9
Framing Workplace Innovation Through an Organisational Psychology Perspective: A Review of Current WPI Studies

Arianna Costantini, Riccardo Sartori and Andrea Ceschi

9.1 Introduction

The European Workplace Innovation Network (EUWIN) has defined workplace innovation (WPI) as a bundle of practices and programs involving changes in business structure, Human Resources management, the relationships with clients and suppliers, or in the work environment itself. From this perspective, what characterizes WPI are the improvements that it is supposed to engender, such as higher motivation at work, better working conditions for employees, increased labour productivity, innovation capability, market resilience, and overall business competitiveness. That is, according to this perspective, all enterprises, irrespective of their size, can benefit from WPI.

Given these potential benefits of WPI for organisational performance and employee well-being, it is not surprising that WPI has received substantial interest from policy-makers, practitioners and scholars from different fields. Indeed, it has come to be seen as a fundamental factor to rely on in order to successfully face intensifying global competition and technological advancement (Boxall and Purcell 2016). In particular, the multidisciplinary perspective to the study of WPI is evidenced in the simultaneous attention to the effects of introduced innovations on individuals (for example in terms of motivation, attitudes, engagement), and on the organisation as a whole.

A. Costantini (✉) · R. Sartori · A. Ceschi
Department of Human Sciences, Verona University,
Lungadige Porta Vittoria, 17, 37129 Verona, Italy
e-mail: arianna.costantini@univr.it

R. Sartori
e-mail: riccardo.sartori@univr.it

A. Ceschi
e-mail: andrea.ceschi@univr.it

P.R.A. Oeij et al. (eds.), *Workplace Innovation*, Aligning Perspectives on Health, Safety and Well-Being, DOI 10.1007/978-3-319-56333-6_9

Nevertheless, this multidisciplinary focus on WPI has led to a variety of different conceptualisations based on the framework of study adopted. It seems that WPI has been investigated at different levels of analysis, for example, in terms of national programs and/or of organisational studies, and within different disciplines, i.e. sociology, management, and organisational behaviour. Even though such a heterogeneous body of literature may be considered to be a necessary multidisciplinary aspect in order to increase the understanding of WPI as a whole, it can create misunderstandings for practitioners who want to foster WPI inside organisations and need evidence of policy outcomes to rely on when making implementation decisions (Beauregard and Henry 2009).

In order to deepen our understanding of the effects of innovations in the workplace, a new integration between developments in theoretical and practical knowledge among different disciplines (Dhondt and Van Hootegem 2015) is needed. In this respect, one area of study, that is, work and organisational psychology (WOP), can provide important insights to promote a deeper understanding of the factors influencing the effective implementation of WPI because it focuses on the understanding of employee motivations, attitudes, and cognitions in the workplace.

In other words, integrating findings from research in WOP with findings from research on WPI could help us better define the linkages and mechanisms through which innovations in the workplace affect employees' attitudes and behaviours, as well as organisational performance. In particular, validated theoretical models and evidence from research in the WOP field provide a wide body of knowledge devoted to understanding how individual motivations, attitudes, and behaviours develop and change according to environmental and contextual influences. In turn, this could represent a relevant and valuable contribution to the understanding of WPI. That is, merging findings from studies investigating WPI and research in WOP could significantly advance both the research and the applied agendas regarding the design, implementation and evaluation of WPI inside organisations.

Therefore, this chapter aims to present an overview of studies on WPI with a specific focus on showing how WOP can contribute to this discussion. Such an approach supports the need for unpacking the mechanisms through which innovation impacts individual and organisational outcomes (Takeuchi et al. 2009).

Before exploring the intersections between WPI and WOP that may contribute to the enrichment of our understanding of WPI, we will present an overview of the concept of WPI, in order to clarify how it has been conceptualised, what it constitutes of and what outcomes are expected from it.

9.2 Domain of WPI

The conceptualisation of WPI has been characterised by definitional variety both among academics and practitioners. Nevertheless, one common feature found across these various conceptualisations has been a tendency to define WPI in terms of its outcomes, specifically, quality of working life and organisational performance

(Pot and Koningsveld 2009). This focus on defining WPI by means of its outcomes rather than by its contents (i.e., practices, policies and initiatives per se) appears tautological because it does not question whether and how WPI interventions are indeed effective in fostering better working experiences and higher organisational performance. On the contrary, it merely assumes that innovations have a positive impact on workers and organisations (Boxall and Macky 2007), a tendency referred to as the maximization fallacy about innovations (Ramos et al. 2016), without considering the specific mechanisms involved in the successful implementation of such practices.

A notable exception to this is the definition of WPI proposed in the report of the third European Company Survey, which focuses on actual practices, as opposed to expected outcomes of WPI. Specifically, it defines WPI as a "developed and implemented practice or combination of practices that either structurally (through division of labour) or culturally (in terms of empowerment of staff) enable employees to participate in organisational change and renewal and hence improve the quality of working life and organisational performance" (Oeij et al. 2015, p. 14). Paying particular attention to the interventions constituting WPI, this conceptualisation identifies two main types of processes pertaining to the introduction of innovations in the workplace: the former concerns structural changes related to production systems and the design of the organisational model; the latter focuses on social aspects fostering positive work behaviours and attitudes, and promoting higher motivation at work.

Even though such a definition is useful to disentangle the different dimensions involved in WPI implementation, it must be noted that the factors constituting both of the proposed dimensions are naturally intertwined, in that institutions are embedded in culture and individuals are embedded in both culture and institutions. That is, the willingness to implement structural changes in the organisation is grounded in values and norms, which are elements of the organisational culture. Likewise, implemented practices aimed at developing and fostering a particular vision of the organisational culture need structural support in order to be implemented. Such an understanding of the interdependent nature of organisational culture and structure is crucial, if WPI implementations are to be effective.[1]

That being said, the clear distinction between processes that are related to structure and processes that are related to culture provides researchers with the opportunity to gain a deeper understanding of the extent to which innovations are associated with positive outcomes for employees and organisations. That is, conceptualising WPI as consisting of these two dimensions, i.e. structure orientation and culture orientation, allows researchers to unpack which specific features of WPI may benefit from being investigated within a specific research field rather than within another. In this context, WOP researchers and practitioners could provide important insights in understanding and supporting the *culture orientation* dimension of WPI, for example, by means of job redesign interventions aimed at fostering

[1]The authors wish to thank Marta Strumińska-Kutra for this precious argumentation.

positive organisational outcomes, such as work engagement and organisational commitment. On the other hand, scholars and practitioners from other research fields, such as management science and sociology, may have more to contribute to our understanding of the *structure orientation* dimension of WPI, providing their knowledge concerning the design of organisational processes and the management of the relationships between the organisation and its stakeholders.

Additionally, in line with the need to clearly focus on the nature of WPI rather than on its outcomes, some authors have suggested that WPI could represent a *special capacity of the organisation* that consists of four sources, namely strategic orientation, smarter organising, flexible working, and product-market improvement (Eeckelaert et al. 2012; Oeij et al. 2014). This perspective, rather than focusing on the nature of the processes constituting WPI (i.e. cultural or structural), highlights the different nature of the sources of WPI strategies. Specifically, it focuses on how external and internal organisational conditions influence the adoption of specific policies and programs. Within this framework, strategic orientation and product-market improvement are policies that focus on conditions external to the organisation that need to be taken into account. Smarter organising and flexible working represent policies that focus on internal organisational issues (Eeckelaert et al. 2012) and are to be considered the main areas in which WOP professionals inside organisations can work on to foster the effectiveness of innovations in the workplace.

In addition, this framing of WPI sources in terms of organisational capacity supports the argument of the bi-dimensional nature of WPI, given that each source may be applied in both a structural and/or a cultural fashion. For example, when the final goal is that of re-organising work in terms of division of labour, a structural orientation will be adopted. However, as previously stated, a structural change in the organisation of work needs to be rooted in a specific cultural dimension. That is, prompting a cultural orientation could represent an alternative or complementary strategy in order to promote the reorganisation of work based on the involvement of employees in the adopted structural changes.

However, although useful for describing and defining WPI, the distinction between these two main dimensions is, in practice, an artifice. Once again, we want to highlight the fact that the factors involved in workplace innovations are interdependent. In other words, the structure and culture orientations are deeply intertwined and result, to different extents, in the aforementioned outcomes: quality of working life (QWL) and organisational performance (OP).

Despite definitional agreement in identifying QWL and OP as the two main expected outcomes of WPI, very limited attempts have been made to clearly define indicators for QWL and OP. In particular, although QWL is a rather old concept (Davis and Cherns 1975), there is still debate regarding its nature (Martel and Dupuis 2006) and no commonly accepted definition has emerged. Studies investigating the QWL-WPI relationship have primarily measured it in terms of organisational commitment or job satisfaction (Dhondt et al. 2014; Oeij et al. 2012). We are unaware of studies investigating how work engagement or work-life balance may be affected by the introduction of WPI. In this respect, taking a WOP perspective on WPI can provide potential advantages because evidence from research in this field offers important explanations of the factors enhancing positive work and organisational attitudes, which are likely to constitute important dimensions of QWL.

Drawing on these premises, the following paragraph will be devoted to explore how the adoption of a WOP framework could contribute to a deeper understanding of factors constituting WPI and its potential effects.

9.3 Intersections Between WOP and WPI

It has been argued that, in order to foster the success of proposed innovation policies in the workplace, it is necessary to consider and imagine how *the pattern of multiple proposed actions* would be linked to the achievement of pursued outcomes (Delery and Doty 1996). That is, given that a policy aimed at fostering innovation in the workplace is introduced within an already established organisation, it is fundamental to consider how the whole range of factors already present in the organisation could impact the effectiveness of the policy, as well as interact with it.

In this regard, WOP research and, in particular, theoretical models developed to investigate how organisational design is related to work attitudes and behaviours, could provide a valuable framework for WPI-policy design. In fact, they would provide not only a strong evidence-based approach to policy design but also the opportunity for being tailored to the unique needs of the organisation. Accordingly, some WPI literature has already adopted a WOP perspective to the study of organisational design aimed at understanding how it can influence employees' health and QWL.

Specifically, current studies have mainly referred to the *Job Demand-Control Model* (JDC) developed by Karasek and Theorell (1992), Karasek (1979). The JDC assumes that work organisation, and, in particular, high control in performing tasks and activities at work, is a key-factor in transforming job demands into opportunities for learning as opposed to risks and stress drivers (Holman et al. 2012). Although the WOP literature has provided considerable support for the hypothesis that the combination of high job demands and low job control is an important predictor of psychological strain and illness (Schnall et al. 1994), support for the hypothesis that control can moderate the negative effects of high demands on well-being is less consistent (de Jonge and Kompier 1997; Van der Doef and Maes 1999). Hence, although the JDC might be useful to gain an understanding of the relevance of organisational design on employee well-being and organisational sustainability, support for the model has been relatively mixed. Moreover, within WOP research more recent organisational models have emerged that factor in a wider range of organisational resources and demands than just job control and work overload (Bakker and Demerouti 2007). However, to our knowledge, it appears that the literature on WPI has never established a connection with these more recent models.

In this regard, the *Job Demands-Resources Model* (JD-R) (Demerouti et al. 2001) could provide an important theoretical basis for the design, implementation and monitoring of WPI practices inside organisations. The JD-R assumes that, whereas every occupation may have its own specific risk factors associated with job stress, these factors can be clustered into two broad categories, i.e. job demands and job resources. Hence, it goes beyond the limits of the JDC, which basically considers only a limited amount of the several factors influencing employees' work outcomes (i.e. work overload, time pressure, and job control) and may be applied to several occupational settings, irrespective of specific professional demands and resources (Bakker and Demerouti 2007). Job demands refer to organisational features requiring employee physical and psychological effort that can result in psychological stress, whilst job resources refer to all those elements in the work environment that help individuals achieve their goals, stimulate personal growth and reduce job demands by facing them. In addition, the JD-R posits that two different underlying psychological processes play a role in the development of negative (job strain) and positive work outcomes. Specifically, it argues that chronic job demands lead to employees' feelings of exhaustion whilst job resources have a motivational potential that foster higher work engagement, individual performance, and work motivation. Research on the JD-R has found strong empirical support for the idea that job demands are predictors of negative work outcomes such as burnout and exhaustion, whilst job resources have been found to predict higher levels of work engagement, extra-role performance, organisational commitment and lower absence duration (Bakker et al. 2003a, b, 2004; Schaufeli and Bakker 2004). Moreover, there is evidence that job resources have positive effects on the relationship between job demands and well-being. Specifically, studies have shown that higher autonomy, feedback, perceived social support and a high-quality relationship with the supervisor can buffer the negative effects of work overload, emotional and physical demands, and work-home interference (Bakker et al. 2005).

These results suggest that the JD-R could represent an important tool for policy-makers and WOP professionals who want to foster WPI inside organisations, because it provides a clear framework for the implementation of innovations in the workplace. Besides providing an evidence-based account for understanding the relationships between resources, demands and work and organisational outcomes, this model provides a paradigmatic approach to the study of organisational variables that influence employees' attitudes and behaviours when introducing WPI. For example, it can represent a reliable means to identify which organisational resources are in specific need of innovations, or, what job demands need to be rethought in order to render them challenging rather than exhausting.

Further, a focus on the positive outcomes related to high job resources permits to shed light on the nature of QWL, which, as previously mentioned, remains a debated concept in need of further clarification. In this respect, applying the JD-R to the study of several organisations operating in a wide range of sectors provided evidence of effects deriving from the demands-resources relationship that are relevant to both QWL and OP. For what concerns the former, the study of organisations through the lens of the JD-R shows that job resources represent one of the most important drivers of work engagement (Bakker and Demerouti 2008), which is defined as a positive, fulfilling, work-related state of mind characterised by vigour, dedication, and absorption (Schaufeli et al. 2002). Moreover, studies have found that engaged employees have high levels of energy, are enthusiastic about their work and are often fully immersed in it (May et al. 2004). Given that QWL constitutes a relatively vague concept related to the well-being of workers, work engagement seems to be a more concrete concept that could constitute one core-dimension of QWL. In addition, work engagement may also potentially affect OP. For instance, research investigating the link between work-engagement and OP, despite the substantial heterogeneity in the way in which performance was measured and conceptualised, found support for the higher engagement-higher performance link (Demerouti and Bakker 2006; Salanova et al. 2005; Xanthopoulou et al. 2007).

Given the above research evidence from a JD-R perspective and given that QWL and OP are defined as the two major outcomes of WPI, the JD-R framework appears to be an effective approach to promote WPI practices that foster high QWL and subsequently, higher OP. Moreover, since policies aiming to foster work engagement must be well-integrated and connected in order to be effective (Gruman and Saks 2011), investigating ways to promote WPI through the JD-R may be effective in not only designing innovative policies aimed at improving job resources but also at harmonising job demands and resources, thereby, promoting higher QWL and OP.

In an attempt to facilitate the integration of research on WOP and WPI and to disentangle the mechanisms underlying the effective implementation of WPI policies, in the following section we present evidence from research on three main concepts: job autonomy, job flexibility and participation in organisational life.

9.4 Examining Current WPI Studies Assuming a WOP Perspective

In this section, we examine findings from WPI studies focusing on three constructs that represent a set of essential, yet not exhaustive, organisational factors that policy makers mainly rely upon when aiming to foster WPI inside organisations: job autonomy, job flexibility, and participation in organisational life. The rationale for sampling these concepts was inspired by the definition of WPI as consisting of a structure and a culture orientation, which, as stated, although useful from a theoretical point of view, must be seen as artificial when it comes to actual practices in organisations. However, assuming this distinction for research purposes helps us understand how interventions specifically aimed at changing organisational design or organisational climate could have an effect on QWL and/or OP.

In identifying relevant literature, we referred to the definitional distinction proposed in the Third European Company Survey and searched for the main psychological constructs representing the contents of the cultural and structural dimensions among current WPI studies. In order to do so, we identified via computer (i.e. PsycINFO, Web of Science, Google Scholar database, which mainly include peer-reviewed journal articles and therefore were considered relevant to our review) WPI publications of the last two decades. Specifically, we searched for studies focused on the relationship between WPI and the three core constructs mentioned above by using the following keywords in relation to (i.e. using "and" as a search option) job autonomy, job flexibility, and participation in organisational life: workplace innovation, quality of working life, high-involvement work systems, organisational innovation, high performance work systems, strategic human resource management, HPWS (the acronym of high performance work systems), organisational climate.

We decided to include only English language sources in order to focus on a transcultural level of analysis. Moreover, these keywords were supposed to appear in the title and/or in the abstract.

We acknowledge that these inclusion/exclusion criteria have factually excluded a large body of WOP research on the three core-constructs considered here (i.e. job autonomy, job flexibility, and participation in organisational life). However, this contribution aims at moving us further along the path of introducing a link between WPI and WOP by focusing on the current state of the art in WPI research. Such an approach aims at fostering future investigations that focus on understanding how WOP could more exhaustively contribute to, and enrich, WPI.

Regarding the papers eventually considered for the analysis (13), it appears that the majority of the identified studies were published in journals in the fields of economics, management, and sociology, with limited references to journals in the field of applied psychology. This limited amount of identified articles is most likely due to our adoption of the proposed definition of WPI as constituted by two main dimensions and the three specific constructs under investigation, which could also be interpreted as a signal of the need for more definitional clarity in the field of WPI.

Table 9.1 reports an overview of the studies considered, specifying the relationships among the different dimensions of WPI, and QWL and OP.

9.4.1 Job Autonomy

Job autonomy is defined as the amount of discretion employees have to carry out tasks, to establish methods of work and the speed or rate of it (Hackman and Oldham 1976; Oldham et al. 1976). Overall, the positive effects of job autonomy on employee well-being, motivation (Karasek 1979; Parker 2003; Singh 2000), and performance have been found to lead to positive organisational outcomes, especially when combined with other organisational practices (Appelbaum et al. 2000). Below, a selected number of studies will be reported that provide an overview of

Table 9.1 Overview of studies included in the review

References	Methodology		WPI dimensions			Outcomes	
			Structure		Culture		
	Quantitative	Qualitative	Job autonomy	Job flexibility	Participation	QWL	OP
Bond and Flaxman (2006)	X		X				+
Dhondt et al. (2014)	X		X		X	+	
Holman et al. (2009)	X		X				+
Humphrey et al. (2007)	X		X			+	+
Oeij and Vaas (2016)	X			X			+
Oeij et al. (2012)	X		X	X		+	+
Oeij et al. (2014)	X	X	X	X	X	+	+
Oeij et al. (2015)		X	X	X	X	+	+
Parker et al. (1997)	X		X				+
Preenen et al. (2015)	X			X			+
Preenen et al. (2016)	X		X				+
Taris et al. (2003)	X		X				+
Zhou et al. (2011)	X			X			+

Note QWL Quality of Working Life; *OP* Organisational Performance; + Positive relationships found findings regarding the relationship between job autonomy and organisational outcomes in terms of QWL and OP.

The first study we will discuss is one conducted by Preenen et al. (2016), which investigated the relationship between employees' job autonomy, self-reported company performance (in terms of revenue and profit), and the moderating role of company maturity among 3311 companies in the Netherlands. They found a main effect for the job autonomy-company revenue relationship. In addition, they found that company maturity moderated the job autonomy-organisational performance link. Specifically, job autonomy was positively associated with employees' perception of company revenue and profit growth, but only for young companies, aged two to five years. Such a moderating role of company age is of particular interest given that, generally, job autonomy is hypothesised to be positively related to organisational outcomes, regardless of the company's maturity. Overall, these findings support the assumption that job autonomy is a key feature to foster positive perceptions of a company's growth (Preenen et al. 2016).

In another study among 2359 call centres in 16 countries, Holman et al. (2009) explored how decisions about work design affect organisational outcomes. They found empirical evidence that job autonomy was negatively associated with voluntary turnover and labour costs, indicating that higher job autonomy enabled employees to better manage and cope with task demands.

Regarding the relationship between job autonomy and indicators of QWL, such as active learning behaviours and higher involvement within the work environment, a longitudinal study conducted within the JDC framework among 876 teachers found a positive effect of job autonomy in promoting high levels of learning (Taris et al. 2003). This finding is in line with arguments made by Parker et al. (1997) whereby job autonomy seems to be a mechanism allowing hands-on learning which gives employees the opportunity to interact with their environment and, at the same time, become more involved in and more knowledgeable about it. Moreover, as noted by the authors, such an experience might then potentially lead to a broader ownership of problems and a more proactive view of performance, interpreted, for example, in terms of the learning process itself (Parker et al. 1997).

In another longitudinal study among call centre workers in the UK, job control, along with individual psychological flexibility and the interaction between these two factors, was shown to predict people's ability to learn a new ITC application, employees' mental health and job performance (Bond and Flaxman 2006).

Overall, research evidence suggests that autonomy, beyond fostering job satisfaction and well-being, could also enhance performance, for example, by enabling quicker responses to problems, due to a more developed understanding of roles (Parker et al. 1997). Finally, job autonomy appears to be an essential element in allowing workers to establish how to pursue their goals and to redefine or optimise paths toward goal accomplishment (Humphrey et al. 2007).

9.4.2 *Job Flexibility*

Defining flexible work as a dimension of WPI directed at optimising personnel availability, Oeij and Vaas (2016) investigated the role of WPI on perceived organisational performance and on sickness absence. In this study, WPI was conceptualised as a special capacity of the organisation consisting of four sources, as previously described (i.e. strategic orientation, smarter organising, flexible working, and product market improvement). Using data collected from a sample of Dutch for profit and non-profit organisations, they found that, among all sources considered, flexible work and organising smarter were those that contributed the least to perceived organisational performance. As an explanation for this surprising finding, the authors suggest that externally oriented resources, such as strategic-orientation and product-market improvement, might be more visible to employees than their counterparts, i.e. flexible work and organising smarter. This, availability bias, in turn could lead to an overestimation of the effects of the externally oriented sources and an underestimation of the effects of the internal sources on performance. Overall, even though reporting a weak effect of flexible work on OP, this study represents an important contribution to the understanding of WPI due to its focus on the differential effects of various sources. Such an approach permits to investigate how different dimensions of WPI do or do not contribute to expected organisational outcomes, i.e. QWL and OP. Moreover, this study found that organisations more active with WPI reported higher perceived organisational performance than organisations less active with WPI. In addition, this relationship was strongest for organisations that were active on more than one of the cited resources simultaneously, confirming the importance of taking a systemic approach to the introduction of WPI.

In another study, using longitudinal firm-level data, Zhou et al. (2011) investigated the role of flexibility on innovation. Specifically, they found that functional flexibility (i.e. the ability of firms to reallocate labour in their internal labour markets, by relying on training that allows personnel to carry out a wider range of tasks) was positively associated with innovation by reducing barriers to knowledge sharing and allowing the building of multiple competencies among employees (Zhou et al. 2011), which may represent elements for improving QWL.

Focusing on internal labour flexibility practices (ILFPs), which reflect the measures that companies take to help their employees in flexibly performing different tasks and roles in their organisation, Preenen et al. (2015) investigated the relationships between these and labour productivity and innovation performance at the company level, in two studies conducted among 4648 companies in the Netherlands. Results showed that ILFPs stimulate labour productivity and company innovation as reported by directors or HR managers.

Taken together, these findings support the value of a deeper investigation of how flexible policies may represent a resource for companies in adjusting to constant dynamic circumstances, by stimulating innovative and creative behaviour along with other positive organisational outcomes such as commitment, learning and knowledge sharing (Preenen et al. 2015), which may constitute dimensions of QWL.

9.4.3 Participation in Organisational Life

Workplace participation, usually defined as the degree to which employees influence decision-making in organisations, is recognized as one of the major drivers of positive outcomes for organisations, such as generic organisational efficiency and workers' well-being and health (Knudsen et al. 2011).

Studies carried out to investigate the role of participation in the workplace identified different ways in which it can be exercised, such as by individual employees, teams and by employee representatives (Hagen and Trygstad 2009; Walters and Nichols 2007). Nowadays, organisations need to find a more dynamic way of conceptualising workplace participation in terms of new job configurations, so as to be able to face the needs and challenges posed by new work and career patterns. That is, more recent HRM practices developed on the basis of the current trends in WOP research tend to promote organisational innovations fostering bottom-up approaches to deal with the needs of the current workforce (Demerouti 2015), which, in turn, improve working conditions by means of higher job control. This type of approach is necessarily based on higher employee participation in organisational life, since individuals are encouraged to adjust their work environment in order to promote and achieve higher QWL and OP.

Knudsen et al. (2011) explored whether employee participation influenced the quality of the work environment and worker well-being at 11 Danish workplaces. Findings from interviews with employees and managers and from questionnaires administered to employees revealed that only democratically governed workplaces led to the experience of a high quality of the psychosocial work environment among employees. Nevertheless, this study also suggests that, when control systems in the workplace systematically demand more from employees than what they can deliver, participation cannot buffer the negative effects of the control system on employees' psychosocial well-being (Busck et al. 2010). These findings suggest that there is a need to take into account the level of job control to allow for the positive effects of participation to emerge.

In addition, the role of organisational level decision latitude on organisational commitment, which may represent a facet of QWL, has been investigated in a study among 2048 employees from six different European countries. Using data from the European Working Condition Survey of 2010, Dhondt et al. (2014) found empirical evidence that among the three different dimensions of job control, i.e. job autonomy, functional support and organisational level decision latitude (OLDL), job autonomy is related to subjective well-being only in combination with OLDL. This suggests that organisations would need to consider all the three dimensions in order to foster QWL. Moreover, they also found that functional support and OLDL are related to organisational commitment more strongly than job autonomy, and that organisational commitment was highest when all the three dimensions were present at the same time (Dhondt et al. 2014). That is, to enhance organisational commitment and well-being, which represent two dimensions of QWL, the different dimensions of job control should be aligned and promoted congruently.

9.5 Conclusions

This contribution aimed at proposing a conceptual integration of the domains of WPI and WOP, in order to deepen our understanding of the potential advantages offered by such an integrative perspective to innovation in organisations. Drawing on the conceptualisation of WPI as composed of two dimensions, i.e. a structural and a cultural orientation, we identified three main constructs that represent organisational practices at the basis of WPI interventions: job autonomy and job flexibility (i.e., the structural orientation of WPI), and participation in organisational life (i.e., the cultural dimension of WPI).

According to Klein et al. (2001), a strong climate for innovation implementation, created by the support of management through a clear and strategic vision for it, represents a fundamental factor in order to create an institutional context informing employees that implementation of innovation is important and even rewarded (Choi and Chang 2009). This reasoning is aligned with the vision of an innovation environment where institutions, organisational cultures and individuals are intertwined and reciprocally influence each other. That is, it suggests that higher QWL and OP are simultaneously achievable when all these different levels of analysis are taken into account as potential factors influencing the introduced innovations. Based on these premises, the first practical issue to be addressed when designing WPI refers to the creation of a supportive implementation context in which management support and encouragement toward innovation can foster employees' positive beliefs in this regard (Purvis et al. 2001; Russell and Hoag 2004). In achieving this, the contribution of WOP research is particularly relevant.

To conclude, evidence from research reported in the cited studies shows the intertwined nature of organisational factors in promoting higher QWL and OP. Indeed, it supports the need to simultaneously consider a multitude of features affecting WPI processes aimed at improving QWL and OP, which, in turn, implies taking a systemic perspective on WPI implementation.

Acknowledgements The authors thank editor Diana Rus who provided insight and comments that greatly improved the manuscript.

The authors and publisher gratefully acknowledge the following permission to use the material in this book: Arianna Costantini, Riccardo Sartori & Andrea Ceschi, Reviewing psychological facets of workplace innovation In: European Work and Organizational Psychology in Practice. Special issue on Workplace Innovation, 2017, Volume 1, 6–18.

References

Appelbaum, E., Bailey, T., Berg, P., & Kalleberg, A. L. (2000). *Manufacturing advantage: Why high-performance work systems pay off.* Ithaca, NY: ILR Press.

Bakker, A., Demerouti, E., & Verbeke, W. (2004). Using the job demands-resources model to predict burnout and performance. *Human Resource Management, 43,* 83–104.

Bakker, A. B., & Demerouti, E. (2007). The job demands-resources model: State of the art. *Journal of Managerial Psychology, 22,* 309–328.

Bakker, A. B., & Demerouti, E. (2008). Towards a model of work engagement. *Career Development International, 13,* 209–223.

Bakker, A. B., Demerouti, E., de Boer, E., & Schaufeli, W. B. (2003a). Job demands and job resources as predictors of absence duration and frequency. *Journal of Vocational Behavior, 62,* 341–356.

Bakker, A. B., Demerouti, E., & Euwema, M. C. (2005). Job resources buffer the impact of job demands on burnout. *Journal of Occupational Health Psychology, 10,* 170–180.

Bakker, A. B., Demerouti, E., Taris, T. W., Schaufeli, W. B., & Schreurs, P. J. (2003b). A multigroup analysis of the job demands-resources model in four home care organizations. *International Journal of Stress Management, 10,* 16–38.

Beauregard, T. A., & Henry, L. C. (2009). Making the link between work-life balance practices and organizational performance. *Human Resource Management Review, 19,* 9–22.

Bond, F. W., & Flaxman, P. E. (2006). The ability of psychological flexibility and job control to predict learning, job performance, and mental health. *Journal of Organizational Behavior Management, 26,* 113–130.

Boxall, P., & Macky, K. (2007). High-performance work systems and organisational performance: Bridging theory and practice. *Asia Pacific Journal of Human Resources, 45,* 261–270.

Boxall, P., & Purcell, J. (2016). *Strategy and human resource management* (4th ed.). Basingstoke: Palgrave Macmillan.

Busck, O., Knudsen, H., & Lind, J. (2010). The transformation of employee participation: Consequences for the work environment. *Economic and Industrial Democracy, 31,* 285–305.

Choi, J. N., & Chang, J. Y. (2009). Innovation implementation in the public sector: An integration of institutional and collective dynamics. *Journal of Applied Psychology, 94,* 245–253.

Davis, L. E., & Cherns, A. (1975). *The quality of working life. Problems, prospects, and the state of the art.* London: Collier Macmillan Publishers.

de Jonge, J., & Kompier, M. A. (1997). A critical examination of the demand-control-support model from a work psychological perspective. *International Journal of Stress Management, 4,* 235–258.

Delery, J. E., & Doty, D. H. (1996). Modes of theorizing in strategic human resource management: Tests of universalistic, contingency, and configurational performance predictions. *Academy of Management Journal, 39,* 802–835.

Demerouti, E. (2015). Design your own job through job crafting. *European Psychologist, 19,* 237–247.

Demerouti, E., & Bakker, A. B. (2006). Employee well-being and job performance: Where we stand and where we should go. In J. Houdmont & S. McIntyre (Eds.), *Occupational health psychology: European perspectives on research, education and practice* (Vol. 1, pp. 83–111). Maia: ISMAI Publications.

Demerouti, E., Bakker, A. B., Nachreiner, F., & Schaufeli, W. B. (2001). The job demands-resources model of burnout. *Journal of Applied Psychology, 86,* 499–512.

Dhondt, S., Pot, F., & Kraan, K. (2014). The importance of organizational level decision latitude for well-being and organizational commitment. *Team Performance Management, 20,* 307–327.

Dhondt, S., & Van Hootegem, G. (2015). Reshaping workplaces: Workplace innovation as designed by scientists and practitioners. *European Journal of Workplace Innovation, 1,* 17–24.

Eeckelaert, L., Dhondt, S., Oeij, P., Pot, F., Nicolescu, G., Webster, J., et al. (2012). *Review of workplace innovation and its relation with occupational safety and health*. Luxembourg: Publications Office of the European Union.

Gruman, J. A., & Saks, A. M. (2011). Performance management and employee engagement. *Human Resource Management Review, 21,* 123–136.

Hackman, J. R., & Oldham, G. R. (1976). Motivation through the design of work: Test of a theory. *Organizational Behavior and Human Performance, 16,* 250–279.

Hagen, I. M., & Trygstad, S. C. (2009). Local flexicurity: Resolving the conflict between direct and representative participation. *Transfer: European Review of Labour and Research, 15,* 557–577.

Holman, D., Frenkel, S., Sørensen, O., & Wood, S. (2009). Work design variation and outcomes in call centers: Strategic choice and institutional explanations. *Industrial and Labor Relations Review, 62,* 510–532.

Holman, D., Totterdell, P., Axtell, C., Stride, C., Port, R., Svensson, R., et al. (2012). Job design and the employee innovation process: The mediating role of learning strategies. *Journal of Business and Psychology, 27,* 177–191.

Humphrey, S. E., Nahrgang, J. D., & Morgeson, F. P. (2007). Integrating motivational, social, and contextual work design features: A meta-analytic summary and theoretical extension of the work design literature. *Journal of Applied Psychology, 92,* 1332–1356.

Karasek, R., & Theorell, T. (1992). *Healthy work: Stress, productivity, and the reconstruction of working life*. New York: Basic Books.

Karasek, R. A. (1979). Job demands, job decision latitude, and mental strain: Implications for job redesign. *Administrative Science Quarterly, 24,* 285–308.

Klein, K. J., Conn, A. B., & Sorra, J. S. (2001). Implementing computerized technology: An organizational analysis. *Journal of Applied Psychology, 86,* 811–824.

Knudsen, H., Busck, O., & Lind, J. (2011). Work environment quality: The role of workplace participation and democracy. *Work, Employment and Society, 25,* 379–396.

Martel, J.-P., & Dupuis, G. (2006). Quality of work life: Theoretical and methodological problems, and presentation of a new model and measuring instrument. *Social Indicators Research, 77,* 333–368.

May, D. R., Gilson, R. L., & Harter, L. M. (2004). The psychological conditions of meaningfulness, safety and availability and the engagement of the human spirit at work. *Journal of Occupational and Organizational Psychology, 77,* 11–37.

Oeij, P., & Vaas, F. (2016). Effect of workplace innovation on organisational performance and sickness absence. *World Review of Entrepreneurship, Management and Sustainable Development, 12,* 101–129.

Oeij, P., de Vroome, E., Bolland, A., Gründemann, R., & van Teeffelen, L. (2014). Investing in workplace innovation pays off for SMEs: A regional innovation initiative from The Netherlands. *International Journal of Social Quality, 4,* 86–106.

Oeij, P., Žiauberytė-Jakštienė, R., Dhondt, S., Corral, A., Totterdill, P., & Preenen, P. (2015). *Workplace innovation in European companies*. Luxembourg: Publication Office of the European Union.

Oeij, P. R., Dhondt, S., Kraan, K., Vergeer, R., & Pot, F. (2012). Workplace innovation and its relations with organisational performance and employee commitment. *LLinE, Lifelong Learning in Europe, 4,* 2012.

Oldham, G. R., Hackman, J. R., & Pearce, J. L. (1976). Conditions under which employees respond positively to enriched work. *Journal of Applied Psychology, 61,* 395–403.

Parker, S. K. (2003). Longitudinal effects of lean production on employee outcomes and the mediating role of work characteristics. *Journal of Applied Psychology, 88,* 620–634.

Parker, S. K., Wall, T. D., & Jackson, P. R. (1997). "That's not my job": Developing flexible employee work orientations. *Academy of Management Journal, 40,* 899–929.

Pot, F. D., & Koningsveld, E. A. (2009). Quality of working life and organizational performance-two sides of the same coin? *Scandinavian Journal of Work, Environment and Health, 35,* 421–428.

Preenen, P. T., Oeij, P. R., Dhondt, S., Kraan, K. O., & Jansen, E. (2016). Why job autonomy matters for young companies' performance: Company maturity as a moderator between job autonomy and company performance. *World Review of Entrepreneurship, Management and Sustainable Development, 12,* 74–100.

Preenen, P. T., Vergeer, R., Kraan, K., & Dhondt, S. (2015). Labour productivity and innovation performance: The importance of internal labour flexibility practices. *Economic and Industrial Democracy,* 1–23.

Purvis, R. L., Sambamurthy, V., & Zmud, R. W. (2001). The assimilation of knowledge platforms in organizations: An empirical investigation. *Organization Science, 12,* 117–135.

Ramos, J., Anderson, N., Peiró, J. M., & Zijlstra, F. (2016). Studying innovation in organizations: A dialectic perspective—Introduction to the special issue. *European Journal of Work and Organizational Psychology, 25,* 477–480.

Russell, D. M., & Hoag, A. M. (2004). People and information technology in the supply chain: Social and organizational influences on adoption. *International Journal of Physical Distribution and Logistics Management, 34,* 102–122.

Salanova, M., Agut, S., & Peiró, J. M. (2005). Linking organizational resources and work engagement to employee performance and customer loyalty: The mediation of service climate. *Journal of Applied Psychology, 90,* 1217–1227.

Schaufeli, W., & Bakker, A. (2004). Job demands, job resources, and their relationship with burnout and engagement: A multi-sample study. *Journal of Organizational Behavior, 25,* 293–315.

Schaufeli, W. B., Salanova, M., González-Romá, V., & Bakker, A. B. (2002). The measurement of engagement and burnout: A two sample confirmatory factor analytic approach. *Journal of Happiness Studies, 3,* 71–92.

Schnall, P., Landsbergis, P., & Baker, D. (1994). Job strain and cardiovascular disease. *Annual Review of Public Health, 15,* 381–411.

Singh, J. (2000). Performance productivity and quality of frontline employees in service organizations. *Journal of Marketing, 64,* 15–34.

Takeuchi, R., Chen, G., & Lepak, D. P. (2009). Through the looking glass of a social system: Cross-level effects of high-performance work systems on employees' attitudes. *Personnel Psychology, 62,* 1–30.

Taris, T. W., Kompier, M. A., De Lange, A. H., Schaufeli, W. B., & Schreurs, P. J. (2003). Learning new behaviour patterns: A longitudinal test of Karasek's active learning hypothesis among Dutch teachers. *Work & Stress, 17,* 1–20.

Van der Doef, M., & Maes, S. (1999). The job demand-control (-support) model and psychological well-being: A review of 20 years of empirical research. *Work and Stress, 13,* 87–114.

Walters, D., & Nichols, T. (2007). *Worker representation and workplace health and safety.* Basingstoke: Palgrave Macmillan.

Xanthopoulou, D., Bakker, A., Demerouti, E., & Schaufeli, W. (2007). How job and personal resources influence work engagement and financial turnover: A diary study in a Greek fast–food company. *International Journal of Stress Management, 14,* 121–141.

Zhou, H., Dekker, R., & Kleinknecht, A. (2011). Flexible labor and innovation performance: Evidence from longitudinal firm-level data. *Industrial and Corporate Change, 20,* 941–968.

Author Biographies

Arianna Costantini is a Ph.D. Candidate in Psychology at Verona University, Department of Human Sciences, Italy. Her field of work is workplace innovation, reward management, and organisational behaviour.

Riccardo Sartori is assistant professor in work and organizational psychology at Verona University (Italy), Department of Human Sciences. His field of work is organisational innovation and the assessment processes linked to this topic, including psychological assessment and human resources management.

Andrea Ceschi is a post-doc in work and organisational psychology at the Department of Human Sciences, Verona University, Italy. His field of work deals with organisational dynamics related to decision-making processes in the workplace.

Chapter 10
Implementing Workplace Innovation Across Europe: Why, How and What?

Peter R.A. Oeij, Steven Dhondt, Rita Žiauberytė-Jakštienė, Antonio Corral and Paul Preenen

10.1 Introduction

Workplace innovation (WPI) is generally beneficial for both business performance and the quality of jobs. Redesigning organisations and work processes matters for better performance and jobs in general (e.g., Bloom and van Reenen 2010; Boxall 2012; Boxall and Macky 2009). The benefits of WPI have been documented for

The authors would like to thank our colleagues of the Eurofound study: IKEI Research & Consultancy (Spain), Workplace Innovation Limited (United Kingdom), ARC Consulting EOOD (Bulgaria), Centre for Working Life Research, Roskilde University (Denmark), Gesellschaft für Empirische Arbeitsforschung und Beratung (Germany), Institute for Modelling and Analysis of Public Policies (Poland), Mykolas Romeris University (Lithuania), University of Piraeus Research Center (Greece).

P.R.A. Oeij (✉) · S. Dhondt · R. Žiauberytė-Jakštienė
TNO, The Netherlands Organisation for Applied Scientific Research, Leiden, The Netherlands
e-mail: peter.oeij@tno.nl

R. Žiauberytė-Jakštienė
e-mail: rita.ziauberytejakstiene@tno.nl

S. Dhondt
KU Leuven, Leuven, Belgium
e-mail: steven.dhondt@tno.nl

A. Corral
IKEI Research & Consultancy, San Sebastián, Spain
e-mail: acorral@ikei.es

P. Preenen
Dutch Tax and Customs Administration, Utrecht, The Netherlands
e-mail: typ.preenen@belastingdienst.nl

P.R.A. Oeij et al. (eds.), *Workplace Innovation*, Aligning Perspectives on Health, Safety and Well-Being, DOI 10.1007/978-3-319-56333-6_10

both employees and organisations across a range of organisational and national contexts. For example, WPI has been linked to both improved individual level outcomes such as indices of quality of working life and improved organisational performance (Eeckelaert et al. 2012; Ramstad 2009), quality of working life (Kalmi and Kauhanen 2008), better organisational performance (Dhondt and van Hootegem 2015; Oeij et al. 2011; Oeij and Vaas 2016), and applicability in SMEs (Oeij et al. 2014). Hence, companies that care about their performance and employees should adopt and implement WPI.

In fact, WPI might be more relevant than ever in the current times of ongoing change and competition, especially given the increasing awareness that many companies focus almost exclusively on technological and business (model) innovation to face today's demands (Dhondt et al. 2015). However, competitiveness is not realised through merely stimulating new technological developments and cost-cutting efficiency policies but needs to go hand in hand with WPI (Pot 2011; Pot et al. 2012, 2016). Despite these positive results of WPI, less is known about how different European companies implement WPI in their own practices and why they do so. In this contribution our central question, therefore, is: *Why and how do companies implement WPI?* To address this question, we will use data from a Eurofound study (Oeij et al. 2015a) that investigated the implementation of WPI in 51 companies across Europe. We will first describe the concept of WPI. Subsequently, the Eurofound study's results are presented, including three company case examples of the implementation of WPI. We end with a conclusion and implications for policy and practice.

10.2 The Concept of WPI

In this chapter we use the following definition of WPI: a developed and implemented practice or combination of practices that structurally (division of labour) and/or culturally (empowerment) enable employees to participate in organisational change and renewal to improve quality of working life and organisational performance (Oeij et al. 2015a: 8, 14). This conceptualisation of WPI implies that one needs to look at the organisation as a whole and consider the reciprocal effects of strategy, structure and culture, if one is to reap the benefits associated with WPI (Howaldt et al. 2016). For instance, hierarchical organisational structures may lead to more directive leadership styles and Human Resource Management (HRM) practices that focus on a clear division of labour and control, whereas less hierarchical structures may lead to leadership styles and HRM practices that are geared at promoting employee involvement, engagement and commitment (MacDuffie 1997; Pot 2011). Therefore, to fully understand WPI, it is essential to not only focus on certain types of HRM practices and their consequences, but to also take into consideration the organisational structure and the management philosophy underlying strategic choices (Dhondt and Oeij 2014; Howaldt et al. 2016; Karanika-Murray and Oeij 2017).

The workplace innovation's 'structure orientation' contains practices that structure work organisation and job design (Oeij et al. 2015a; de Sitter et al. 1997). As described, these practices concern the division of labour, the division of controlling ('managing') and executing tasks, and they provide employees with structural decision latitude or control capacity (Dhondt et al. 2014a). This means that employees are structurally given influence over their work, management, production system, and organisation through, for example, co-creation in work design, employee budget or planning control, self-organizing and self-steering teams. Such an approach goes beyond HR-dominated streams (such as High Performance Work Practices and High Involvement Work Practices—discussed in Oeij et al. 2015a), as it is rooted in the choices made about how to design the production system and work organisation. Structure-oriented practices can stimulate employee-control or autonomy, and provide a ground for employee (and employee representatives') voice. These are crucial for individual level motivation and innovative behaviour (Preenen et al. 2015, 2016).

The workplace innovation's 'culture orientation' contains practices that provide opportunities for employees to participate in various ways, for example, in organisational decision-making through dialogue (Oeij et al. 2015a) and are focused on enhancing employee engagement and participation. An example of such a practice would be higher management visits to the shop floor in order to engage in dialogue with the employees. These culture-oriented practices do not only concern employees, but they could also include employee representatives, as in the case of social dialogue and collective bargaining. Culture-oriented practices can stimulate commitment and provide employees (and employee representatives) with voice (Totterdill and Exton 2014).

10.3 The Eurofound Study

The Eurofound study '*Workplace innovation in European companies*' (Oeij et al. 2015a, b) is a multiple case study among 51 companies from 10 EU Member States. Its purpose, among others, was to explore why and how companies apply WPI in order to offer policy makers in Europe recommendations for how to pursue and stimulate WPI across Europe.

10.3.1 Sampling

The companies were selected from the European Company Survey 2013 (ECS survey) database comprising about 30,000 companies (Eurofound 2015). For this purpose, a WPI-index score[1] was constructed to rank all companies in terms of their WPI-features (Dhondt et al. 2014b; Oeij et al. 2015b). The top 5% of the companies in the ranking were selected; this means that, according to the ECS-survey data, these companies are mature in terms of WPI-features. WPI-mature companies were selected because they have already undergone the whole WPI implementation process, therefore, enabling the analysis of leverage factors and barriers, as well as results of WPI. In order to achieve some variation across Europe the companies were divided according to the following regional breakdown: Continental and Western Europe (Denmark, Germany, Ireland, The Netherlands, United Kingdom—102 companies were approached, 22 companies participated), Southern Europe (Greece, Spain—105 companies were approached, 12 participated), and Central and Eastern Europe (Bulgaria, Lithuania, Poland—17 companies participated from the 154 that were approached). The final cases were selected via direct contacts with the companies explaining the purposes of the project. Our final sample varied in sector, and size as follows: *Company size*: SMEs 50–249 employees (27 companies) and large companies with 250 employees or more (24); *Branch*: industry/manufacturing (21); commercial services (14); social services (16).

10.3.2 Methods and Fieldwork

In each company, face-to-face or group interviews were intended with a manager, a group of employees, and employee representatives to get an extensive and broad picture. These were always persons who were involved and knowledgeable of the WPI practices to be studied. However, for diverse reasons, usually operational difficulties during the fieldwork, in five companies it was impossible to talk to employees and in 16 companies no employee representatives were available.

The fieldwork was carried out by nine European research institutes using a standardized methodology and formalized questionnaires about the how and why of

[1]The WPI-Index score (Dhondt et al. 2014; see also Oeij et al. 2015b) consists of separate items derived from the ECS Management Questionnaire, which are linked to the theory of high performance work systems. Using Principal Component Analysis (PCA), a latent structure in the data was found, which consisted of 7 factors: (1) innovation (product and organisational innovation), (2) voice (employees/employee representatives having a say in decisions and changes), (3) learning and reflection (training and feedback), (4) structure and system (variable pay), (5) work organisation autonomy (autonomy), (6) work organisation career (long-term career plans), (7) hierarchy. The WPI-Index score was calculated as the sum of the scores of these separate factors, implying that each of these elements was given the same weight in the WPI-index.

WPI use (see Oeij et al. 2015b[2]). All in all, about 200 people were interviewed (for exact numbers see Tables 10.3 and 10.4), following specific questionnaires for each interviewee category (in total, 3 questionnaires per firm). The information gathered was inputted into a data file and each case was described in a mini-case study report (2–3 pages).[3] In each company, specific WPI practices were identified (up to 168 practices in total).

Subsequently, the questionnaires were analysed using Qualitative Comparative Analysis (QCA). Analyses revealed the 'conditions' within companies that explain the presence of substantial WPI practices. Together, these conditions constitute successful routes ('configurational paths') that can be regarded as implicit strategies employed to become a WPI company. Case study reports were used to assess whether different types of WPI practices could be distinguished. Qualitative information from interviews was used to get a richer description of contextual factors, drivers and motivations, ways of developing and implementing WPI, and the impacts of WPI. This combined approach, leveraging information from different data sources, enabled an in-depth analysis of the companies and their WPI practices.[4]

10.3.3 Results of WPI Implementation in Practice

10.3.3.1 Why Do Companies Want to Implement WPI?

To see why companies introduced WPI-practices a distinction was made in the questionnaires between two drivers or targets, namely, to improve the quality of performance of the organisation or to improve the quality of working life and employee engagement. The analysis of the questionnaires revealed the existence of a third category that essentially combined both drivers. Although economic reasons drive the decision to introduce WPI, most practices identified in the case studies (69%) are targeted at both goals (i.e., the enhancement of the company's performance and quality of working life), while the remaining practices are approximately equally divided between those that focus on quality of working life (18%) and quality of performance (14%) (Table 10.1).

Table 10.1 presents a total of 168 WPI-practices that where identified in the selected companies. Half of these practices (53%) are either focusing on WPI-structure elements (14%), WPI-culture elements (20%) or are a mixture of

[2]Full technical report with the methodology can be found on the Eurofound website at https://www.eurofound.europa.eu/publications/report/2015/working-conditions/third-european-company-survey-workplace-innovation-in-european-companies.

[3]All cases can be found on the Eurofound website at: https://www.eurofound.europa.eu/workplace-innovation-in-european-companies-case-studies.

[4]An elaborate description of the fieldwork and methodology can be found in Oeij et al. (2015b) which can be downloaded from the Eurofound website.

Table 10.1 Types of practices applied and drivers (percentages)

Drivers	WPI			Total WPI	HR	Other	Total
	WPI-structure	WPI-culture	WPI-mixed				
Quality of performance	1.8	1.2	3.6	6.5	3.6	3.6	13.7
Quality of work	3.0	3.6	4.2	10.7	6.5	0.6	17.9
Both: quality of work and performance	8.9	15.5	11.3	35.7	28.6	4.2	68.5
Total	13.7	20.2	19.0	53.0	38.7	8.3	100.0
Number of practices	23	34	32	89	65	14	168

Source Oeij et al. (2015b: 21)

structure and culture practices (19%). Quite a high proportion of identified practices were assessed as being exclusively HR-practices (39%), which we see as too limited to qualify as a genuine WPI-practice. The practices in this category are 'typical' or 'traditional' HR-practices in the fields of, for example, personnel recruitment, training, competence development, performance appraisal, working conditions, remuneration, flexibility and health, risk and safety measures. The category 'other' (8%) consists of practices such as cost-effectiveness, efficiency improvement and ICT-practices that also do not qualify as WPI.

Table 10.2 provides some concrete examples of the 168 practices identified. The complete list of practices (including HR-practices) can be found in the Annex to the report (Oeij et al. 2015b).

Given that most WPI-practices in our sample seem to be aimed at both achieving economic goals and better work, it is likely that they not only lead to better company performance but also to increased employee engagement and a better quality of working life. WPI practices, such as the ones mentioned above, tend to be aligned with employee interests and, as we will show later on, lead to agreement among managers, employees and employee representatives as to what is more or less important.

Regarding the different paths or routes employed by companies on their way to becoming mature WPI-companies, we found that organisations followed their own unique routes. Indeed, most of them applied more than one WPI-practice, often a combination of structure oriented, culture oriented and HR measures. This may indicate that a 'bundling' of measures might be beneficial, as is proposed in the High Performance Work Practices (HPWP) literature. However, we cannot draw any definite conclusions regarding the ideal combination of WPI-practices, given the wide variety of WPI-practices combinations we found in our cases (Oeij et al. 2015a).

Before we turn to the 'how' question, we will take a closer look at the motives, leverage factors and impacts of WPI. The following sections are based on both quantitative data from the questionnaires as well as interview data from managers,

Table 10.2 Examples of WPI-practices

Types of practices	Examples[a]
WPI: structure orientation	Educational organisation (BG-EDUC-UNI-S): Self-managing teams were introduced as a system for organising day-to-day duties and activities. This approach ensures that the team members have sufficient flexibility to decide how to implement their tasks taking into account their own capacities and time schedule
	Research organisation (ES-SCI-ENVIRONM-L): Minimising organisational levels and enhancing autonomous teams is done by ensuring that there are no more than two hierarchical levels between the lowest and the highest levels. This also facilitates the existence of self-managed working teams that have the freedom to organise themselves
	News and Journalism organisation (NL-INFO-NEWS-L): Job enlargement by expanding sales jobs with account management tasks; also cross-functional teams were installed to realise innovation projects across departments
WPI: culture orientation	Museum (DK-ART-MUSEUM-S): Partnership with unions. New projects and organisational changes are debated in a joint committee with union representatives, OSH representatives and management representatives. This committee is initiating new practices such as training and support for new employees
	Energy company (BG-ENER-GAS-S): The Knowledge Management System, OGpedia, is a voluntarily developed IT-based information sharing system. All employees can share and gain new knowledge
	Postal organisation (LT-SERV-POST-L): "Loyalty Day" aims to enhance communication and knowledge sharing between managers and first line workers. Managers voluntarily visit workers at their work site and gather information about specific processes and possible issues. This increases sustainability, efficiency and good organisational communication
WPI: mixed	Research organisation (ES-SCI-WORK-L): Flexitime practices allow workers to have a say regarding their working times: they can adjust their starting and exiting hours, also ad hoc exits (with manager's permission) are allowed
	Financial service company (EL-FIN-BANK-L): An initiative for personal development: every year teams of 1–2 people take part in a challenge defined by the top leaders. In this way ideas can be passed from young talents to the top management. Young talents are supported by coaching sessions and assessment tools and gain experience
	Pet food processor (DE-AGRO-PETFOOD-S): Overall Competences: Ready to do any job in the production line, an overall qualification was given to the production staff, enabling the employees to take over every job in the production. After the mechanisation of production most of the employees had the chance to upskill and take over a skilled workers' task

[a]Company codes are indicative of country, branch of activity and size (small 50–250 and large 250+) and used to ensure anonymity (*Source* Oeij et al. 2015a: 25–26)

employees and employee representatives. Generally, the level of agreement between these stakeholder groups was high.[5]

10.3.3.2 What Are Motives to Implement WPI?

Although companies did choose varying paths to WPI and selected different (combinations of) WPI-practices, their reasons for initiating WPI reflect much commonality. Our quantitative and qualitative analyses show that, from an organisational perspective, economic motives for initiating WPI are dominant (Table 10.3). In this sense, from the viewpoint of the 'organisation as a whole', the most prominent three general motives identified by the three groups of interviewees for initiating WPI implementation, were efficiency improvement, to gain competitive advantage and to enhance innovative capability. However, many companies understand that achieving economic goals largely depends on the role that employees play, as reflected in motives such as becoming an attractive employer and increasing acceptance by employees (Table 10.3).

Apart from looking at motives for 'the organisation as a whole', the investigation of 'motives' was also approached as possible desired impacts for each group of stakeholders separately (management, employees and employee representatives) (Table 10.3). We found that motives for WPI implementation from both the managers' and employees' perspectives overlap,[6] and, moreover, are aligned with the general reasons to initiate WPI.[7] The three most salient motives are economic and business goals, learning and development opportunities, and performance.

Interestingly, all three actor groups saw motives related to the quality of organisational performance as more important than those related to the quality of working life.

---

[5]The drivers are derived from the WPI-practices that are implemented by the companies; the motives are part of the interview checklists that were applied (for more details see the Technical report, Oeij et al. 2015b).

[6]Statistical results (based on McNemar tests) indicated that there are no significant differences in how frequently managers, employee groups and employee representatives selected the top-3 motives (economic and business goals, learning and development opportunities, performance).

[7]All three groups indicated the improvement of efficiency as the most important motive for the "organisation as a whole", while gaining competitive advantage and enhancing innovative capability were the second two most important motives. No differences were found among the three groups in the frequency of selecting the improvement of efficiency. Managers' selected gaining competitive advantage more often than groups of employees and employee representatives (respectively $p = 0.049$ and $p = 0.039$); managers also selected enhancing innovative capability more often than employee representatives ($p = 0.039$). Here and in other comparisons attention should be paid to data missing from employee representatives (>30%).

Table 10.3 General motives for the implementation of WPI

	Manager	Group of employees	Employee representatives
	Percentage of companies		
For the 'organisation as a whole'			
To improve efficiency	80	80	74
To gain competitive advantage	78	58	65
To enhance innovative capability	75	58	65
To become an attractive employer	57	53	44
To enable the acceptance by employees	37	31	47
To enable the embedment of new technology and ICT	37	33	35
To improve industrial relations with unions	18	9	47
From managers' and employees' perspective			
Economic and business goals	94	89	88
Learning and development opportunities	78	71	74
Performance	61	62	59
Public goals	31	33	32
Flexibility	31	42	38
Shareholder interests	25	24	29
Labour market position	25	18	35
Balance private-work life situation	25	24	32
N of respondents	51	45	34

Source Oeij et al. (2015b: 27)

10.3.3.3 What Are Important Leverage Factors for the Implementation of WPI?

Leverage factors are actions, measures or means that drive the successful implementation of WPI-practices. The most important three leverage factors for WPI implementation are employee involvement, top management commitment, and, at a distance, leadership or the involvement of a powerful person. Again this was reported by all three groups of interviewees (Table 10.4).[8] While reasons and motives to start WPI point to business-related arguments, employee involvement seems a sine qua non when it comes to adoption and implementation.

[8]Statistical (McNemar) test indicated that there are no significant differences in how frequently managers, employee groups and employee representatives selected the top-3 leverage factors (employee involvement, top management commitment, and leadership, powerful person).

Table 10.4 Leverage factors for WPI implementation

	Manager	Group of employees	Employee representatives
	Percentage of companies		
Employee involvement	82	84	88
Top management commitment	80	69	68
Leadership, powerful person	67	56	65
Organisational, non-conflictive climate	49	42	50
Resources, enough money and people	33	38	29
Time, no interference from reorganisation	18	20	24
N of respondents	51	45	34

Source Oeij et al. (2015b: 27)

10.3.3.4 What Are the Impacts or Expected Impacts of WPI?

Impacts of WPI-practices, like drivers, can be divided into effects on organisational performance and on employees. Four types of impacts have been researched: impacts on the organisation, management, employees, and employee representatives. Table 10.5 presents the top three-to-five effects for each type.

For the organisation, according to all three groups of interviewees, employee engagement was the most important outcome of WPI, followed by long term sustainability,[9] and, at some distance, high performance, better customer focus/client focus, efficiency, and profitability. For employee representatives, notable outcomes were also the establishment of good work and more positive employment relations. Remarkable maybe, but according to employees, establishing good work was ranked lower than efficiency, profitability and high performance (not visible in Table 5, see Oeij et al. 2015b; 28–29).

The most important impacts of WPI for managers' interests are efficiency and sustainability; for employees' interests learning opportunities, voice/participation, and challenging and active jobs; and for employee representatives'/union interests employee voice.

[9]McNemar Tests also indicated that there are no significant differences in how frequently managers, employees and employee representatives selected the two top outcomes (employee engagement and longer term sustainability), hence, there is agreement among groups.

Table 10.5 Impacts of WPI

	Manager	Group of employees	Employee representatives
	Percentage of companies		
For the organisation			
Employee engagement	82	78	85
Longer term sustainability	73	64	62
High performance	67	56	59
Establishing good work	63	47	62
For managers/managers' interests			
Efficiency	73	64	74
More sustainability	71	60	62
Competitiveness	65	53	59
Innovation/innovation capability	61	47	59
Satisfied client, customer	61	53	59
For employees/employees interests			
Learning opportunities	71	67	59
Voice, participation	59	56	59
Challenging, active jobs	57	64	44
Healthy work	43	49	56
For employee representatives/union interests			
Employees voice	79	67	85
Sustainable organisation	56	33	50
Equality, fairness	35	33	41

Source Oeij et al. (2015b: 28–29)

10.3.4 Summary

In sum, when we consider why companies implement WPI, what they see as the most important leverage factors, and which effects or impacts of WPI they expect for the organisation, managers, employees and employee representatives, it becomes clear that the three different respondent groups tend to largely agree with each other. Given that economic goals are triggering the initiation of WPI and that employee involvement is a key factor in the introduction of WPI, it is intriguing to see how much agreement emerges among all stakeholders. All three actors regard:

- employee engagement, longer term sustainability and high performance as the most important impacts for the organisation;
- efficiency, increased sustainability and competitiveness as the most important impacts for managers;

- learning opportunities, voice/participation and challenging and active jobs as the most important impacts for employees;
- employee voice as the most important impact for employee representatives.

Hence, in many instances, the process of introducing WPI-practices tends to improve not only economic performance, but also employee engagement and quality of working life. Figure 10.1 captures these findings.

10.3.5 How Do Companies Implement WPI?

Now that we have outlined the motives for introducing WPI and their associated leverage factors, we will discuss 'how' WPI is being implemented. The process of initiation, adoption and implementation of WPI-practices reveals a common pattern across companies. As previously described, companies choose paths that differ among companies, yet within companies there is agreement among managers, employees and employee representatives regarding why WPI should be introduced, how to do it, and what impacts are desired. Our research suggests that, often, it is management that initiates WPI, and that the main motive is economic. Once this decision has been taken, employees are involved to help design and implement the intervention. Moreover, consulting employee representatives is common among those companies who advocate communication and employee interests. Be reminded that our sample is from the companies who score the highest on WPI. Many of these companies are WPI-mature and from the case studies we learned that they have come this far after many years. The way that WPI-practices get implemented seems to reveal a generally applied pattern (Fig. 10.1):

1. **The initiative of a WPI** often has an economic purpose and very often this is dominant (see 1 in Fig. 10.2). However, in many cases WPI-practices are not solely targeted at economic goals. Often they are combined with or embedded in organisational, job and HR-related measures.
2. Once the WPI-initiative has been refined into a measure or set of measures, employees (and often employee representatives) play an important role in **(co-)designing and developing the WPI-practice** and its **implementation** (see 2a in Fig. 10.2). This happens because management tends to realise that it is impossible to implement WPI without the engagement of employees. Given that employee participation in the design and implementation phase is inextricably linked to employee engagement and possibly improved quality of working life, this can result in the achievement of employee-favourable targets (see 2b in Fig. 10.2).
3. The target of improved economic performance is often not only a direct effect of the implemented WPI-practice but, is in most cases, also indirectly influenced by employees and employee representatives. When economic targets are achieved, they may well coincide with the targets of improved quality of

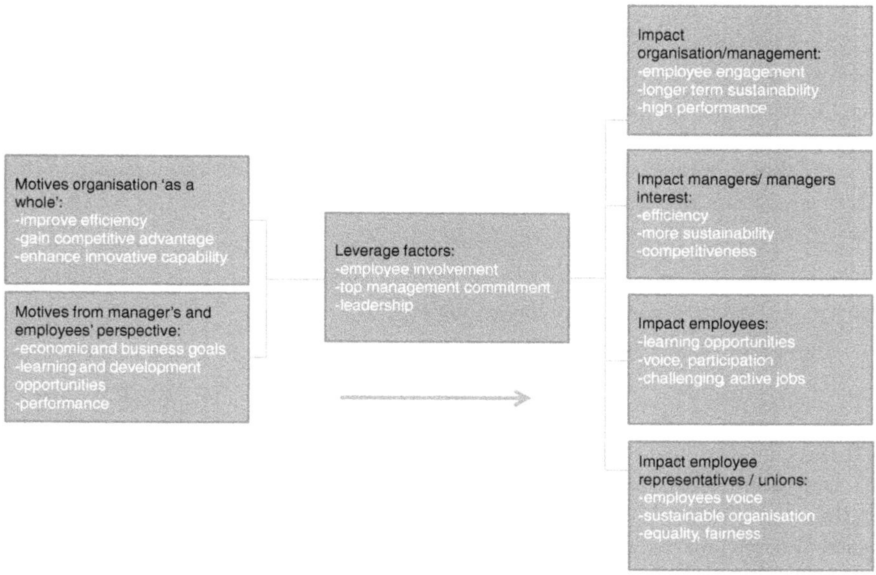

Fig. 10.1 Agreement about the main motives, leverage factors and impacts of WPI according to three respondent groups (managers, employees, employee representatives)

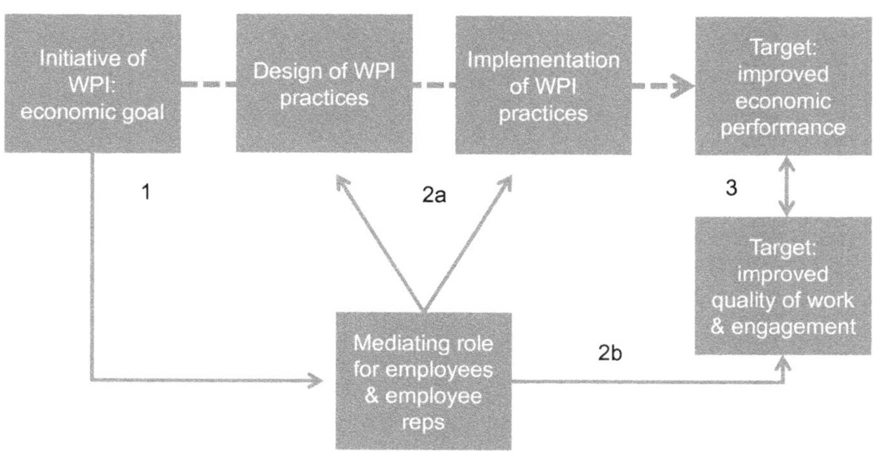

Fig. 10.2 Pattern of implementing WPI-practices (Oeij et al. 2015a: 59)

working life and employee engagement. Vice versa, an improved quality of working life and employee engagement can contribute to improved economic targets (see 3 in Fig. 10.2).

In summary, it appears that (initial) reasons and motives to initiate WPI are mainly economic. In the next phase, concrete WPI-practices are designed and implemented. Here, it becomes apparent that employees get to play a major role, especially in light of the fact that the most important leverage factor for adoption and implementation is employee involvement. Interestingly, managers, employees and employee representatives seem to agree that employee engagement in the whole process is a necessary condition for WPI.

10.3.6 Three Company Examples of the WPI Implementation Process

In our research we found that companies adopt and implement WPI in their own specific way. Below, we will present three examples of WPI implementation from the United Kingdom, Denmark and Lithuania in order to highlight the uniqueness of the WPI process (Oeij et al. 2015a: 53–54). The UK example shows how leadership enables employee participation, while the Danish example mirrors a stepwise approach of management engaging in partnership with unions. The Lithuanian case exemplifies the taking up of dialogue between management and employees, which is relatively new to the region. We chose these examples due to their variety, distinctiveness, and richness in terms of implemented WPI practices.

10.3.6.1 Examples of the WPI Implementation Process

UK Example: Leadership

Energy company (UK-ENER-ELEC-L): "We want this to be a business where views are listened to and where communications are open and honest. We also want this to be a workplace where positive ideas are encouraged and where achievements are celebrated" says the Head of HR. The introduction of Open Forums replaced the previous company-wide meetings and suggestion schemes which had struggled to stimulate open and constructive dialogue and feedback. The CEO's open leadership creates trust and employees feel confident about the future. According to one employee: "It is interesting isn't it, you go to the Open Forums and people will say what they think and absolutely nobody will turn round and go, I can't believe he said that… They might not agree with you but nobody will actually knock anyone for having a view because we are encouraged to have a view. That's really empowering I think."

Danish Example: Partnership with Unions
Service organisation (DK-SERV-PARK-S): Organisational changes are discussed by the manager and the union representatives. They have a partnership and value each other's opinions. The manager explains: "It is nice to have representatives who are not afraid to step up against me in a constructive dialogue". The implementation approach consisted of (1) management took initiative, (2) external consultants supported the process, (3) experiments were conducted (a work team tested new meeting practices or the like), (4) 'invitation' to the same knowledge for all training, and (5) implementation of the practices, but not necessarily in the same way everywhere. No evaluation was done but adjustments were made along the way. Both management and employees believe that it is important to design the process in a manner that creates 'enthusiasts' amongst the employees. The union representative explains: "It gives a huge boost to the company that we work together to create a great workplace. … That's what made us 'the best workplace' (a Danish award) in 2004". The employees believe that, even though management determines the direction, they have to have the trust to be able to discuss it: "It should be perfectly legal to say our outspoken opinion to our manager—and it is. There may well be disagreement, but you have to be able to discuss things" (employee).

Lithuanian Example: Dialogue with Personnel
Hotel (LT-ACCOM-HOTELS-S): The WPI practice, Think Guest Feedback, consists of regular middle management meetings where middle managers from all departments (Front Office, Reservations, Conference Hall, Lobby, Restaurant, Sky Restaurant, Room Service, Marketing and others) regularly meet and review Hotel ratings on dedicated social media platforms. They discuss particular guest feedback cases and joint actions that could improve guest stay experience (and feedback as a result), brainstorm on how guest feedback could be stimulated and collectively addressed, take important information back to the teams of their departments for further action, produce minutes of their observations and recommendations to top management on improvement of various hotel operational aspects and share experiences with each other. Think Guest Feedback involves, for example, prompt reaction to guest feedback (especially when negative) before they leave the hotel, and constant organisational learning from any mistakes made. It implies staff empowerment, not only in the sense that they could solve emerging problems straight away, but also that each of them could feel like owners of the business and be pro-active in preventing negative guest experiences. Mutual trust, goodwill and respect across departments (not to solve your own issues at other's costs) and between all levels of organisational management were stimulated. According to the General Director, the initiative is still very new, but after a few months, it is already showing benefits.

Whereas these examples are different in terms of the interplay between management, employees and their representatives, they are all similar in the sense that cooperation between actors is fundamental to improve the business.

10.4 Conclusion and Lessons for Practice

As mentioned in the introduction, WPI practices have been associated with better organisational performance as well as with softer outcomes such as employee engagement. Yet, to date, we know relatively little about why and how companies implement WPI across Europe.

The general conclusions that emerge from our research can be captured in a few lines (Oeij et al. 2015a: 62). The initiative to start WPI practices usually comes from company owners or managers. However, these managers/owners have understood that, for WPI to be successful and to help them reach benefits in terms of company performance and sustainability, employees' and employee representatives' involvement and participation are crucial. Typically, the reasons driving management's decision to implement WPI practices are related to efficiency, competitiveness and innovation enhancement. In a number of cases, management's decision to implement WPI is triggered by other factors such as:

- a situation of crisis or difficulties in the company's performance that requires significant changes to survive and remain competitive in a changing and globalised market, where the traditional products/services and ways of working need to be revised and adapted in order to satisfy the requirements of increasingly exigent and sophisticated customers;
- sometimes, the former is also combined with a take-over from (or merger with) another (multinational) company which brings in new forms of work organisation and new work practices, systems, etc. that involve workplace innovation. In these cases, there is a kind of 'WPI know-how transfer' from the headquarters to the subsidiary.
- In several of the Eastern European case studies, the privatisation of public enterprises and the associated reorganisation processes have served as a background to the implementation of WPI, seeking greater efficiency and employee involvement that were previously lacking.

Interestingly, factors related to job quality and good working conditions do not emerge as primary reasons or motives for WPI, but rather as either a pre-condition or a result of WPI. This means, that the objective of WPI introduction is not to improve the working conditions or the working environment as such, but that, in order to enhance employee involvement and their contribution to the company's performance and innovation processes, a good set of working conditions is required.

The companies in our sample used different paths to become WPI-mature organisations, meaning they applied different combinations of WPI-practices and stressed different organisational choices (Oeij et al. 2015a). Given that our sample consisted of only WPI-mature companies, it is possible that, for not (yet) WPI-mature organisations other success paths may emerge in future research. We found that organisations can choose different production systems that enable WPI implementation, such as flow structures and teams (Achterbergh and Vriens 2010; Christis 2010). This implies that WPI is related to organisational changes at the 'root' of the production process of making products and delivering services (MacDuffie 1997).

Although companies differ in their implementation strategies, constructive cooperation between management and employees seems to be a key factor for successful WPI, as our three case descriptions imply. Additionally, the way that WPI-practices get implemented seems to reveal a general pattern across companies (Fig. 10.2). Importantly, it appears that organisations can make strategic choices with their organisational structure. Moreover, it seems beneficial for the emergence of WPI to strengthen the position of employees and employee representatives. This can help boost WPI-practices, which, in turn, may improve both economic performance and quality of working life.

10.4.1 Some Pointers for Practitioners

In our sample, WPI is supported by all organisational players: managers, employees and employee representatives. This clearly indicates good employment relationships and industrial relations among the company stakeholders, which result in constructive cooperation, communication and collaboration, the containment of labour conflicts and a striving for common interests and goals. To achieve this, management and leadership behaviours are required that not only bring the business forward but simultaneously stimulate trust and employee engagement. In turn, it is necessary that employees engage in entrepreneurial and intrapreneurial behaviours and apply their talents in support of innovation and co-creating change. In addition, it is crucial that employee representatives and unions balance their interests with those of the employees and companies.

Choosing the right WPI-practices to implement is not easy. It requires linking an organisation's strategy to its management philosophy, and subsequently, to its structure and culture. Moreover, this has to be done by taking into account the viewpoints of management, employees and external stakeholders—unions, customers, etc. What does this mean? First, the management philosophy determines the strategy (i.e., whether the organisation's goals are to be achieved top down or bottom up). That same philosophy serves as the basis for the design of the working processes (i.e., the structure). In this respect, one can choose a more top down or a more bottom up approach regarding the division of labour and the way in which these working processes are managed. In turn, this will be reflected in the culture of an organisation, especially if one looks at leadership styles and the manner in which employees are engaged. In this respect, organisations can show more of a control-orientation versus more of a commitment-orientation, depending on former choices. The established culture will determine whether employees will be more pro-active or risk avoidant. Strategy, structure and culture together constitute a unitary system and need to be treated as a whole. Merely implementing employee-friendly HR-measures, like innovation competitions, job performance interviews or company suggestion boxes while leaving a top down structure intact, will therefore, result in disappointment rather than satisfaction, in the long run.

When developing WPI-practices one should take a whole system approach and understand that the root causes of behaviour need to be addressed, and that these can often be found in the interplay between strategy, structure and culture. It is nonetheless difficult to assess the potential impacts of WPI-practice at the outset. This is partly because it is difficult to predict the outcomes, partly because of the complex manner in which WPI-practices interact with other organisational factors, and partly because some effects can be quantified, but many others cannot and remain 'qualitative' evaluations. To be able to build a proper 'business case' for such WPI-practices in which a trade-off can be made between quantitative and qualitative aspects, and which reflects the viewpoints of different stakeholders, employers and employees can collaboratively engage in a stepwise approach (Oeij et al. 2012). This approach implies that employers and employees analyse the (future) productivity challenge and strategy of the organisation, and link this to possible workplace innovation practices, and their effects on performance and quality of work. By making a trade-off between the advantages and disadvantages, and by applying a dialogue approach (Oeij et al. 2006: 258–259), they can build a solid business case for their final choice. The dialogue approach means that viewpoints from different perspectives are taken into account.

10.4.2 Coda

This contribution tried to make clear why leading WPI-companies apply WPI and how they implement it. WPI-mature companies have mature relationships between management, employees, and employee representatives. These stakeholders understand that they need each other and that strategy, structure, and culture can best be seen as a system that should be balanced (see also Chap. 20). Workplace innovation thus requires a holistic or integral view on change. Moreover, companies can take different paths and employ a variety of practices for effective WPI implementation.

Acknowledgements The authors and publisher gratefully acknowledge the following permission to use the material in this book: Peter Oeij, Steven Dhondt, Rita Žiauberytė-Jakštienė, Antonio Corral & Paul Preenen, Implementing workplace innovation across Europe: Why, how and what? In: European Work and Organizational Psychology in Practice. Special issue on Workplace Innovation, 2017, Volume 1, 46–60.

References

Achterbergh, J., & Vriens, D. (2010). *Organizations. Social systems conducting experiments* (2nd ed.). Berlin: Springer.

Bloom, N., & van Reenen, J. (2010). Why do management practices differ across firms and countries? *Journal of Economic Perspectives, 24*(1), 203–224.

Boxall, P. (2012). High-performance work systems: What, why, how and for whom? *Asia Pacific Journal of Human Resources, 50,* 169–186.

Boxall, P., & Macky, K. (2009). Research and theory on high-performance work systems: Progressing the high-involvement stream. *Human Resource Management Journal, 19,* 3–23.

Christis, J. H. P. (2010). Organization and job design: What is smart organizing? In H. A. M. van Lieshout, L. Polstra, J. H. P. Christis, & B. J. M. Emans (Eds.), *Management of labour. Societal and managerial perspectives* (pp. 39–71). Groningen: Hanzehogeschool Groningen University of Applied Sciences.

de Sitter, L. U., den Hertog, J. F., & Dankbaar, B. (1997). From complex organizations with simple jobs to simple organizations with complex jobs. *Human Relations, 50*(6), 497–534.

Dhondt, S. & Oeij, P. R. A. (2014, September). Social innovation related to innovation in management studies. In J. Howaldt, A. Butzin, D. Domanski, & C. Kaletka (Eds.), *Theoretical approaches to social innovation—A critical literature review* (pp. 122–150). Dortmund: SI-Drive [EU Seventh Framework Programme]. https://webgate.ec.europa.eu/socialinnovationeurope/en/magazine/methods-and-tools/articles-reports/theoretical-approaches-social-innovation

Dhondt, S., Oeij, P. R. A., & Preenen, P. T. Y. (2015). Working in the digitalized world: The meaning of the link between advanced manufacturing and workplace innovation in the EU growth strategy. In C. M. Schlick (Ed.), *Arbeit in der digitalisierten Welt* (pp. 155–170). Frankfurt/New York: Campus Verlag.

Dhondt, S., Pot, F. D., & Kraan, K. O. (2014a). The importance of organizational level decision latitude for well-being and organizational commitment. *Team Performance Management, 20* (7/8), 307–327.

Dhondt, S., Preenen, P. T. Y., Oeij, P. R. A., Corral, A., Isusi, I., & Totterdill, P. (2014b, February). *European company survey: construction of the workplace innovation index and selection of companies*. Leiden: TNO.

Dhondt, S., & van Hootegem, G. (2015). Reshaping workplaces: Workplace innovation as designed by scientists and practitioners. *European Journal of Workplace Innovation, 1*(1), 17–25.

Eeckelaert, L., Dhondt, S., Oeij, P. R. A., Pot, F., Nicolescu, G. I., Webster, J., et al. (2012). *Review of workplace innovation and its relation with occupational safety and health: Literature review*. Bilbao: European Agency for Safety and Health at Work.

Eurofound (2015), Kankaraš, M., & van Houten, G. *Third European company survey 2013—Overview report: Workplace practices—Patterns, performance and well-being*. Luxembourg: Publications Office of the European Union.

Howaldt, J., Oeij, P. R. A., Dhondt, S., & Fruytier, B. (2016). Workplace innovation and social innovation: An introduction. *World Review of Entrepreneurship, Management and Sustainable Development, 12*(1), 1–12.

Kalmi, P., & Kauhanen, A. (2008). Workplace innovations and employee outcomes: Evidence from Finland. *Industrial Relations: A Journal of Economy and Society, 47*(3), 430–459.

Karanika-Murray, M. & Oeij, P. R. A. (2017). How can work and organizational psychology fortify the practice of workplace innovation?—Chapter 20 in this book.

MacDuffie, J. P. (1997). The road to "root cause": Shop-floor problem-solving at three auto assembly plants. *Management Science, 43*(4), 479–502.

Oeij, P. R. A., de Looze, M., ten Have, K., van Rhijn, G., & de Graaf, B. (2012). From productivity strategy to business case: Choosing a cost-effective intervention for workplace innovations. *JCC: The Business and Economics Research Journal, 5*(2), 171–184.

Oeij, P. R. A., de Vroome, E., Bolland, A., Gründemann, R., & van Teeffelen, L. (2014). Investing in workplace innovation pays off for SMEs: A regional innovation initiative from The Netherlands. *International Journal of Social Quality, 4*(2), 86–106.

Oeij, P. R. A., Dhondt, S., & Korver, T. (2011). Social innovation, workplace innovation and social quality. *International Journal of Social Quality, 1*(2), 31–49.

Oeij, P. R. A., & Vaas, F. (2016). Effect of workplace innovation on organisational performance and sickness absence. *World Review of Entrepreneurship, Management and Sustainable Development, 12*(1), 101–129.

Oeij, P. R. A., Wiezer, N. M., Elo, A.-L., Nielsen, K., Vega, S., Wetzstein, A., et al. (2006). Combating psychosocial risks in work organisations: Some European practices. In S. McIntyre & J. Houdmont (Eds.), *Occupational health psychology: European perspectives on research, education and practice* (Vol. 1, pp. 233–263). Maia, Portugal: ISMAI Publishing.

Oeij, P. R. A., Žiauberytė-Jakštienė, R., Dhondt, S., Corral, A., Totterdill, P., & Preenen, P. T. Y. (2015a). *workplace innovation in European companies*. Luxemburg: Office for Official Publications of the European Communities. Study commissioned by Eurofound.

Oeij, P. R. A., Žiauberytė-Jakštienė, R., Dhondt, S., Corral, A., Totterdill, P., & Preenen, P. T. Y. (2015b). *Workplace innovation in European companies. Technical annex.* Dublin: Eurofound. https://www.eurofound.europa.eu/sites/default/files/ef1540_-_technical_annex_-_third_european_company_survey_workplace_innovation_in_european_companies.pdf

Preenen, P. T. Y., Oeij, P. R. A., Dhondt, S., Kraan, K. O., & Jansen, E. (2016). Job autonomy matters for young companies' performance growth. *World Review of Entrepreneurship Management and Sustainable Development, 12*(1), 74–100.

Preenen, P. T. Y., Vergeer, R. A., Kraan, K. O., & Dhondt, S. (2015). Labour productivity and innovation performance: The importance of internal labour flexibility practices. *Economic and Industrial Democracy*. Advance online publication, March 24, 2015.

Pot, F. D. (2011). Workplace innovation for better jobs and performance. *International Journal of Productivity and Performance Management, 60*(4), 404–415.

Pot, F. D., Dhondt, S., & Oeij, P. R. A. (2012). Social innovation of work and employment. In H.-W. Franz & J. Hochgerner (Eds.), *Challenge social innovation* (pp. 261–274). Berlin: Springer.

Pot, F. D., Totterdill, P., & Dhondt, S. (2016). Workplace innovation: European policy and theoretical foundation. *World Review of Entrepreneurship, Management and Sustainable Development, 12*(1), 13–32.

Ramstad, E. (2009). Promoting performance and the quality of working life simultaneously. *International Journal of Productivity and Performance Management, 58*(5), 423–436.

Totterdill, P., & Exton, R. (2014). Defining workplace innovation: The Fifth Element. *Strategic Direction, 30*(9), 12–16.

Author Biographies

Peter R.A. Oeij is senior researcher/consultant at TNO, The Netherlands Organisation for Applied Scientific Research (Leiden, The Netherlands). The main topics of his work are innovation management, workplace innovation, social innovation, productivity, flexibility, working smarter and team work.

Steven Dhondt is visiting professor at the KU Leuven (Belgium; Chair Social Innovation) and senior research scientist at TNO, The Netherlands Organisation for Applied Scientific Research (Leiden, The Netherlands). He is the network coordinator for the European learning network on Workplace Innovation (2013-2017) for the DG Enterprise & Industry.

Rita Žiauberytė-Jakštienė is a junior researcher at the Netherlands Organisation for Applied Scientific Research (TNO). Formerly she has been a recruitment consultant in personnel selection agencies in Lithuania. Her research interests include social innovation, workplace innovation, and goal-setting topics.

Antonio Corral is Unit Director at IKEI Research & Consultancy (Spain). He is an economist and MBA with more than 25 year experience in social and economic research and public policy projects. He is national correspondent in Spain for the European Observatory of Working Life (EurWORK) of Eurofound.

Paul Preenen is a researcher at the Dutch Tax and Customs Administration, Centre for Knowledge and Communication. He holds a Ph.D. in Work and Organisational Psychology from the University of Amsterdam. Previously, he has worked for the Netherlands Organisation for Applied Scientific Research (TNO).

Chapter 11
Workplace Innovation as Institutional Entrepreneurship

Helge Hvid and Vibeke Kristine Scheller

11.1 Introduction

Supported by the EU-Commission, a campaign for Workplace Innovation (WPI) was launched, supported by EUWIN.[1] The objective of this campaign is to promote the 'Columbus egg' laid by the 'Sociotechnical Movement' back in the 1950s (Trist and Bamforth 1951). Through experiments in British coal mines, Trist and Bamforth developed their principles of work organisation and social relationships in the workplace. According to Trist and Bamforth, following these principles enables both companies and workers to (a) increase productivity through better resource utilisation and greater innovation and (b) improve health and quality of working life, more learning opportunities and maintaining workers' capacity within the labour market (Trist and Bamforth 1951). Furthermore, it can be argued that the welfare state gains through increased tax revenues and decreased costs related to social care and health. Following the principles of sociotechnical systems thus creates a win–win–win situation: employers, employees and the state all have something to gain.

Nevertheless, the British mining companies rejected Trist and Bamforth's findings, choosing to manage their work organisation according to traditional Taylorist principles (Pasmore 1995).

However, despite Trist and Bamforth's limited success in implementing their proposals, their ideas remained alive. They were the point of departure for what

[1]European Workplace Innovation Network; see http://www.portal.ukwon.eu.

H. Hvid (✉)
Roskilde University, Roskilde, Denmark
e-mail: hh@ruc.dk

V.K. Scheller
Copenhagen Business School, Frederiksberg, Denmark
e-mail: vks.ioa@cbs.dk

P.R.A. Oeij et al. (eds.), *Workplace Innovation*, Aligning Perspectives on Health, Safety and Well-Being, DOI 10.1007/978-3-319-56333-6_11

might be called a social movement, supported by leaders, trade unionists, consultants, teachers and researchers at business schools, technical universities and departments of psychology and sociology (Fricke 2000; Gustavsen 2007; Pasmore 1995). As a social movement it has made a difference, but has never been allowed to dominate management and organisational thinking.

Over the years the sociotechnical movement has prevailed under different names: quality of working life in the Anglo-Saxon world in the 1970s and 1980s; industrial democracy in the Scandinavian countries; humanising work in Germany; and as modern sociotechnology in the lowlands (de Sitter et al. 1997; van Eijnatten 1993, 2007). In the last 10–15 years new concepts largely based on the basic assumptions of the sociotechnical movement have been launched as 'high performance work systems', 'high involvement work systems', 'sustainable work systems' and 'workplace innovation' (Boxall and Winterton 2015; Gittell et al. 2012; Oeij et al. 2015). Thus, the sociotechnical approaches have developed and changed throughout history. However, we still find that Trist and Bamforth's (1951) core concept of 'responsible autonomy' (as opposed to individualised helplessness) is the essential element of the vision underlying the many new concepts (Boxall and Winterton 2015).

The sociotechnical movement has been driven by great enthusiasm, trying to convince managers and employees that the principles of the different approaches are obviously sensible. However, despite the solid arguments for following the principles of the sociotechnical movement, dissemination of the underlying ideas is very slow. Actually, it seems as if the development of work organisation is moving away from the principles of the sociotechnical movement (Lopes et al. 2014; Oinas et al. 2012). This paradox will be explored in this article by answering this research question: What is the reason for this 'non-spread' of sociotechnical approaches to work organisation (i.e. workplace innovation) and can neo-institutional theory (i.e. institutional entrepreneurship) provide us with an explanation?

Critically oriented researchers studying changes in working life have presented hundreds of case studies documenting how standardisation and Taylorism are increasing (see, for instance, the journals *Work, Employment and Society, Human Relations, Nordic Journal of Working Life Studies*). Researchers have documented how precariously employment is linked to degradation of working conditions (Standing 2011). In the tradition of critical management studies, it has been documented how managerial power is becoming increasingly sophisticated (Alvesson et al. 2009). Also, studies that are closely linked to the sociotechnical movement provide evidence that development is going in the wrong direction. Eurofound conducts studies of European working conditions every five years (European Working Conditions Survey). These studies include sections that are largely based on Karasek's famous job demand–control model (Karasek 1979; Karasek and Theorell 1990), which is inspired by sociotechnical thinking. According to these studies, the general trend in the evolution of work in Europe since 1995 has been increased quantitative demands, less influence at work and fewer opportunities to learn. This is despite the fact that the level of education has largely been elevated and traditional industrial work has vanished. The picture is slightly better in the Nordic countries than for Europe as a whole (Lopes et al. 2014; Oinas et al. 2012).

The research thus clearly indicates that principles of work and management that create unhealthy working conditions, loss of qualification, reduced employability and increased welfare costs for governments are gaining increased support, while the logic of the sociotechnical movement retains only a marginal position. This is in denial of the fact that following the principles pays off, which is well documented. Appelbaum et al. confirmed this when they renewed the concept of sociotechnical systems under the name 'high performance work systems' (Appelbaum et al. 2000). Their study involves 44 industrial companies and a survey of more than 4000 employees. Several reviews of existing studies of the effects of sociotechnical systems document a positive association between sociotechnical systems (including high performance work systems) and productivity (Boxall and Macky 2009; Brödner and Latniak 2003; Totterdill et al. 2009). Despite the methodological difficulties of getting studies of concepts to impact on production (Boxall and Macky 2009), and although it might be overstated to claim that the principles are superior when it comes to productivity in all cases, there are strong indications for productivity gain through following the principles of sociotechnical systems.

From the history of the sociotechnical movement we can thus conclude that rational economic arguments are not sufficient to make sociotechnical systems the dominant principle of work organisation. To understand the conditions for disseminating sociotechnical principles (currently expressed through the concept of workplace innovation) we have to understand management and companies as actors that are not guided by rational economic arguments alone. We therefore turn to neo-institutional theory for inspiration. The main point here is that companies follow economically rational terms only to an extent. Organisations are increasingly guided by the desire to gain legitimacy in their relationships to other organisations (DiMaggio and Powell 1983; Scott 2001). Proponents of sociotechnical principles, currently in the form of WPI, must therefore understand themselves as institutional entrepreneurs, who, on the basis of the existing institutional settings, must create legitimacy around WPI. We will therefore examine how initiatives to implement WPI coexist and are in dialogue with the existing institutional setting of work.

11.2 WPI from a Neo-Institutional Perspective

Neo-institutional theory does not deny that market conditions, technology and governmental regulations set the framework for companies and their development. However, either the market, or technology, or legislation gives companies precise indications of what rational behaviour is on the market, or for publicly owned companies; what the right strategy is to secure survival and growth. The future is first of all characterised by uncertainty. Therefore, it seems appropriate for organisations to imitate other organisations in the same position–that is what DiMaggio and Powell (1983) call institutional isomorphism–and to gain legitimacy in relation to the institutions they interact with. These institutions can be authorities, financial institutions, customers, citizens, unions, and so on. On this basis, the assumption is that organisations will be more motivated to incorporate WPI if there are organisations in their surroundings from which they can copy it and if they can achieve greater legitimacy by introducing WPI than, for example, by implementing lean, key performance management or other fashionable management principles.

Scott (2001, p. 52) states that institutions are founded on three pillars. We will examine in the following section how the sociotechnical movement has influenced organisations through:

(1) the regulatory pillar, based on rules and sanctions;
(2) the normative pillar, based on expectations and social obligations; and
(3) the cultural-cognitive pillar, based on common beliefs and shared logics of actions.

The sociotechnical movement has for decades aimed to institutionalise the principles of sociotechnical systems through the cultural-cognitive pillar. Efforts have been made to reshape the dominant institutional logics with arguments about 'responsible autonomy' instead of 'individualised helplessness' (Trist and Bamforth 1951). The arguments have been legitimated with references to research and best cases where the principles of sociotechnical systems have worked and led to success. EUWIN, who are launching the WPI in a European context, is using exactly these means to gain legitimacy for WPI (http://portal.ukwon.eu/).

However, as we described in the earlier sections of this article, good arguments (affecting pillar 3) are often not sufficient to change institutional logics, which frequently are reproduced in institutional norms (pillar 2) or through established rules (pillar 3).

Efforts to institutionalise the principles of the sociotechnical movement have not been solely cultural-cognitive (through pillar 3). In the North European countries there has been a certain normative institutionalisation of these principles. Working conditions in for instance the Nordic countries are moderately regulated through laws (pillar 1) and much more regulated by norms (pillar 2) based on negotiation between the social partners. Kettunen (2012) argues that social citizenship in the Nordic countries is based to a limited extent on rights and much more on interests. Working conditions in the Nordic countries have more than elsewhere evolved in a constant negotiation between employers and employees supported by tripartite systems (Gustavsen 2007). Within these negotiations, sociotechnical principles have been deemed suitable because they promise something attractive to both parties.

Although working conditions are largely regulated through negotiations, there is also statutory regulation in the Nordic countries that supports the principles of the sociotechnical movement. The Working Environmental Acts to some extent support the principles (in Norway, Sweden and the Netherlands). Further, the Nordic countries (as well as several other countries) have encouraged the spread of sociotechnical principles through legislation by supporting companies and networks of companies in their implementation of sociotechnical principles (Alasoini 2016). Alasoini suggests that the experience of these schemes confirms the fruitfulness of a neo-institutional horizon of understanding the proliferation of sociotechnical principles. According to Alasoini, supporting individual companies in their implementation of sociotechnical principles has little effect, because the regulatory and normative support is missing, and the legitimacy of the principles is weak. However, when it is possible to support networks of companies who can copy each other and establish a broader legitimacy around their activities, the impact of the support is much greater.

Neo-institutional theory provides a framework for understanding the institutionalisation of organisations. However, neo-institutional theory has been criticised for not understanding what creates organisational change, and for lacking an actor perspective. Neo-institutional theory has a weak understanding of power and change (Hardy and Maguire 2008; Hirsch and Lounsbury 2015).

To meet this criticism, recent years' interest in institutional entrepreneurship has been growing. The main focus is identifying the institutional prerequisites for creating change. Battilana et al. (2009) used existing research to develop a very simple model for understanding institutional entrepreneurship:

1. *Enabling conditions for institutional entrepreneurship*: All institutions operate in a field with established cultural-cognitive beliefs and shared logics, norms and rules. One such field may be quite heterogeneous, which can create openings for institutional entrepreneurs. Based on the openings offered in the field, the institutional entrepreneur can formulate new visions or modify existing visions.
2. *Creating alliances*: Through alliances with different actors it can be possible to experiment with the vision in various contexts, to achieve the necessary legitimacy behind the vision.
3. *Institutional changes*: Permanent changes in the shared logics and norms are established (between e.g. employees and managers, different professions, the organisation and its stakeholders). These institutional changes further influence the institutional field.

The first step in Battilana, Leca and Boxenbaum's model has been taken: institutional entrepreneurs (among all those creating EUWIN) modified existing sociotechnical visions and convinced the European Commission to promote workplace innovation: '*[...] the Commission sees WPI as a motor for innovation and competitiveness, and as a way to transform workplaces to make better use of human talents and skills*' (Oeij et al. 2015, p. 5).

In our study based on six case studies, we will examine how WPI as an idea is embedded in the six institutions, which alliances are created concerning WPI, and whether the WPI initiatives have created changes in the shared logics of the institutions. That will help us to answer our research question (see above), which we now can further specify: can the theoretical approach to institutional entrepreneurship provide us with insight in the dissemination and maintenance of sociotechnical approaches to work organisation?

11.3 Methodology

This chapter is based on six case studies included in a project about WPI funded by Eurofound (Oeij et al. 2015). The project used quantitative methods to identify contextual drivers for WPI and qualitative methods to understand the motives among employers and employees for implementing WPI initiatives, to describe WPI practices and to map the consequences of WPI activities for employees and company performance. The project comprised 51 case studies conducted in 10 European countries, with six of the case studies conducted in Denmark.

Case selection was based on quantitative data from the 'Third European Company Survey' (ECS, 2013), which is a telephone survey of companies in Europe with 10 or more employees. Of the 30,000 companies involved, 1284 were pre-selected. The pre-selected companies had a high score on the WPI index, constructed on the basis of the 'Third European Company Survey'. The researchers in each country got a list of potential companies they could invite to participate in the study.[2]

The common denominator for participating companies was their participation in the ECS, which nominated them as being particularly innovative. This nomination rendered them 'critical cases' (Flyvbjerg 2004) as they had made special efforts towards developing WPI practices, making them model companies for studying benefits and challenges linked to WPI. Working with critical case studies suggests that these benefits and challenges would apply to a broader selection of companies (Flyvbjerg 2004). If these companies had not incorporated WPI fundamentals into the shared logic of their organisation (institution), how could we expect other companies to do so? If these companies met great challenges moving towards WPI, how would more ordinary companies fare?

Recruiting case companies, however, proved to be more difficult than expected. Researchers asked Danish companies that had participated in the ECS and had high scores on WPI parameters if they wanted to participate in the follow-up study. The initial contact took place over the phone and included a short structured interview about WPI. Of the 64 companies contacted, only six agreed to participate. At each of these six companies individual interviews were conducted with a manager and one or more employee representatives. In addition, focus group interviews were conducted with employees involved in WPI activities at the company. The interview guides supported the creation of a coherent 'story' about the implementation

[2]For more information, see the 'Technical annex' of the 'Third European Company Survey–Workplace innovation in European companies': http://www.eurofound.europa.eu/sites/default/files/ef1540_-_technical_annex_-_third_european_company_survey_workplace_innovation_in_european_companies.pdf.

and benefits of WPI at each company. There were only a few instances of different assessments of WPI. The researcher attended a full day at each company. In Denmark 18 interviews were conducted (individual and focus group interviews). All interviews were transcribed and translated into English.

To make a coherent analysis of the 51 cases possible, the project needed a clear interview guide and standardised coding matrixes (see Chap. 10 in this publication). For this article we chose to return to the interview data and devise a new, open coding. Thus, we coded the interviews according to the following four categories:

(1) *WPI activity*
(2) *institutional foundation (pillars)*
(3) *institutional alliances (shared logics)*
(4) *institutional change–organisational changes and changes in shared logics.*

11.4 Institutional Embedding of WPI Activities in Six Danish Companies

In this section we will present the context and WPI activities of each company. Following these descriptions each activity will be analysed in terms of its 'institutional embeddedness', alliances and changed institutional logic.

1. **The Design Company: Organising Open Innovation and 'Multidisciplinary Customer Segment Teams'**

Company: The Design Company is an international company that produces designs for various customer segment groups (retail, hotels, furniture manufacturers, etc.). It has 400+ employees, of whom 145 currently work in the headquarters in Denmark, in product development, marketing, finance, sales, HR, IT, logistics, customer service, sample and order expedition and quality. The products are manufactured at sites in several European countries.

WPI activity: In 2012, The Design Company introduced new 'multidisciplinary customer segment teams' (MCST) with representatives from subsidiary companies, in-company product developers, designers, customer service assistants, logisticians and representatives from the departments of quality and the environment. These teams meet physically only rarely, because members live and work in different countries. However, the rare face-to-face meetings make the everyday communication between the team members much easier. According to both managers and employees, this teamwork practice provides the company with a 'multi-perspective' understanding of its customers' needs, enabling the company to provide better service. Open innovation and discussion in the 'multidisciplinary customer segment teams' provide a platform for 'disturbing business as usual' and exploring new possibilities for providing customer-oriented services.

Teamwork, knowledge-sharing and employee initiatives are regarded as supportive of company success at The Design Company. Hence, the product development department conducts regular innovation meetings that are conducted with a very open agenda: '*We formulate all sorts of utopian and "crazy" ideas about how to organise work. Afterwards, we work systematically on how to transform them into specific solutions*' (employee). Interdisciplinary cooperation and knowledge-sharing provide The Design Company with a competitive advantage.

Institutional context: The Design Company is a non-hierarchical organisation with an informal management style and few formalised institutional setups: no trade union representatives, no collective agreements, no collaborative committee and no well-established professions. Only when it comes to occupational health and safety does a formal employee representation exist because it is obligatory under the law.

Institutional allies for the WPI vision and activities: The institutional foundation of the MCST seems to be the management group. It was from here that both the idea and the decision came. This seems to be a vulnerable anchoring of the MCST. There are a few follow-ups on the activities in the MCST and the necessary renewal of the MCST is not secured. Through the interviews we got the impression that there is quite a positive attitude towards the MCST within the organisation, but the MCST does not play a prominent role in everyday life. We had to remind the interviewees repeatedly that it was the MCST that was the reason for our visit to the company.

WPI initiated institutional change: The MCSTs are a result of institutional entrepreneurship conducted primarily by the management group; however, we doubt whether the MCST will last, owing to weak institutionalisation and the lack of an alliance supporting the MCST. Nevertheless, the MCST will probably have a lasting impact on The Design Company because the MCST confirm the cultural-cognitive pillar of the company's institutional setting.

2. **The Union: 'Internal University' Practices within a 'Grass-Roots Organisation'**

Company: The Union, with 17,000 members and 75 employees, is divided into eight geographical districts, a secretariat with administrative tasks and a group of approximately 25 consultants based in a large city in Denmark. The Union provides guidance services, conducts courses and takes part in the political processes of collective agreements. The management structure in The Union is flat, and it has always been this way as the organisation started out as a 'grass-roots movement'. Employee representatives' perspectives are valued: '*We speak our minds when change happens. We are able to raise critical questions that management has forgotten to consider*' (employee representative). According to the employees, WPI enables the organisation to 'absorb' ideas on how to improve services. Sharing knowledge also leverages the ability to adapt quickly to legislative and labour market changes.

WPI activity: The employees introduced an 'internal university' practice in 2008 with the aim of strengthening their ability to share knowledge internally. At 'internal university' meetings, which happen approximately every other month and normally last for half a day, employees present legislation amendments and changed practices. 'Internal university' is an opportunity to discuss interesting cases with co-workers and to disseminate specialised knowledge. The employees describe the practice as very dynamic and well-functioning.

Employees, managers and representatives agree that sharing knowledge reduces organisational vulnerability. The internal university approach, where the employees decide form, frequency and themes, seems to be a strength because it creates employee motivation. However, it is also a weakness because the daily operations can take over, when managerial prioritisation of the practice (e.g. sending out formal calendar invitations) is lacking.

Institutional context: The Union is organised quite informally, based largely on the cultural-cognitive pillar. The cultural-cognitive foundation in The Union is tied to the organisation's activities within the collective agreement system, which promote learning opportunities and job quality for its members.

Institutional allies for the WPI vision and activities: It is very difficult or impossible to make institutional alliances in The Union because the institutional constitution is very weak and most decisions are taken informally.

WPI initiated institutional change: The 'internal university' is a result of institutional entrepreneurship conducted by some of the employees. We doubt that the initiative will continue because it is weak in terms of institutionalisation and there is no alliance supporting it. Nevertheless, the 'internal university' might have a lasting impact on The Union because it confirms the institution's cultural-cognitive pillar.

3. **The City Museums: Introducing Autonomous Work Organisation**

Company: The City Museums is a Danish knowledge-based company with 140 employees–half academic personnel and half unskilled and skilled employees (e.g. craftsmen working at museum sites). In recent years, the pressure to compete with other attractions (e.g. cinemas, amusement parks) has increased, thus the museum realises the importance of employee involvement and initiative in creating appealing experiences for museum guests.

WPI activity: A new teamwork structure was introduced in 2012 at one of the city museums, the Village Museum, which is an open-air museum with antique buildings. Previously, the craftsmen's work was limited to performing maintenance tasks that took up 80% of their workdays. However, the Village Museum is a 'living' workplace, where guests watch them working on the antique buildings. Therefore, the craftsmen wanted more flexible work organisation with the possibility of conveying their maintenance tasks to the audience. Now the team self-organises work, task sequences, work modes, duty rosters, budgets, and so on. It improves their ability to plan, so they can perform tasks that are interesting for the guests during business hours. Beforehand, a functional manager devised the schedules then distributed tasks to the various professions. The new model implies that management has confidence in the employees' abilities to prioritise, which is very much appreciated by the craftsmen.

Institutional context: The City Museums has many strong institutional connections to do with the managerial systems of the municipality that owns the museum, collective agreements, trade union representatives and formalised collaboration between management and employees. The museum occupies employees belonging to strongly institutionalised professions: academics and specialised handicraft. However, there are also employees with weaker professionalisation such as administrative staff.

Institutional allies for the WPI vision and activities: WPI activity has been discussed and negotiated in institutional settings inside and outside the Village Museum. The craftsmen's strong professionality makes a WPI project possible for them. Attempts were made to include the administrative personnel in another WPI project. However, this proved more difficult, perhaps because their professionality is not as strongly institutionalised.

WPI initiated institutional change: The new obligations for the maintenance staff have been institutionalised. Management, representatives and employees agree that the new practices provide greater job satisfaction for employees while simultaneously enabling the company to provide better service: '*You can get a completely different organisation if you focus on autonomy and job satisfaction*' (employee). The WPI activity has somewhat changed the institutional logic of the organisation.

4. **The Municipal School: Meeting Challenges with Collaborative Practices**

Company: The Municipal School, with 90 employees and 700 students, is the result of a merger between two municipal schools in 2011. The merger was a big challenge at the same time as extensive public reforms were calling for new practices. The reforms meant new requirements for quality assessment, inclusion of children with special needs and new agreements on work hours for teachers. It was then a matter of 'making-sense' of the organisational changes while simultaneously gaining competitive advantage.

WPI activity: Originally teachers were individually responsible for teaching specific subjects across year groups. Now they are collectively responsible for an entire year group within a team of teachers and pedagogues. This requires a different approach to teaching, because the teachers within the team should be able to teach a broader set of subjects. '*It is important to strengthen teamwork because cooperation is crucial in order to implement the necessary pedagogical and didactic tools. That is why we thought that we [had] better create some strong teams, who, of course, have the academic competences, but [also] have to cooperate*' (manager).

Institutional context: Teamwork among the teachers is, in different forms, implemented in most Danish public schools. This school, however, prioritise its employees' involvement very highly. Employees, union representatives and managers participated in joint training sessions after the merger to ensure that they were all involved as 'partners' in the process. Organisational changes are discussed in forums with both direct and indirect participation from employees. The 'educational development forum' (a representative committee) plays a significant role, but also more formal cooperation takes place with union representatives, following a framework agreement made in the region. '*They [management] are very open to ideas and projects; if, for example, you ask, "Is it okay to spend a few hours on this idea?" I have never heard a "No." Innovative ideas run freely and they never interfere with them*' (employee).

Institutional allies for the WPI vision and activities: The comprehensive school reform, including changes to work hours, has created deep conflict between teachers and government. This conflict took place in most schools, although this one succeeded in institutionalising employee involvement in the change and maintenance of the work organisation.

WPI initiated institutional change: As already mentioned, the work organisation was re-institutionalised and the collaboration between management and employees was strengthened. The institutional logic of the organisation was affected.

5. **Road and Park Services: Introducing 'Joint Management' and Partnerships**

Company: Road and Park Services is a company with 118 employees that carries out technical construction services and projects relating to roads, parks and forests on Danish island. It is an independent private enterprise in the island's Regional Municipality and primarily performs tasks commissioned by the municipality. The company focuses on employee participation, partnerships with unions and job autonomy.

WPI activities: In 2003, the company participated in a development project about 'joint management' together with Local Government Denmark (employers association) and 3F (United Federation of Danish Workers). The project aimed to inspire workplaces to engage in partnerships and support more job autonomy for employees. It was an opportunity to learn how to utilise and develop competencies and there were funds for extensive training programmes for participating companies. The partnership between the management group and the union representatives began to flourish during this process.

Organisational changes (restructurings, collective agreements and the like) are discussed by the manager and union representatives at a monthly meeting. As an union representative explains, they have gone from a 'trade-off' relationship between management and representatives to a partnership in which they value each other's opinions and respect each other as equal partners, which provides a good basis for negotiations: '*It is nice to have representatives who are not afraid to step up against me in a constructive dialogue*' (manager). The employees believe that trust allows the company to respond more quickly to new opportunities.

The employees at Road and Park Services work in self-managing work gangs with a high degree of autonomy. They do not just organise their own work; in most cases the gangs price the tenders, too. '*This means that our prices are much more rooted in the gangs. It is not just something a "silly manager" has come up with by calculating costs*' (manager).

Whenever a gang wins a tender, they have already considered how to carry it out: '*We won the tender of maintenance of all the public toilets. The gang had decided on a tight time schedule of seven minutes and twelve seconds to clean each toilet. They had come up with that by themselves*' (manager). The manager is convinced that you can get the best organisational results by involving the employees, as they are 'way better' at developing their own work than he is.

Institutional context: Road and Park Services, like The Waste Management Company (below), is a private company owned by the municipality. That gives the organisation both relative freedom (compared to regular municipal institutions) and a relative sense of security (compared to privately owned companies in the field).

Institutional allies for the WPI vision and activities: Road and Park Services has created strong alliances with the trade union, the employer's association and the municipality.

WPI initiated institutional change: Work organisation has been changed, autonomy has increased significantly and workers are involved in the management of the company. However, there are dark clouds on the horizon. Recently, municipal politicians have begun to interfere and impose top-down management on the technical areas. Perhaps Road and Park Services will in the near future be subject to 'new public management' principles, like the rest of the municipality. The dominant institutional logic of Road and Park Services will perhaps be overruled by another strong logic.

6. The Waste Management Company: Supporting Employee Initiative

Company: The Waste Management Company (170 employees) collects and manages waste from a large Danish municipality. Collaboration between management and union representatives is highly developed within the company, which has a long tradition of employee participation and has had employee members of the executive board since 1994. The company prioritises safety for its employees and supports employee initiative. Representatives from the company participate in open forums, before sharing new ideas and innovations with other companies. The management group stresses the importance of involving employees in developing successful innovations, as they '*know everything about their jobs*' (manager).

WPI activity: Lately, union representatives, employees and managers have developed two trucks that are safer for employees and that enable faster and easier handling of chemical and heavy duty waste.

Recently the employees have used their initiative to create another product innovation, involving extracting methane gas from shredder waste. The idea came up when the employees working in one of the repositories wanted to measure the temperature in the waste repository. They drilled holes in the waste, wherein they placed some temperature tubes. They chose to add a filtering air tube and extract some air, mainly out of curiosity. It turned out that the waste produced methane gas. After discovering this, the company developed a way of extracting and utilising this gas. The gas extraction is an important part of a sustainable business plan, as it has a positive climate effect and at the same time makes profit that evens out increasing costs in other departments:

'*Clearly, in the project with the gas, the money and the revenue that could come out of it are not insignificant. There will be revenue [from] it and, even though we are self-reliant, these revenues will probably be used for paying our future expenses [...]. If we have to pay for cleaning the water that seeps from the repository for the next 30 years, such revenue [will be] very welcome*' (manager).

Institutional context: The Waste Management Company is a private company owned by the municipality (i.e. Private Public Partnership). The formal construction of the company has given the organisation a great deal of freedom (compared with other municipal institutions) and the ownership has given the organisation a relatively high sense of security (compared with other private institutions). In a number of years the company has used this freedom and security to develop an involving work organisation and a democratic management style.

Institutional allies for the WPI vision and activities: The institutionalisation of a democratic managerial style, where employee representatives are involved in daily management, supports the vision of WPI. Further, both the union and the municipality are using the company to profile themselves. The union uses the company as a showcase for employee involvement, and the municipality uses the company for branding itself as a 'sustainable organisation'.

WPI initiated institutional change: The institutional relationship between employers and workers' representatives has extended its scope. Product development has been introduced as an issue included in local industrial relations. Gradually the institutional logic of the organisation has changed.

11.5 Analysis

To analyse the enablers of and threats to the implementation of WPI, we will use the previously presented neo-institutional theory framework. We will analyse the institutional embeddedness of the WPI initiatives in the case companies by using Scott's theory on institutions, based on three pillars: (1) the regulatory pillar, (2) the normative pillar and (3) the cultural-cognitive pillar. Institutional embeddedness is presented in the table below in terms of which of the three pillars have been involved in WPI activity. Further, we will use the previously presented theories on institutional entrepreneurship to analyse alliances to the vision of WPI, and whether the WPI initiatives have created permanent changes in the shared logics, enabling further development of WPI (Table 11.1).

In the two first mentioned cases, WPI activity is weakly embedded, alliances relating to WPI activity are almost non-existent and the activity will probably not change the institutional logics of the organisations.

In the two last mentioned cases, WPI activities are strongly embedded, alliances relating to WPI activity are many and have impact, and the WPI activities are

Table 11.1 Enabling or threatening factors relating to institutional embeddedness of WPI

Company	Institutional embeddedness	Alliances	Changes in shared logics
The Design Company	Cultural-cognitive patterns	Management only. No internal or external alliances	Logic for international cooperation has changed. However, basic institutional logic has not
The Union	Cultural-cognitive patterns	Single innovative employees only. No internal or external alliances	Existing institutional logics are maintained, not changed
The City Museums	Regulatory, normative and cultural-cognitive patterns	Alliance between professions, management and employee representatives	Institutional logics in the village museum have been changed, although they could be threatened by institutional logics imposed by the municipality
The Municipal School	Regulative, normative and cultural-cognitive patterns	Alliance between management, employee representatives and employees. Supported by the general school reform	Institutional logics in the school have been changed, although they could be threatened by institutional logics imposed by the municipality as well as the national conflict between teachers and the association of municipalities
Road and Park Services	Regulatory, normative and cultural-cognitive patterns. WPI activity has a long history, and is strongly embedded in the organisation, having created a strong company branding	Alliances created internally between management and employees and externally with labour market institutions and funds	Institutional logics have changed dramatically over the years. WPI inspired logics could, however, be 'too exotic' in a business world where companies are imitating each other. In this case logics inspired by new public management, imposed by the municipality, could change the institutional logics supporting WPI
The Waste Management Company	Regulatory, normative and cultural-cognitive patterns. WPI activity has a long history, and is strongly embedded in the organization, having created a strong company branding	Alliances created internally between management and employees and externally with labour market institutions and funds	Institutional logics have changed dramatically over the years. WPI inspired logics could, however, be 'too exotic' in a business world where companies are imitating each other

changing the institutional logics of the organisations. This does not necessarily mean that the WPI activities in these institutions will survive, because they are challenged by other institutional logics–those imposed by new public management. However, the institutional logics, influenced by ideas connected to WPI, are actually in a kind of battle with the logic of new public management, and because of that will have an impact beyond the single organisation.

The two institutions mentioned in the middle have adopted institutional logics included in WPI. However, their changes are more incremental and less radical than in the two last cases, and the possible clash with other institutional logics, that is, those imposed by new public management, will not be as confrontational.

11.6 Conclusion

The central premise of this chapter is that the convincing argument that the introduction of WPI creates a win–win–win situation where both employers, employees and the state gain advantages is not sufficient to encourage mass dissemination of WPI. Therefore we asked: What is the reason for this 'non-spread' of sociotechnical approaches to work organisation (i.e. Workplace Innovation) and can neo-institutional theory (i.e. institutional entrepreneurship) provide us with a frame for an explanation?

On the basis of our study we conclude that the dissemination of WPI can be seen as a matter of institutional entrepreneurship, moving through three steps.

The *first step* is to find openings in the field that create opportunities for the dissemination of new visions. The 'entrepreneurs' behind WPI (mostly those behind EUWIN) have succeeded in getting support for their vision from the European Commission. However, our study indicates that we have still not found the right openings for the vision of WPI among ordinary companies. Perhaps WPI should be better adjusted to the institutional realities of different kinds of companies, and perhaps also to national, regional and sectoral conditions. Could WPI be included in instruments monitoring psychosocial working environment or quality of management? Could WPI be included as an issue in the collective agreements? Can managers improve their reputation if they have participated in WPI activities? Can trade unions gain support from their members by supporting WPI?

The *second step* in the process of institutional entrepreneurship is to create alliances around the vision. Alliances around WPI activities in the six cases are very different in content, strength and scope. We note that where 'institutional entrepreneurs' have succeeded in creating extensive alliances both within the company and externally, WPI activities are most comprehensive, and create changes in the shared logics forming the work organisation. These alliances were most often supported by public institutions, funds, or labour market institutions. It seems as if a local institutionalisation supporting WPI is essential.

The *third step* is the creation of permanent changes in the shared logics of the work organisation with lasting institutional implications. In four of our cases that has happened. These new logics are however, challenged in two of our cases by other strong logics related to new public management. Perhaps we will find that these new organisational logics will be overturned by the logics of new public management. However, even though the single case after a period of time perhaps will go through a development from extraordinary to ordinary, the case company has contributed to the development of the shared logic of public management.

We can hardly generalise on the basis of our six case studies, as the number and character of the interviews limit their depth. However, our study emphasises the relevance of an institutional perspective on the implementation of WPI, and stresses that certain institutional conditions may be essential for this implementation.

References

Alasoini, T. (2016). *Workplace development programmes as institutional entrepreneurs: Why they produce change and why they do not*. Aalto University, Espoo.

Alvesson, M., Bridgman, T., & Willmott, H. (2009). *The Oxford handbook of critical management studies*. Oxford, UK: Oxford University Press.

Appelbaum, E., Bailey, T., Berg, P., & Kalleberg, A. E. (2000). *Manufacturing advantage: Why high-performance work systems pay off*. London: Cornell University Press.

Battilana, J., Leca, B., & Boxenbaum, E. (2009). How actors change institutions: Towards a theory of institutional entrepreneurship. *The Academy of Management Annals, 3*(1), 65–107.

Boxall, P., & Macky, K. (2009). Research and theory on high-performance work systems: Progressing the high-involvement stream. *Human Resource Management Journal, 19*(1), 3–23.

Boxall, P., & Winterton, J. (2015). Which conditions foster high-involvement work processes? A synthesis of the literature and agenda for research. *Economic and Industrial Democracy*, Epub ahead of print (September 2, 2015). doi:10.1177/0143831X15599584

Brödner, P., & Latniak, E. (2003). *Sources of innovation and competitiveness: National programmes supporting the development of work organisation*. Gelsenkirchen: Report to DG Employment and Social Affairs.

de Sitter, L. U., den Hertog, J. F., & Dankbaar, B. (1997). From complex organizations with simple jobs to simple organizations with complex jobs. *Human Relations, 50*(5), 497–534.

DiMaggio, P. J., & Powell, W. W. (1983). The iron cage revisited: Institutional isomorphism and collective rationality in organizational fields. *American Sociological Review, 48*(2), 147–160.

Flyvbjerg, B. (2004). Five misunderstandings about case-study research. In C. Seale, G. Gobo, J. F. Gubrium, & D. Silverman (Eds.), *Qualitative research practice* (pp. 390–405). London: SAGE Publications.

Fricke, W. (2000). Twenty five years of German research and development programs "Humanization of work/work and technology". *Concepts and Transformation, 5*(1), 133–138.

Gittell, J. H., Seidner, R., & Wimbush, J. (2012). A relational model of how high-performance work systems work. *Organization Science, 21*(2), 490–506.

Gustavsen, B. (2007). Work organization and 'the Scandinavian model'. *Economic and Industrial Democracy, 28*(4), 650–671.

Hardy, C., & Maguire, S. (2008). Institutional entrepreneurship. In R. Greenwood, C. Oliver, K. Sahlin, & R. Suddaby (Eds.), *The SAGE handbook of organizational institutionalism* (pp. 198–217). Los Angeles: Sage.

Hirsch, P., & Lounsbury, M. (2015). Toward a more critical and "powerful" institutionalism. *Journal of Management Inquiry, 24*(1), 96–99.

Karasek, R. A. J. (1979). Job demands, job decision latitude, and mental strain: Implications for job redesign. *Administrative Science Quarterly, 24*(2), 285–308.

Karasek, R. A. J., & Theorell, T. (1990). *Healthy work: Stress, productivity and the reconstruction of working life*. New York: BasicBooks.
Kettunen, P. (2012). Reinterpreting the historicity of the Nordic model. *Nordic Journal of Working Life Studies, 2*(4), 21–43.
Lopes, H., Lagoa, S., & Calapez, T. (2014). Declining autonomy at work in the EU and its effect on civic behavior. *Economic and Industrial Democracy, 35*(2), 341–366.
Oeij, P., Žiauberytė-Jakštienė, R., Dhondt, S., Corral, A., Totterdill, P., & Preenen, P. (2015). *Workplace innovation in European companies*. Luxemburg: Publications Office of the European Union.
Oinas, T., Anttila, T., Mustosmäki, A., & Nätti, J. (2012). The Nordic difference: Job quality in Europe 1995-2010. *Nordic Journal of Working Life Studies, 2*(4), 135–152.
Pasmore, W. A. (1995). Social science transformer: The socio-technical perspective. *Human Relations, 48*(1), 1–21.
Scott, R. W. (2001). *Institutions and organizations*. Thousand Oaks, CA: SAGE Publications.
Standing, G. (2011). *The Precariat: The new dangerous class*. London: Bloomsbury Academic.
Totterdill, P., Exton, O., Exton, R., & Sherrin, J. (2009). *Workplace innovation policies in European countries*. Nottingham: UKWON.
Trist, E., & Bamforth, K. (1951). Some social and psychological consequences of the longwall method. *Human Relations, 4*(3), 3–38.
van Eijnatten, F. M. (1993). *The paradigm that changed the work place*. Assen: Van Gorcum.
van Eijnatten, F. M. (2007). Socio-technical systems. In S. Clegg & J. R. Bailey (Eds.), *International encyclopedia of organization studies* (pp. 1448–1451). London: Sage.

Author Biographies

Helge Hvid is a Professor, Department of People and Technology, Roskilde University, Head of Centre for Working Life Research. Chief editor of Nordic Journal of Working Life Studies. Current research: Temporal structures at work, jobcrafting, psychosocial working environment.

Vibeke Kristine Scheller b. 1987, is PhD fellow at Department of Organization, Copenhagen Business School. Main research focuses: Healthcare innovation, everyday innovation in organizations, working life studies, temporal structuring and practice based research.

Chapter 12
Creating the Bottom-up Organisation from the Top: Leaders as Enablers of Workplace Innovation

Peter Totterdill and Rosemary Exton

12.1 Introduction

The evidence is clear. Workplace innovation enhances both economic performance and quality of working life, presenting a compelling case for change in organisations characterised by traditional cultures and ways of working (Pot et al. 2016; Ramstad 2009; Totterdill 2015).

There is a growing number of case studies of companies such as Innocent and Red Gate Software created by a new generation of entrepreneurs who attribute their success to empowering and non-hierarchical workplace practices and who are not burdened with conventional management attitudes and organisational practices. Such cases, inspiring though they can be, should not lead us to underestimate the challenge faced by those seeking to drive change in more traditional and longer established organisations, often in the face of resistance or inertia. A systemic and transformational approach is required if established practices are to be uprooted and replaced with active workforce involvement and participation (Totterdill 2015).

The term "workplace innovation" is now widely used to describe such an approach. Based on their meta-analysis of 120 case studies across ten European countries, Totterdill et al. (2002) argue that it takes diverse forms but is always characterised by the adoption of workplace practices that enable employees at every level of the organisation to use and develop their full range of skills, knowledge, experience and creativity in their day-to-day work (Totterdill et al. 2002). Workplace innovation stresses the importance of individual job autonomy, self-managed teams, permeable organisational walls and ceilings, systems and

P. Totterdill (✉) · R. Exton
Workplace Innovation Europe CLG, Dublin, Ireland
e-mail: peter.totterdill@workplaceinnovation.eu

R. Exton
e-mail: rosemary.exton@workplaceinnovation.eu

P.R.A. Oeij et al. (eds.), *Workplace Innovation*, Aligning Perspectives on Health, Safety and Well-Being, DOI 10.1007/978-3-319-56333-6_12

procedures based on trust and empowerment, systematic opportunities for employee-led improvement and innovation, shared and distributed leadership, and employee voice in strategic decision-making.

There seems to be little evidence to show that evidence about the benefits of workplace innovation motivates or stimulates its actual adoption in companies. While much is known about the practices and behaviours associated with workplace innovation, less is known about the drivers and situational settings which stimulate and sustain it. We need a much better understanding of the factors which shape workplace practices at enterprise level not least because successive surveys demonstrate a substantial gap between the evidence from research of "what works" and what actually happens in European enterprises (Eurofound 1997, 2012, 2015a; OECD 2010).

These contextual factors can reflect national characteristics as well as those at the workplace, and a "varieties of capitalism" (Hall and Soskice 2001) approach might be used to explain differences in the distribution and nature of workplace innovation in Europe (Totterdill et al. 2016). Hall and Soskice distinguish two distinct types of capitalist economy based on the ways in which firms operate in the market and interact both with each other and with actors such as government and trade unions. In Liberal Market Economies (LMEs) such as the UK, firms interact principally through market mechanisms; in Co-ordinated Market Economies (CMEs) such as those in Northern Europe, non-market forms of interaction such as social partnership and regulation play a more prominent role in shaping patterns of interaction.

Constraints on the development of workplace innovation are likely to be greater in liberal market economies than in the co-ordinated market economies. Godard (2004) argues that high performance workplace practices challenge the LME paradigm in which distrust and adversarial relations often underpin the employment relationship. In the CME paradigm, levels of trust are higher and provide a sounder basis for ways of working based on autonomy and self-management. For example, enterprise level "Co-operation Committees" long-established within the national social partnership arrangements in Denmark, can play a significant role in stimulating and resourcing workplace innovation (Mulvad Reksten and Dragsbæk 2014). Likewise, public policy intervention to promote and support the adoption of workplace innovation is widespread within CME countries, and actively involves both labour market partners, but is rare elsewhere (Totterdill et al. 2016). In LME countries such as the UK, explanations for the adoption of workplace innovation are therefore more likely to be found at enterprise level, with a much greater emphasis on voluntarism and managerial discretion.

Analysis of fifty-one case studies undertaken as part of the Third European Company Survey provides a rare insight into the "paths to workplace innovation" found at enterprise level (Eurofound 2015b). In particular, the study explores the variations and similarities between organisations in which workplace innovation:

- is instigated by top management or by employees;
- involves employees in the design and implementation of workplace practices, or not;

- is driven by the strategic goal of securing a sustainable future (the "burning platform") or by specifically targeting performance or quality of working life indicators;
- reflects the existence or absence of organisational freedom from control by a parent company or by other stakeholders.

While these variations are seen to shape the specific characteristics of workplace innovation in each organisation, common outcomes are found in terms of innovative behaviour based on enhanced employee autonomy.

Each of the six case studies conducted in the UK describes the instigation of workplace innovation practices from the top down, perhaps reflecting the relative absence of the workplace social partnership arrangements associated with CME countries. In four of the cases, workplace innovation resulted directly from the appointment of a new Chief Executive committed to distributed forms of leadership and with a transformational agenda. Three of these cases are examined below.

It is equally important to understand the nature of the journeys that followed. While some people in shared leadership roles have a clear vision of the systemic change they are trying to establish, others begin the journey from one of several specific issue-related angles such as industrial relations, a changing competitive environment, concordance with a parent company's corporate values or the need to sustain competitiveness. At its best, this begins an incremental though sustained journey of experimentation, trial and error, and learning as workplace innovations are established, consolidated and extended, characterised by 'bottom up' employee empowerment and initiative. As the Eurofound case study report argues, the resulting practices "do not often constitute a coherent programme, and in many cases they reflect developments over a number of years" (Eurofound 2015b, p. 41).

12.2 Workplace Innovation and "the British Disease"

The UK has a multi-faceted and well documented problem with productivity, skills development and skills utilisation. According to the UK Commission for Skills and Employment (UKCES 2009, p. i):

"Our stock of skills and their optimal deployment fare relatively poorly when compared internationally according to skills utilisation measures such as labour productivity and levels of qualifications among different workforce groups. Access to opportunities for skills acquisition is uneven as are their impacts."

UKCES and other commentators describe this as a "low skill equilibrium". The economy is seen to be trapped in a vicious spiral of low value-added and low skills; enterprises are staffed by low skilled staff producing low quality goods and services to which the training market responds rationally by providing training aimed at the demand for low skills.

This "British disease" has a long clinical history, achieving cyclical but transitory public policy prominence over several decades though seemingly without reaching solutions capable of closing the productivity gap with competitor nations (Wright and Sissons 2012). Survey findings (UKCES 2009; LLAKES 2012) point to:

- a widening gap in the labour market between the number of workers with qualifications at various levels and the number of jobs that require those qualifications;
- the tendency for UK employers to require lower educational qualifications for otherwise similar jobs than their counterparts in many other developed countries;
- 35–45% of workers with qualifications that are not fully utilised in their current jobs (Wright and Sissons 2012) but which would be of economic value if they could be put to better use in more demanding roles;
- the slow pace at which UK employers have adopted workplace innovation practices despite long-established evidence that they are associated with enhanced levels of productivity and performance.

The two latter points are of particular relevance to this article as well as providing a partial explanation for the “British disease”. The fundamental premise underlying workplace innovation is that traditional ways of organising and managing work limit the ability of employees at all levels to use and develop their full range of skills, knowledge, experience and creativity, both in performing their functional tasks and in contributing to improvement and innovation, thereby weakening productivity, competitiveness and quality of working life. Workplace innovation seeks to broaden job roles and employee discretion at both individual and team levels, transcend vertical and horizontal demarcations, enable employee-led improvement and innovation, and engage the tacit knowledge of frontline workers as a resource for all levels of decision making. It therefore embraces the concern with skills utilisation and development in the workplace (Totterdill 2016).

These workplace practices enhance the ability of employers to secure a full return on their investments in training and technology as a result of improvements in performance, innovation and quality of working life. Moreover, as a 2015 study by CEDEFOP (the EU’s Centre for the Development of Vocational Training) shows, increasing the complexity of jobs leads to enhanced opportunities for workplace learning and development.

“Workplace innovation” emerged during the early years of the century as a unifying concept which brought together work organisation, human resource management and other antecedents. Pot (2011, p. 404) describes workplace innovation in terms of “new and combined interventions in work organisation, human resource management and supportive technologies” which are strategically informed and highly participative in nature.

In defining workplace innovation, it is important to recognise both process and outcomes. The term describes the participatory *process* of innovation which leads to *outcomes* in the form of participatory workplace practices. Such participatory practices grounded in continuing reflection, learning and improvement sustain the process of innovation in management, work organisation and the deployment of technologies.

Workplace innovation is fuelled by open dialogue, knowledge sharing, experimentation and learning in which diverse stakeholders including employees, trade unions, managers and customers are given a voice in the creation of new models of collaboration and new social relationships. It seeks to build bridges between the strategic knowledge of the leadership, the professional and tacit knowledge of frontline employees, and the organisational design knowledge of experts. It seeks to engage all stakeholders in dialogue in which the force of the better argument prevails (Gustavsen 1992; Pot et al. 2016).

Research highlights the importance of internal consistency in such policies and practices (Huselid et al. 1997). As Teague (2005) suggests, organisations with mutually reinforcing employment practices achieve superior performance as their collective impact is greater than the sum of individual measures.

Such insights led to the creation of *The Fifth Element*[1] as a means of providing practical and actionable insights into the vast body of research evidence relating to workplace innovation in order to enhance productivity, performance and employee health and well-being (Totterdill 2015). Grounded in a cross-cutting analysis of research studies and documented cases, *The Fifth Element* identifies four basic building blocks which, in concert, create a system of mutually reinforcing practices:

Element 1	Empowering jobs and self-managed teams
Element 2	Flexible organisational structures, people-centred management practices and streamlined systems and procedures based on trust
Element 3	Systematic opportunities for employee-driven improvement and innovation
Element 4	Co-created and distributed leadership combined with 'employee voice' in strategic decision-making

Element 5, *The Fifth Element*, represents the enterprising behaviour, the culture of innovation, the high levels of employee engagement, and the organisational and individual resilience which flourish when the other four combine to shape experience and practice across the whole organisation.

If workplace innovation produces tangible economic and employee benefits at enterprise level it is also likely to have wider impacts on the labour market and economy. Skills demand is enhanced because employers need individual workers to embrace wider technical functions and, critically, to enhance generic skills including problem solving, communication and team working. Product and service quality are enhanced while the rate of innovation grows, thereby breaking out of the low skills equilibrium trap (CEDEFOP 2015; OECD 2010; UKCES 2009).

[1]http://uk.ukwon.eu/the-fifth-element-new.

12.3 The Gap

Even though evidence about the effectiveness of workplace innovation has been around for a long time (Totterdill 2015), successive surveys show that the vast majority of UK companies do not make systematic use of it. One UK survey estimated that less than 10% of employees work in self-managing teams, a basic building block of good work organisation, while less than 30% have a say in how their work is organised. The UK compares unfavourably with several other Northern European countries against many such indicators of employee involvement and participation (LLAKES 2012).

At enterprise level the limited spread of workplace innovation practices can be understood in terms of several interwoven factors (Totterdill et al. 2002; Business Decisions Limited 2002) including:

- an excessive tendency to see innovation purely in terms of technology;
- low levels of awareness of innovative practice and its benefits amongst managers, social partners and business support organisations;
- poor access to robust methods and resources capable of supporting organisational learning and innovation;
- barriers to the market for knowledge-based business services and the absence of publicly provided forms of support;
- the failure of vocational education and training to provide knowledge and skills relevant to new forms of work organisation.

Resistance to high involvement work practices can also be explained in terms of the embedded structures that shape management behaviour. Power can be seen as a zero-sum game: to empower workers, managers may wrongly perceive that they have to lose it, potentially challenging their self-identity and status within the organisation (Hardy and Leiba-O'Sullivan 1998).

12.4 Driving Workplace Innovation in UK Organisations

As with other liberal market economies, the UK lacks both the social partnership structures and the public policy frameworks that help to stimulate the adoption of workplace innovation in the CME countries. UK governments have relied on a market-driven approach to workplace innovation and instigated no policies or programmes to close the gap in productivity caused by the very long tail of companies who fail to respond to evidence. This stands in stark contrast to France, Germany and some Nordic countries where national and regional workplace development programmes have existed for some decades (Totterdill et al. 2009).

In this context it is unsurprising that enlightened leadership should play an important role in driving workplace innovation within public and private enterprises, though the systemic adoption of highly participative working practices remains relatively rare in UK enterprises. Leadership theory is a highly contested field but leadership development has nonetheless gained increasing prominence through business school curricula, professional institutions and consultant offerings. Early theories were primarily focused on the distinction between "task focus" and "people orientation" (for example Vroom and Yetton 1988) but the emergence of "New Leadership Theories" led to the celebration of "transformational", "charismatic", "visionary" and "inspirational" leadership (Storey and Holti 2013), often drawing on the burgeoning hagiographies of business leaders such as Jack Welch and Steve Jobs.

The dark side of such leadership approaches soon began to emerge including the potential for abuse of power, narcissism, destabilisation, blind obedience and fear of questioning. It can even be argued that the extraordinary trust in the power of charismatic CEOs displayed in these leadership approaches "resembles less a mature faith than it does a belief in magic" (Khurana 2002, p. 8).

Alternative approaches focused on leadership as a creative and collective process (Senge 1990) were less concerned with the central, charismatic individual and more with the creation of opportunities for employees to seize the initiative and contribute to decision making. Such shared and distributed leadership relates to a concern with empowerment (Caldwell 2005) and change agency, a phenomenon characterised by dispersed responsibility for change and not to be confused with the more heroic or charismatic models of change leadership (Buchanan et al. 2007). For Gronn (2002) the principle advantage of distributed leadership is that it builds organisational capability, and is, therefore, a key element of workplace innovation in that it helps to release the full range of employee knowledge, skills, experience and creativity.

Recent developments in the UK add extra weight to the focus on leadership. On the same day in June 2015 two public agencies coincidentally published guidelines on productivity (Acas 2015) and employee well-being (National Institute for Clinical Excellence 2015), both referring to the need for positive leadership styles which encourage creativity, new ways of doing things and opportunities to learn, and for leaders to be open, honest and visible.

12.5 The Case Studies

The authors were part of a team of researchers commissioned by the European Foundation for the Improvement of Living and Working Conditions (Eurofound) to interview a sample of companies identified from their responses to the earlier European Company Survey (Eurofound 2015a) as scoring highly in relation to indicators of workplace innovation.

During 2014–2015, interviews were held in fifty-one selected companies covering ten EU member states. In each company interviews were held wherever possible with a manager (mostly an operational director or HR manager), a group of employees and an employee representative. These provided in-depth, qualitative information about workplace innovation practices, the motivations and actions of different actors and their contributions to their development and implementation, and their outcomes and effects including company performance and quality of work. Interviews were recorded and transcribed, and for each case, a coding matrix was completed in which all the answers were incorporated in a systematic manner (Eurofound 2015b).

Of the six UK case studies, we selected the three which most clearly illustrate the choices that leaders make in shaping working practices and culture within their organisations. While in all three cases workplace innovation was instigated by a newly appointed Chief Executive and led to greater employee empowerment and innovation, their journeys of transformation were very different, reflecting the varied contexts within which each operated. In the following section both the divergences and commonalities of the three journeys are explored.

12.6 Drax Power: Transforming Industrial Relations and Leadership

Drax Power is the owner and operator of Drax Power Station in North Yorkshire, the largest power station in the UK. It is a wholly owned subsidiary of Drax Group plc. Some 850 people are directly employed by Drax Power at the power station site, around 600 of whom operate the plant. The remainder are in corporate and other Drax Group functions. Union density amongst plant workers is around 80%, though much lower in senior roles and corporate services.

Drax Group has been listed on the London Stock Exchange since 2005. AES, the previous US owners of the power station, transferred ownership to its lenders following a period of financial hardship for the business triggered by a period of all-time low power prices. Drax Group recognised that coal-fired electricity generation would have little future beyond 2020 due to national and EU regulatory regimes. Lobbying by the company, led by its Chief Executive Dorothy Thompson, persuaded government of the environmental and strategic benefits of conversion to biomass, reducing greenhouse gas emissions typically by 80% compared with coal-fired generation. A massive programme of investment which is still underway led to consequent changes in working practices and the need to engage the workforce in the process.

Improved industrial relations and greater staff engagement in innovation and improvement were central to the future vision for Drax. Between 2010 and 2012 the company developed a twin-pronged strategy based on both representative and direct participation in order to build the culture of openness and participation which it saw as vital to its future success.

12.6.1 Working Together with Trade Unions

Prior to 2005 industrial relations were adversarial and this legacy persisted even after the formation of the Drax Group. This was a context not conducive to the transformation facing the company, limiting prospects for employee engagement in the change process. Richard Neville's arrival as Head of HR in 2009 created an opportunity to break with the past, not least because of his previous track record in securing effective partnerships with unions in other utility companies.

Overcoming their initial scepticism, Richard worked closely with union representatives to design 'Working Together', Drax Power's partnership agreement. Working Together doesn't just focus on communication and consultation at corporate level but seeks to shape line management behaviour at all levels of the organisation. It commits managers to a process of "open, honest, clear and accurate" communication and two-way dialogue; the agreement has been reinforced by a programme of management training and development in partnership behaviours, from supervisors upwards.

The emphasis is also on early stage trade union involvement rather than consulting on a finalised proposal. In addition to the quarterly Joint Consultative Committee, senior managers and union representatives meet weekly in informal "Point of Contact" sessions, providing an off-the-record sounding board for ideas and allowing specific issues to be addressed flexibly.

Replacing adversarial industrial relations with management-union partnership has played a vital role in Drax Power's technological and cultural transformation. In particular, it has engaged unions as active and knowledgeable partners in the programme of technological innovation and prevented adversarial industrial relations from becoming an obstacle to change. Of equal importance it has helped to build trust and engagement in change throughout the wider workforce.

12.6.2 Leading a Culture of Communication and Consultation at Drax

Drax's flagship communication channel is the twice-yearly Open Forum instigated by Dorothy Thompson and involving the senior executive team in face-to-face, hour-long dialogue sessions with the whole workforce. In view of very strict rules about how many people can leave their posts for reasons of safety, and to keep the plant going, Forums are conducted in sessions ranging from one or two people to groups of forty.

This very significant investment of time by the Chief Executive and her senior team exposes them to the concerns, ideas and tacit knowledge of employees at all levels of the organisation. While part of each session is based on a presentation to employees, the emphasis is firmly on two-way communication. Open Forums replaced the previous company-wide meetings and suggestion schemes which had struggled to stimulate open and constructive dialogue and feedback. Outcomes and suggestions from the different Forum sessions are assimilated and disseminated throughout the company using various media channels.

Employees at all levels report that the Forums symbolise Drax's visible and approachable leadership, helping to drive a more open culture characterised by improved communication through both formal and informal channels.

12.6.3 *Who Benefits?*

12.6.3.1 Improved Understanding

A 2010 workforce survey asked whether employees understood the company's future direction. Only 11% of respondents strongly agreed that they did. When the same question was asked in 2011, 96% of respondents agreed (and 63% *strongly* agreed), representing a major shift in engagement levels.

12.6.3.2 Partnership Working from a Trade Union Perspective

Union representatives stressed during the interview that partnership working hasn't yet shaped the behaviour of managers at all levels and there are still blockages. But they acknowledged that the agreement provides a tangible focus: "[Management] will discuss more things with you, more things about the business." It is a safeguard: "But when all else fails you can always refer back and say, this is what we have signed, why aren't you following this procedure?" And informal contact plays an important role: "Every fortnight we have a Point of Contact meeting so there is more chance of us getting across things that we find are not right and we can sit down and discuss those things."

12.6.3.3 Supporting People Through Change

One person summarised the general view expressed during the employee peer group interview: "the workforce is very well supported … the first thing the business did before implementing any of those changes was roll out the Vision and Values … and basically let everybody know what was going on and … that we would be supported through it … The way the changes are implemented, it is a very inclusive process."

12.6.3.4 Employee Voice

One employee commented: "It is interesting isn't it, you go to the Open Forums and people will say what they think and absolutely nobody will turn round and go, I can't believe he said that … they might not agree with you but nobody will actually knock anyone for having a view because we are encouraged to have a view. That's really empowering I think."

12.6.4 Prognosis

Drax Power has succeeded in improving industrial relations, securing union commitment to a common vision and creating a more open and entrepreneurial culture. This has helped in the company's transformation and is releasing staff ideas for innovation and improvement. Dorothy Thompson's open leadership creates trust and employees feel confident about the future. Remaining obstacles such as instances of unsupportive management behaviour need a consistent approach if the new culture is to become fully embedded.

12.7 Skanska: Living Corporate Values

Skanska, a Swedish-owned company established in 1887, is one of the world's leading project development and construction groups. The UK operation was established with the acquisition of Kvaerner Construction in 2000, McNicholas Utilities in 2006 and Atkins Highways Services in 2013. It is now one of the country's largest contractors, employing more than 5000 people.

The UK construction industry traditionally had a poor image tarnished by corruption, quality issues, insufficient regard to safety, a poor environmental record and the blacklisting of union activists. When Skanska acquired Kvaerner it inherited many of these practices, at odds with the Swedish parent company's open and transparent culture and values. Mike Putnam's appointment as CEO in 2009 demonstrated a clear commitment to break with the past. Visibility and accessibility is characteristic of his leadership style and in a 2012 article (Oliver 2012), he explained that:

"One of the big transformations has been to adopt the Swedish approach to openness and transparency."

"Everybody in the UK talks about it but in the Skanska Group it is at a completely different level. When you have your values you need to be visibly seen to follow them. The behaviour that backs up that leadership is absolutely crucial."

Although the Swedish parent's values provided a powerful driver, the interviews suggested that Putnam is known in the company to be motivated by "gut feel" rather than spending time preparing an elaborated business case for culture change.

"Respect" is a word that recurred frequently in our interviews with managers, union representatives and frontline staff alike. It is reflected in the company's core values which aspire to:

- zero accidents;
- zero environmental incidents;
- zero tolerance of bribery and corruption;
- zero defects.

Respect forms part of a culture which, according to a senior manager interviewed as part of the study, is intended to "penetrate the company's DNA", releasing employee voice at every level of the business. While hierarchical structures remain in the business, he argued, hierarchical management behaviour is increasingly out of place.

12.7.1 Mutually Reinforcing Practices: The Journey to Culture Change

In Skanska, gradual improvements in culture have resulted from a series of incremental workplace innovation initiatives that took place between Putnam's appointment in 2009 and the interviews in 2014. Interviewees in each group were clear that the new CEO was personally responsible for stimulating and sustaining each initiative. While each initiative is relatively modest in itself, they consistently reflect Putnam's commitment to Skanska's core values and build on each other to create a sustainable momentum of change:

- Breaking down silos: Skanska UK started with the acquisition by the Swedish parent of three businesses with separate cultures and practices. The first part of the journey was to demolish the walls between different parts of the organisation, centralising and transforming "Enabling Functions" such as HR and Finance. This helped embed Skanska's core values in each function and allowed consistent, company-wide initiatives to develop, addressing management development, healthy working and environmental sustainability.
- Changing management behaviour: Management development programmes focus on culture, not basic procedures. Skanska's Great Boss initiative defines expected management competencies and behaviours which are reinforced through management training programmes and appraisals. Progress is measured through an annual working climate survey. Union representatives interviewed for the study agreed that it is working and that there has been a steady change in management culture.
- Reinforcing ethical behaviour: At least once a quarter—and often monthly—management team meetings will spend half an hour exploring an ethical dilemma relevant to the business, either hypothetical or real. This reinforces the message to all managers that Skanska seeks to be driven by its core values.

- Trade union partnership: Senior management teams work closely with the trade unions which are seen as vital partners in reinforcing company values and in ensuring health and safety. Early-stage involvement and informal dialogue between union representatives and management at all levels play a critical role in dealing with potential issues quickly and collaboratively.
- An "Injury Free Environment": Safe, healthy working is driven from the top. It is seen as part of a shared learning culture rather than a regulatory stick. A rare death on any Skanska site in the world leads to a Global Safety Stand Down - a one minute silence plus an opportunity to learn from the experience and improve practice. Employee voice and union safety representatives play a critical role in highlighting risks and identifying better ways of working. This extends to the supply chain, ensuring consistent practices on site.
- A learning culture, not a blame culture: "If things go wrong and you tell us you get nothing but support."
- Engaging employees in improvement and innovation: Skanska recognises that there is no single way of stimulating employee initiative. An open and enabling management culture is the starting point, supported by specific initiatives including local Consultation Forums for frontline workers and union representatives, the Skanska Way Week—a focus for dialogue on important issues such as wellbeing and mental health, a You Said/We Did board, and even an Innovation App that enables employees to take a picture to illustrate ideas for improvement.
- Open access to learning: Prominently positioned in the main entrance hall of Skanska's headquarters, the Learning Centre hosts diverse personal development opportunities from on-line courses to tailored one-to-one support.

12.7.2 Who Benefits?

Skanska has one of the best health and safety records in the UK construction sector. Likewise, levels of employee engagement are consistently high. Annual engagement surveys typically secure a 75–85% favourable response to every question set.

Active engagement with trade union representatives in working towards common goals has secured positive employment relations.

A clear strategy has been adopted for integrating new acquisitions. Skanska acquired Atkins Highway Services in 2013, and these practices are ensuring the integration of its staff into the company's culture.

12.7.3 Prognosis

This case study does not describe a dramatic example of workplace innovation. It is about a series of incremental innovations sustained by consistent leadership and slowly generating a strategic change in culture, working practices and employee engagement.

It is still work in progress, and the company recognises that it will always be so. An open and innovative culture is not something you create just once: rather it needs to be maintained and renewed continuously.

Asked if there was one message he would like to give to Mike Putnam, one trade union representative said that his would be: "fantastic job, keep going!"

12.8 Bristan: Positioning for Sustained Competitiveness

Bristan was founded in 1977 in Birmingham (UK) as a family owned business. It is now a leading supplier of showers, taps and bathrooms in the UK with exports to Europe and Russia. It has been awarded one star status in the Sunday Times *Top 100 Best Companies To Work For*.

12.8.1 Leadership by Behaviour

The company is led by Chief Executive Jeremy Ling. He joined Bristan Group in 2009 when it was a family-owned "can do" company but with "a lack of clear focus and inconsistent objectives". One of Jeremy's first tasks was to guide the sale of the company to Masco, a US-based group of companies. Each company within the group operates autonomously and Bristan has retained its family atmosphere.

Employees interviewed as part of the study tended to see Jeremy's arrival as transformative. A series of measures instigated by him since his appointment in 2009 and our visit in 2014 have placed continuous improvement at the heart of the company's practice and this is reflected in the experiences of several of the employees interviewed. Jeremy himself describes the Bristan Group as "always moving forward".

Jeremy's approach is informed in part by his past frustrations when working for larger and less empowering organisations. He set out to create a culture based on shared leadership, values and behaviours: "empowerment of my top team and feedback on performance are essential for us to be a learning organisation". He and his senior team are guided by "boundaries" rather than strict role definitions and are trusted and empowered to be entrepreneurial within those boundaries. Jeremy points out that all employees have two duties: to develop themselves and to change and develop their roles.

One person summarised the view of Jeremy's leadership held by members of the employee peer group interviewed as part of the study: "This senior management team are the best ever, they know all our names." He argued that Jeremy is passionate about bringing people with him. The "Big Briefing", a quarterly meeting to keep everyone aware of company progress, results and updates, enables employees to ask questions and give feedback. Everyone attends in prearranged hourly slots and people are encouraged to send Jeremy emails on the topic he presents.

Jeremy argues that line managers drive the organisation and they take their own decisions about how to lead their teams. Self-managed teamworking is an important part of the culture of non-hierarchical behaviour that Jeremy is creating in Bristan. Teams are empowered to address issues from customers directly, working within boundaries set to give room for entrepreneurial behaviour. Open plan offices and working groups enable cross functional collaboration throughout the company, and job swaps provide opportunities to work in different areas and support career progression.

Training, skills development, apprenticeships and career progression are a priority for Jeremy, and the majority of the senior team have been promoted from within the company. The "Leadership by Behaviour" training programme based on the empowerment-based values that Jeremy introduced to Bristan is cascaded from the senior team to all managers.

Employees value the company's commitment to their learning. Individual development goals, learning needs and SMART objectives are discussed at conversational one to one meetings with line managers each month; together with mid-year reviews these lead to a "no surprise" year-end appraisal.

The senior team makes sure that everyone works within Bristan Group's SHINE values—Straightforward, Helpful, Innovation, No Limits to Customer Service, and Empower and Engage. Each quarter, three finalists from each of the five categories are chosen by the company's Employee Forum, which then decides on an overall winner. These winners are then entered into the Star Performer category, with the overall winner being announced at an award ceremony in January. At the end of the year, the company holds an awards ceremony where the SHINE Star of the Year is named.

Continuous improvement (CI) is part of the everyday vocabulary at every level. CI champions organise improvement projects and teams organise cross-functional working groups involving people from all areas of the company to improve the way they work. At the time of interview, in 2014, there were 146 CI initiatives in progress, each based on an opportunity identified by an employee. Bringing people together from different parts of the company enables problems to be seen "with a fresh pair of eyes." Every employee is also encouraged to sign up for a Kaizen project, taking time out for a week to join a cross-functional team. Between four and six Kaizen events are held every year and outcomes are presented to the senior team as an important means of learning, engagement and networking.

For example, a warehouse-led "fast despatch" group recruited from across the company met for a week and reviewed the despatches, cutting 7.5 miles of walking per week. Everyone in the warehouse is involved in continuously improving the integration of technology for "Just in Time" with their team working practices and USP (Unique Selling Point) of "next day delivery". The warehouse people rotate their roles and tasks regularly, but individuals can self-organise if they want to stay to complete something.

An Employee Forum offers further opportunities to discuss issues or ideas for improvement. Thirteen Employee Representatives are nominated by their peers and attend four formal meetings a year with a member of the senior management team. The Forum helps run 'PeopleFest' which is an annual event encompassing sessions on well-being, career, financial and health advice, cycle to work, benefits, discount vouchers and child care information. The Forum is given a budget to organise the annual family open day which is the climax to PeopleFest.

Trust and open communication are reflected in the employee survey with an 89% return rate. Managers reflect on what to do differently based on survey results. As part of their action plan they write individual 'I promise to…' commitments on the Management Pledge Board in the canteen.

12.8.2 Who Benefits?

Open leadership, communication and trust have enabled Bristan Group to undertake a potentially difficult 'journey of efficiencies' with a reduction in overall staff numbers and the closure of sites, while retaining high levels of staff engagement. Profits have risen as a result of these actions and employees have gained yearly pay rises and even bonuses throughout the recession. Employees are kept fully informed of plans to expand and are upbeat about the future. One employee summed up the feeling of the interview group: "Bristan Group are investing in their people so people will stay - those who leave come back."

12.9 Conclusion

These three organisations undertook major journeys of transformation during the period between 2009 and 2014, stimulated by the appointment of new Chief Executives. In each case, the CEO-driven momentum of change led to an increase in bottom-up initiative including the greater empowerment of frontline workers in their day-to-day jobs and to the amplification of employee voice in decision-making. This reinforces the view of Gronn (2002) and other writers that distributed leadership builds organisational capability and helps to release the full range of employee knowledge, skills, experience and creativity. Since the adoption of workplace innovation practices in LME economies relies more on managerial discretion alone than in CME countries, distributed leadership models have a particularly important role to play.

12.9.1 Positioning the Organisation for a Better Future, not Responding to Crisis

None of the three companies was motivated to change by short term problems or issues. In each case, the Chief Executive was driven by a 'gut feel' that responsiveness to future opportunities and resilience against future challenges needed to be strengthened by making better use of the skills, experience and creativity of the company's entire workforce.

12.9.2 Diverse Entry Points Leading to Systemic Approach

The journeys of each organisation began from very different starting points but each CEO emphasised consistency of values, aligning work organisation, organisational structures, systems and procedures, high involvement innovation and improvement, and employee voice in order to create a sustainable body of mutually reinforcing practices based on participation and empowerment.

12.9.3 Workplace Innovation as 'the Sum of the Parts'

Each of the three companies achieved profound improvements in culture and working practices through the cumulative impact of small, incremental changes rather than by big leaps. None of the CEOs began with a masterplan: rather they created opportunities for inclusive reflection and dialogue that led to the identification of ideas for improvement and innovation. These transformations reflect the *process* of generating and implementing these ideas as much as the ideas themselves.

12.9.4 Consistent Values and Behaviours

Workplace innovation can be deflected by short-term problems and contingencies, especially when senior management becomes overly involved in crisis management (Business Decisions Limited 2002). Jeremy Ling, Mike Putnam and Dorothy Thompson share an approach to leadership based on delegation and the rejection of micro-management. This enables them to focus on strategic matters including consistency in the design and implementation of workplace innovation initiatives.

Acknowledgements The authors and publisher gratefully acknowledge the following permission to use the material in this book: Peter Totterdill & Rosemary Exton, Creating the bottom-up organization from the top: Leaders as enablers of workplace innovation. In: European Work and Organizational Psychology in Practice. Special issue on Workplace Innovation, 2017, Volume 1, 73–87.

References

Acas. (2015). *Building productivity in the UK*. London: Acas Strategy Unit.

Buchanan, D., Addicott, R., Fitzgerald, L., Ferlie, E., & Baeza, J. (2007). Nobody in charge: Distributed change agency in healthcare. *Human Relations, 60,* 1065–1090. doi:10.1177/0018726707081158

Business Decisions Ltd. (2002). *New forms of work organisation: The obstacles to wider diffusion.* KE-47-02-115-EN-C, DG Employment and Social Affairs, European Commission, Brussels. Retrieved from http://uk.ukwon.eu/document-store/new-forms-of-work-organisation-the-obstacles-to-wider-diffusion-final-report

Caldwell, R. (2005). *Agency and change: Rethinking change agency in organizations*. Abingdon: Routledge.

CEDEFOP. (2015). *Matching skills and jobs in Europe: Insights from Cedefop's European skills and jobs survey.* Retrieved January 3, 2016 from http://www.cedefop.europa.eu/en/publications-and-resources/publications/8088

Eurofound. (1997). *Employee participation and organisational change.* EPOC survey of 6000 workplaces in Europe. Dublin: European Foundation.

Eurofound. (2012). *5th European working conditions survey*. Overview report. Dublin: Eurofound.

Eurofound. (2015a). Kankaraš, M., & van Houten, G. *Third European company survey 2013—Overview report*. Publications Office of the European Union.

Eurofound. (2015b). Oeij, P., Žiauberytė-Jakštienė, R., Dhondt, S., Corral, A., Totterdill, P., & Preenen, P. *Workplace innovation in European companies*. Luxembourg: Publications Office of the European Union.

Godard, J. (2004). A critical assessment of the high-performance paradigm. *British Journal of Industrial Relations, 42*(2), 349–378. doi:10.1111/j.1467-8543.2004.00318.x

Gronn, P. (2002). Distributed leadership as a unit of analysis. *Leadership Quarterly, 13,* 423–451. doi:10.1016/S1048-9843(02)00120-0

Gustavsen, B. (1992). *Dialogue and development*. Maastricht: Van Gorcum.

Hall, P. A., & Soskice, D. (Eds.). (2001). *Varieties of capitalism. The institutional foundations of comparative advantage.* Oxford: Oxford University Press.

Hardy, C., & Leiba-O'Sullivan, S. (1998). The power behind empowerment: Implications for research and practice. *Human Relations, 51,* 451. doi:10.1177/001872679805100402

Huselid, M. A., Jackson, S. E., & Schuler, R. S. (1997). Technical and strategic human resource management effectiveness as determinants of firm performance. *Academy of Management Journal, 40*(1), 171–188. doi:10.2307/257025

Khurana, R. (2002). The curse of the superstar CEO. *Harvard Business Review, 80*(9), 60–66.

LLAKES. (2012). *Skills and employment survey.* Retrieved January 3, 2016 from www.llakes.org.

Mulvad Reksten, P., & Dragsbæk, P. (2014). How trade unions can stimulate, resource and sustain workplace innovation. Experiences from Denmark. *EUWIN Bulletin.* Retrieved June from http://uk.ukwon.eu/tu-in-wi-denmark

National Institute for Clinical Excellence. (2015). *Workplace policy and management practices to improve the health and wellbeing of employees.* NICE Guideline NG13. London: NICE.

OECD. (2010). *Innovative workplaces: Making better use of skills within organisations*. Paris: OECD Publishing.

Oliver, A. (2012). Mike Putnam: Recession riding. *New Civil Engineer*. Retrieved from http://www.newcivilengineer.com/mike-putnam-recession-riding/8627486.fullarticle

Pot, F. (2011). Workplace innovation for better jobs and performance. *International Journal of Productivity and Performance Management, 60*(4), 404–415. doi:10.1108/17410401111123562

Pot, F., Totterdill, P., & Dhondt, S. (2016). Workplace innovation: European policy and theoretical foundations. *World Review of Entrepreneurship, Management and Sustainable Development, 12*(1), 13–32. doi:10.1504/WREMSD.2016.073428

Ramstad, E. (2009). Promoting performance and the quality of working life simultaneously. *International Journal of Productivity and Performance Management, 58*(5), 423–436. doi:10.1108/17410400910965706

Senge, P. (1990). *The fifth discipline: The art and practice of the learning organization*. London: Century Business.

Storey, J., & Holti, R. (2013). *Towards a new model of leadership for the NHS*. London: NHS Leadership Academy.

Teague, P. (2005). What is enterprise partnership? *Organization, 12,* 567. doi:10.1177/1350508405052759

Totterdill, P. (2015). Closing the gap: The fifth element and workplace innovation. *European Journal of Workplace Innovation, 1*(1). Retrieved from http://journal.uia.no/index.php/EJWI/article/view/166

Totterdill, P. (2016). *Innovative workplaces: Enhancing skills utilisation through workplace innovation*. Paris: OECD.

Totterdill, P., Dhondt, S., & Milsome, S. (2002). *Partners at work? A report to Europe's policy makers and social partners*. Nottingham: The Work Institute. Retrieved from http://uk.ukwon.eu/document-store/partners-at-work

Totterdill, P., Exton, O., Exton, R., & Gold, M. (2016). High-performance work practices in Europe: Challenges of diffusion. *European Journal of Workplace Innovation, 2*(1). Retrieved from http://journal.uia.no/index.php/EJWI

Totterdill, P., Exton, O., Exton, R., & Sherrin, J. (2009). *Workplace innovation in European Countries*. Report to KOWIN (Korean Ministry of Labour). Nottingham: UKWON. Retrieved from http://uk.ukwon.eu/document-store/workplace-innovation-policies-in-european-countries

UKCES. (2009). *High performance working: Developing a survey tool*. Evidence Report 5. Wath-on-Dearne: UK Commission for Employment and Skills.

Vroom, V. H., & Yetton, P. W. (1988). *The new leadership: Managing participation in organizations*. Englewood Cliffs, NJ: Prentice Hall.

Wright, J., & Sissons, P. (2012). *The skills dilemma: Skills utilisation and low-wage work*. London: The Work Foundation.

Author Biographies

Peter Totterdill is a Director of Workplace Innovation Europe CLG, an international not-for-profit company committed to stimulating and resourcing better ways of working. He is also a Visiting Professor at two universities.

Rosemary Exton is also a Director of Workplace Innovation Europe CLG. Her principal field of expertise is in employee empowerment and management-union partnership. She brings practical experience as a manager, clinician and trade union representative.

Chapter 13
Lean as a Tool for Local Workplace Innovation in Hospitals

Liv Starheim and Peter Hasle

13.1 Introduction

Hospitals are under severe economic pressure and the staff experience constant pressure to deliver more. Hospitals use lean management to meet these growing performance demands. However, the productivity results of lean management are mixed (D'Andreamatteo et al. 2015; Hasle et al. 2016; Prætorius et al. 2015), the consequences from an employee's perspective are debatable (Hasle et al. 2012), and while there is a growing literature on lean in hospitals, there is little research on the consequences for the employees (See an example in Ulhassan et al. 2014).

Employee participation is emphasized in the lean philosophy (Womack 1996), yet it may not always be stressed in practice. As the work pressure on staff originates largely from productivity demands (Westgaard and Winkel 2011), to date, it is not clear as to whether lean can be used not only to improve productivity, but also to ease pressure on staff via a focus on participation and, thereby, develop local possibilities for workplace innovation (WPI). In this chapter, we pursue an answer to this question. Our use of the term workplace innovation is in line with the following definition of workplace innovation applied in the area of health and safety: WPI are "*strategy induced and participatory adopted changes in an organisation's practice of managing, organising and deploying human and non-human resources that lead to simultaneously improved organisational performance and improved quality of working life*" .

L. Starheim (✉)
Implementations and Performance Management,
Technical University of Denmark, Kongens Lyngby, Denmark
e-mail: lst@crecea.dk

P. Hasle
Center for Industrial Production, Aalborg University Copenhagen,
Copenhagen, Denmark
e-mail: hasle@business.aau.dk

P.R.A. Oeij et al. (eds.), *Workplace Innovation*, Aligning Perspectives on Health, Safety and Well-Being, DOI 10.1007/978-3-319-56333-6_13

Our starting point is Value Stream Mapping (VSM), which is a participatory tool for process improvement (Rother and Shook 2009). VSM is a systematic process for mapping all the steps in a production process (in this case a particular part of the treatment of a patient) and identifying problems (in lean terminology: waste) in the process. It has proven to be useful in the context of hospitals (Henrique et al. 2015). Insights into the work process can potentially also be used to identify and improve the work environment (Jarebrant et al. 2016). We have developed psychosocial lean (P-lean), which is a VSM-based methodology aimed at the improvement of the psychosocial work environment and have tested the methodology in a qualitative case study in a large university hospital in Denmark. We start with a short presentation of the results of the case study, which is presented in detail in other publications (Hasle et al. 2016; Starheim et al. 2016), followed by a broad discussion of the dilemmas related to the use of a contested rationalisation methodology (Harrison 1994; Stanton et al. 2014; Stewart et al. 2009).

Employee involvement is a fundamental principle for workplace innovation but in order to be effective, tools are necessary. They are needed to help employees get new insights into their work and to enable participation. We argue that P-lean constitutes such a tool, and that the power of lean management, normally used for rationalisation, can be turned around and be used to benefit the employees.

13.2 The Development and Testing of P-Lean

The design of the intervention was based on the literature on lean management and VSM (Jarebrant et al. 2016; Rother and Shook 2009), participation (Van Eerd et al. 2010), as well as the particular context of our case hospital. The intervention, named psychosocial lean (P-lean), aimed to improve the psychosocial work environment (Leka et al. 2008), were we follow the demand control model of Karasek and Theorell (1990) and effort-reward model of Siegrist (1996).

The testing of the intervention methodology included a pilot and five test wards (Table 13.1) in a large Danish university hospital covering different specialties.

The design was tested in a pilot project in a ward specializing in lung diseases and was, subsequently, adjusted based on feedback from the pilot test. The adapted intervention was implemented in five wards. It was introduced to the departments by internal lean and work environment consultants who, as facilitators, also supported the application of the methodology.

In each ward, the intervention consisted of:

- A small planning group, meeting 8–10 times with their appointed facilitator throughout the project period of six months.
- Three VSM workshops which aimed at improving challenges in the psychosocial work environment for selected work processes. Six to eight employees participated as representatives of all involved staff groups together with the

Table 13.1 Participating departments

	Specialty	No. of employees	Professions	Structure
Z	Department of Clinical Physiology (urology examinations)	84	• Doctors • Lab technicians	• Management: a lab technician and a doctor
			• Physicists • Nurses • Nurse assistants	• The department consists of several wards and the ward studied is directly controlled by the management of the department
B	Department of Nephrology	153	• Doctors • Nurses	• Management: a doctor and a nurse
			• Nurse assistants	• The department consists of two wards, each managed by one doctor and a nurse
O	Department of Medicine	277	• Doctors • Nurses	• Management: a doctor and a nurse
			• Nurse assistants • Physio- and occupational therapists • Secretaries	• The department consists of five wards; each is managed by one doctor and a nurse
G	Department of Gynaecology and Obstetrics (gynaecological examination)	335	• Doctors • Midwives	• Management: a doctor, a nurse, and a midwife
			• Nurses • Nurse assistants	• The department consists of 11 wards and the ward studied has a doctor and midwife as local management
E	Department of Paediatrics	237	• Doctors • Nurses	• Management: a doctor and a nurse
			• Nurse assistants • Secretaries	• The department consists of two wards; each is managed by a doctor and a nurse

planning group. The workshops were facilitated by internal lean and work environment consultants from the hospital.
- Observations and registrations of the wards' daily work routines between the three VSM workshops.
- Action plan and follow-ups integrated in the structure of the ward meetings and in the planning group meetings.

The results were evaluated based on a qualitative research design. Data collection included shadowing of activities in the wards, interviews with key stakeholders and written information. Quantitative indicators collected by the wards were included and used when possible.

The data from each ward were analysed by extracting elements that supported and hampered the process and the outcome of the process. Subsequently, we compared these elements with the intended mechanisms (Pawson and Tilley 1997) in the intervention design. The mechanisms consisted of well-known improvement factors, such as management support, employee participation and facilitator support, combined with a well-structured intervention design which, by directly targeting work activities, indirectly aimed at improving the employees' psychosocial work environment.

13.2.1 Findings

Table 13.2 shows an oversight of the results in the five departments.

Below, we will discuss in more detail the results from Urology examinations in the Department of Clinical Physiology (case Z in Tables 13.1 and 13.2), in order to provide a more tangible understanding of our results. Overall, we will show how the intervention did not only improve the psychosocial environment, but also included considerable improvements in productivity.

The first VSM workshop in case Z revealed a particular morning problem. In theory, the list of acute patients from the emergency ward should be available on the coordinators' fax machine at 7.45, but this rarely happened. At 8.00 every morning, the personnel removed the curtains from the reception window and opened for contact with waiting patients. Typically, the assistant nurses experienced problems, given that not all waiting patients were on the list for the day. Some patients came on the wrong day, at the wrong time, or had not been added to the list, due to the delayed information from the acute ward. In addition, some patients were not properly hydrated, which made it impossible to go ahead with the examination at the scheduled time. In their attempt to solve these problems, the assistant nurses would call the ward coordinator, trying to schedule an examination time for the waiting patient. However, they often received a gruff answer from (especially one of) the coordinators, who tried to make the schedule for the day, with a delayed or missing list from the acute ward. Frustration started growing, and the result was that the assistant nurses gave up calling and just hoped for a solution to appear later.

Table 13.2 Results in the five participating departments

	The processes selected for VSM	Project activities	Improvements
Z	The good reception of patients	• Three VSM workshops	• A redesign of the reception process
		• Descriptions of morning situations	• A new position as team coordinator
		• Observations • Registrations	• Disturbance of other colleagues during morning planning stopped
		• Patient questionnaires	• Cooperation and climate improved considerably between receptionist and coordinators
			• All patients examined–even those not prepared properly on arrival; no patients are rejected and sent back for a new appointment
			• Time for four extra examinations daily identified in VSM
B	Information in shifts and delegation of job tasks in accordance to qualifications	• One whole day VSM workshop	• Clear morning procedure for reading patients' information
		• Consultant observed morning procedures	• Delegation through qualifications made transparent
O	Disruptions in medicine dispensing room	• Registration of interruptions in three dispensing rooms	• The importance of informal communication recognized and moved outside the dispensing room
			• Film produced by the employees, showing disturbances in the dispensing room
G	A good balanced workday. The nuchal scan was selected as the process which most often disturbed the balance of the work day	• Three VSM workshops • Interview with patients	• Tangible solutions regarding towels, PC's, parking, letters to patients, information signs
		• Registrations of time used for the particular scanning methodology	• Action plan for different scenarios for reorganizing the whole scan procedure, including the demand for resources and competences
E	Well-structured workday beginning with well-structured meetings	• Three VSM workshops	• Well-structured and well prepared morning meetings
		• Consultant observed and evaluated morning meetings	• Reorganizing teams
			• Time outs to coordinate
			• Small changes such as moving the coffee table, local medical supply depots etc.

Table 13.3 Indicators on the effect of the P-lean intervention in Department Z, urology examinations

Situation before P-lean intervention	Situation after P-lean intervention
2–3 patients denied their scheduled examination on a weekly basis	No denial of service, all patients are examined on the scheduled day
26 examinations each day	30 examinations each day
Problem identification: personality	Problem identification: disturbance in a work process that needs concentration
Avoidance and lack of respect	Helpfulness and mutual recognition
The scheduling process consists of 11 steps involving 4 employees	The process consists of 4 steps involving typically 2–3 employees.
Diuretics are used—inconvenience for patient	No diuretics are used
2–3 disturbing phone calls from the reception during the morning scheduling work	No disturbing phone calls from the reception during the scheduling work

The staff shared an important experience through VSM as they shifted the problem understanding from blaming people (i.e., the ward coordinator) to a focus on the organisational disturbances during the preparation of the daily plan. This new understanding was the point of departure for the development of a number of different solutions which were subsequently implemented. The results of the whole intervention are summarized as Table 13.3.

However, the department also chose to not focus on solving problems that derived from relations with other departments, in particular the emergency department, which often was the cause of missing information regarding incoming patients. The management and the staff considered it too difficult to intervene in the work of other departments. The same constraint was experienced in the other involved departments.

Another important feature was the importance of the external support from the internal facilitators. Although the VSM was the main tool for the eye-opening effect, the facilitators played an important role in making sure that the procedures were followed and the VSM was applied in a qualified manner.

13.2.2 Concluding Remarks on the Case Presentation

Hospitals are described as a typical example of a professional bureaucracy (Glouberman and Mintzberg 2001a). Although there are many different groups of professionals within hospitals, doctors and nurses are the two largest groups. Both groups consist of professionals who have traditionally had high autonomy in the execution of their work tasks. As a result, distinct silos have developed within hospitals, where the professionals only engage in limited communication and coordination across professions even though they share the common task of treating patients.

It is therefore well in line with the problem of professional silos that communication and coordination proved to be the focus points for all the intervention wards in our case study. The different professional groups tend to focus on their individual tasks and, subsequently, to lose the overview of the whole workflow. Indeed, when constraints for their work appear, they are inclined to attribute the problems to persons and personalities rather than to the organisation of the work. It is therefore key to understand and solve the real underlying problems (i.e., in this case, the way the work was organised), so as to foster better social relations in communication and coordination.

The involvement of employees via the P-lean intervention methodology seems to be a feasible way to work with the problems related to both the psychosocial factors at work and the inefficiencies related to constrained communication. The point is that VSM opens new insights into the work processes and the flow connecting different processes, which, in turn, facilitates a reframing of the problem from one of personal conflicts to one pertaining to the organisation of work. The extensive involvement of the employees in the VSM analysis has, at least, two distinct advantages. First, the employees are the only ones with the necessary practical in-depth knowledge of the process that would allow for a VSM intervention, which digs sufficiently deep to create new insights. Second, solutions can only be identified, accepted, and implemented in practice if they are meaningful for the employees.

The experience from the VSM workshops shows that staff was highly motivated to make improvements and that they also had the necessary competencies both as a result of their education and their daily practice. Such an approach is in line with Appelbaum et al. (2000) who focused on the learning possibilities in lean. The authors suggest that ability, motivation, and opportunity (AMO) are necessary components of using lean as a learning process. Ability and motivation are already very high in hospitals, and the problematic issue is the "opportunity to perform." The complexity of the context and many different, partially conflicting demands make it difficult for staff to take action. It is here, that P-lean, with the VSM workshops, provided an opportunity to achieve a new understanding of the daily work tasks. The workshops provided a new frame for the opportunity to perform, by providing insights into the impact all the different professions' activities had on the overall treatment and care tasks. Even though participants may have had preconceived opinions about other staff members and, in particular, other professions as the main reason for certain problems, the workshops changed this perspective. As a result, participants came to respect others' contributions to these tasks and to see their interconnectedness. What also contributed to this process of mutual learning was that they had fun doing it. Typically, the energy in the room was high during analysis. Sometimes the jokes and pointing out the absurdities in their work made them laugh and they felt relieved that they could share these problems. This, in turn, led to growing trust in handling these problems together.

This local embeddedness, whereas beneficial on the one hand, is perhaps also the reason for one of the limitations of P-lean. What we found was that, the VSM discussions produced not only the necessary knowledge across professions to improve the daily work in the involved wards, but also provided knowledge and ideas for improvement activities across groups. However, these cross-departmental challenges that emerged in the workshops were not selected for improvement activities. This leads us to believe that the absence of representatives from other departments limited the VSM groups' trust in possible changes across departments. If P-lean were to be applied for cross-departmental activities, it would demand representation of the staff involved in the daily relations between departments. Earlier experience with VSM has, for instance, shown that it is possible to cross departmental borders. It happened, for instance, in a hospital where the pharmacy staff and nurses providing chemotherapy medication for outpatient clinics, together, solved problems related to the late delivery of the medication. Importantly, this was a problem that management had been fighting for years without any clear results (Hasle et al. 2016).

13.3 Possibilities and Barriers for Lean as a Tool for Local Workplace Innovation in Hospitals

We use our experience with P-lean as the point of departure for a discussion of the broader possibilities and barriers for the use of lean as a tool for local workplace innovation in hospitals. We do this by highlighting important elements in the design and the practical experience with P-lean and use the literature to discuss these elements from a broader perspective of the consequences of applying lean. We focus on three main issues:

- The difference between the approach in P-lean and the traditional drive for productivity increase in hospitals.
- The relation between the concept of value in lean, professional healthcare values as well as employee well-being as a value.
- Lean is typically criticised for standardisation and management control and involves a risk of reducing employee control and autonomy.

13.4 The Difference Between the Approach in P-Lean and the Traditional Drive for Productivity Increase in Hospitals

The most common question we get when we present our study to workplaces and researchers is: How can you claim that a lean-inspired method can be used to improve the work environment when the general experience is that lean tends to have a negative impact upon the work environment?

There are good reasons for asking this question. Budgets in hospitals are severely restricted and the staff is continuously asked to deliver more. In Danish hospitals there has been, for the last ten years, a yearly productivity increase of more than 2% (Regioner 2014). Lean is considered to be a part of this productivity pressure and is often identified by hospital staff as being the symbol of 'productivity above patient care'. The literature on lean has reported this experience of hospital staff, with lean being equated to 'lean and mean' (Harrison 1994; Landsbergis et al. 1999). In addition, lean has been heavily criticised in the manufacturing industry (Stewart et al. 2009) and in the context of healthcare, scholars have also highlighted potentially negative impacts of lean, such as the intensification of work (Hasle 2010; Stanton et al. 2014; Trägårdh and Linberg 2004).

However, more recent research developments are painting another picture. Reviews are showing that lean in simple manufacturing work may have a negative impact, but not necessarily in more complex work (Brännmark and Håkansson 2012; Hasle et al. 2012). Especially in hospitals, several studies indicate that the impact often is more positive than negative (Dellve et al. 2015; Ulhassan et al. 2014). One possible explanation for these conflicting findings could be found in the manner in which lean is applied in practice and the extent to which employees are involved. Therefore, the important question is probably how lean is applied in practice and to what extent the employees are involved (Hasle 2014; Holden et al. 2015).

First of all, in our case, the question of potential efficiency gains came up in the initial meeting with the overall steering group of the hospital. However, during discussions, the emphasis was placed upon the work environment as a core value for improvement. In fact, the chair of the steering group, the hospital vice CEO, agreed that the wards could make use of the potentially gained time resources themselves. This was a strong signal to the wards that improvement of the work environment was given priority and would not be used as an opportunity to remove resources.

Overall, P-lean differs from traditional lean processes by explicitly focusing on the psychosocial work environment. The results from the test ward's risk assessment of the psychosocial work environment constituted the starting point for P-lean activities. The organising committee in the departments, therefore, also included occupational health and safety (OHS) representatives. Moreover, the selection of priority problems and processes to be dealt with was based on the employees' wishes for improvement of the psychosocial work environment. Hence, the productivity increases which resulted from the P-lean implementation can be considered unintended yet positive side-effects.

The VSM is designed in such a way that the priority remains the improvement of the work environment. In our case, it was the local planning group who decided which work processes to include in the VSM workshops, such as, for instance: "Improving the reception of the patient." In the VSM workshops, we followed the lean specifications for VSM: i.e., putting the brown paper on the wall, describing work activities with big yellow post-its, adding red post-its, when the production flow was disturbed, and green post-its describing the possible solutions. In contrast to the typical VSM process, in our version of the VSM process, we did not adopt a higher management definition of what type of value should be added. The workshop groups themselves defined what kind of value this VSM process should add, who the beneficiaries would be, and what kind of value should be created in the improvement process. In this respect, we parted from other lean VSM methods.

Another important difference in our methodology as compared to other lean VSM methods is the addition of a good work environment as a value in itself into the process. A post-it with an orange colour was used in the value stream mapping both for identification of problems with the work environment and for discussing the value of the proposed improvement for the work environment. The facilitator explicitly asked the workshop group members to describe which impact the improvement of the work process would have on staff well-being. Moreover, in the analysis of the activities, the orange post-its were used to make problems with the experiences of stress, frustration, and irritation explicit and tied them to the processes which created these feelings. In those cases, where the employees expressed that they experienced stress or frustration, the facilitator helped by asking what improvement goals for the processes could be agreed on for that particular process. Interestingly, the general experience was that the employees most often phrased these goals in terms of patient outcome. For instance, one such goal was: "To feel secure that the patient is taken properly care of when we send the patient home."

The third difference in the P-Lean methodology to traditional lean thinking is that waste is not regarded as the core enemy. For instance, time used for the purpose of coordinating and strengthening necessary relations is welcomed and included in the improved work processes. In our example from the urology examination, the introduction of one more coordinator into the examination rooms to answer telephone calls from the reception, is an example of an additional activity that made it possible to use the existing four examination rooms more efficiently. The new coordinator located in the examination rooms had the information on how to place an acute patient in a free examination room, and how to use the four rooms efficiently. Time waste was also eliminated, for example, by reducing disturbances to the coordinator's work who made the daily lists, both in terms of avoiding telephone calls during the busiest time of the day for her, the mornings, and by reducing the need for disturbing visits by moving the work plan to another location. Yet, the purpose of this elimination was to strengthen coordination, and improve the coordinator's work environment - not to reduce time as a cost.

However, the important point is that P-lean tends to also lead to increased productivity. In the case of urology, the efficiency in terms of patients examined per day was improved as well as the patient experience, given that patients are not turned away any more. By using a reasonably different point of departure from the traditional lean approach, P-lean provides productivity, but avoids the common staff experience that the productivity changes are enforced from above without consideration for patients and staff.

13.5 The Relation Between the Concept of Value in Lean, Professional Healthcare Values as Well as Employee Well-Being as a Value

A major issue in this context is the question of value. In the traditional use of lean in the manufacturing industry, the question of value is quite simple: Value is created by delivering the best possible product at the lowest possible cost. In hospitals, however, the concept of value is more complicated and sometimes contradictory. Young and McClean (2008) suggest at least three different dimensions of the value concept in hospitals:

- Clinical value—the best treatment of the patient
- Experiential value—patients' experience of care
- Operational value—the best cost effectiveness

The professional staff in hospitals has a strong focus on doing the best for the patient although the focus may vary slightly between doctors and nurses. That is, for the doctor, the focus is primarily on the clinical value; whereas for the nurse, it is primarily on the experiential value. The operational value is considered to be a question for the management and is most often seen as contradictory to clinical and experiential value (Hasle et al. 2016). Whereas this may be so in some cases, in many other cases, there will be alignment between the three dimensions of value, even though the staff may initially think otherwise. In fact, this is exactly what we experienced during the testing of P-lean.

A clear example of this can be found in Department G, who chose to work with the nuchal fold scanning of pregnant women early in their pregnancy. This type of scan is carried out in order to discover genetic variance in the foetus and assess whether there is a risk of malformations. In a 30-min scan offered to women with risk for genetic mutation, sonographers perform measurements that would indicate the level of genetic variations. The sonographers have a midwife background, and they regarded the short time allotted for the scan as a problem, as they felt a professional obligation to advise the pregnant women based on their pregnancy history and have a dialogue about lifestyle. The department management had tried to convince the sonographers that priority should only be given to the identification of genetic variation and that the general dialogue should be left to the normal midwife consultation later in the pregnancy. However, the midwifes doing sonography experienced the examination as stressful because they tried, during the short time allocated, not only to perform the scan but also to deliver a full midwife consultation. This particular problem was identified during the first VSM, and alignment between the clinical and operational value began. The department carried out interviews with patients and they learned, to their surprise, that the patients at the nuchal fold scan did not expect more than the information about having a healthy baby and maybe the sex of the baby. Hence the experiential value was lower than the sonographers assumed in their desire to provide a full midwife consultation.

The conflict was not resolved during the VSM process, but the participation of the concerned staff and the presentation of the data from the interviews with the patients opened a new dialogue about the sonograph examinations, and an action plan was developed with scenarios of how to use a combination of the sonographers' professional knowledge already in this early stage of the pregnancy. Thereby, the P-lean process helped remove the sonographers' frustrations caused by their clinical values. P-lean helped change the situation by shifting it from being a conflict between staff and management to a professional discussion about the treatment of the patients.

13.6 Lean and Standardisation as a Risk of Reducing Employee Autonomy and Control of Work

Hospitals constitute a standard example of professional bureaucracies (Glouberman and Mintzberg 2001a, b). In this sense, they have what Parsons (1969) calls a contract with society, where the professions (doctors, nurses and others) have established a monopoly on their profession and, as a consequence, have committed themselves to provide quality services of a high ethical standard. Hospitals are hierarchical organisations where high status doctors are at the top followed by the other professions. Traditionally, it has been the doctors who solely took decisions regarding patient treatment. Later on, nurses have also enjoyed quite large autonomy in the care processes.

This traditional autonomy has come, in several ways, under pressure these days. First, hospitals have become large organisations delivering key services to society. The complexity is growing due to more different diagnoses, more medicines, more specialisations, and more chronic patients with multiple diagnoses (Hasle et al. 2016). This, in turn, increases the need for coordination, which renders the individual health professional more dependent on others. At the same time, the budget constraints imposed by New Public Management (NPM) (Hood 1995) as the adopted dominating strategy is attacking professional autonomy. NPM focuses on restricting the professionals' autonomy as they are considered to focus too much on their own interests and too little on the public interests (O'Flynn 2007). Interestingly, NPM has been criticised for not delivering on its efficiency promises (Hood and Dixon 2015; Simonet 2013).

The introduction of NPM results in an experience of reduced control by hospital staff (Bolton 2004; Worrall et al. 2010). Given that lean has, to a great extent, been equated with NPM, it is often resisted, even though the negative outcomes for hospital staff are not particularly evident (see the discussion earlier in the chapter). Nonetheless, this experience of reduced autonomy and control is obviously bad for the employees' health and well-being (Karasek and Theorell 1990; Siegrist 1996).

At the same time, hospitals are also organisations where the employees are well educated and have a strong dedication to fulfil the professional goal of caring and treating patients. They are therefore fully qualified to take decisions about life and death and they also have the qualifications and routines for implementing new treatments and new medical technology. The employees therefore feel fully qualified to take all necessary decisions.

However, the solution to this perceived loss of autonomy and control cannot be a return to "the good old times" where doctors are in charge of the hospitals and every doctor, on an individual basis, makes decisions about what to do. Hospitals have grown past the traditional organic organisations into large bureaucracies with high complexity where each professional is dependent on the decisions and actions of other professionals. Hence, even though the health professional might be naturally inclined to prioritize every individual patient, there is a need to ensure that there are sufficient resources for all patients.

Paradoxically, in spite of the bad reputation of lean, it may provide the possibility for increasing control simultaneously with securing the efficient delivery of service. The point is to move organisations/hospitals from a coercive to an enabling bureaucracy. Adler and Borys (1996) have pointed out that autonomy and control over work cannot be provided as simple bottom-up solutions in larger organisations. They have therefore suggested the concept of an enabling bureaucracy as an alternative to a coercive bureaucracy, as would be exemplified in NPM where the focus is on controlling the employees (Fig. 13.1).

The simple structure is most often found in small enterprises where it can be either organic if employees collaborate on an equal basis or autocratic if the owner

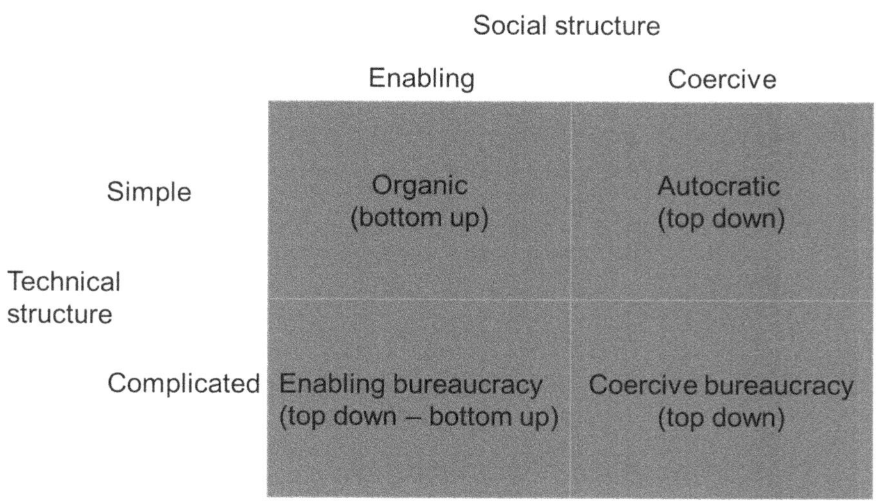

Fig. 13.1 Enabling and coercive organisations (Adler and Borys 1996)

Coercive bureaucracy	Enabling bureaucracy
Focus on performance standards to highlight poor performance	Focus on best practice supported by performance measures
Standards to minimize game playing and cost monitoring	Customization of standards to different experience and room for improvisation
Design system to keep employees out of control loop	System provides for control of own work – transparency to understand interconnections
Standards are instructions to follow and not to be challenged	Standards are best practice templates to be improved

Fig. 13.2 The difference between coercive and enabling bureaucracy (adapted from Adler and Borys 1996)

is the one who decides. The complicated structure is found in larger organisations where the coercive bureaucracy constitutes an example of management trying to control decisions top-down. In contrast, the enabling bureaucracy would, on the one hand secure the necessary top-down coordination and, on the other hand, provide employees with the best possibilities for controlling the bureaucracy. Simultaneously, it provides the opportunity of using the bureaucracy as a learning platform given that employees have possibilities for adjusting and changing standards (Fig. 13.2).

This understanding of an enabling bureaucracy fits very well with our P-lean approach. Surely, there need to be standards on how to coordinate and communicate and these were addressed in the P-lean projects. However, these standards are developed and controlled by the involved local staff.

An interesting point is that the solutions the staff developed through P-lean can be characterised as standard operating procedure (SOP). They developed, for instance, standards for how to organise meetings (length, agenda, and distribution of tasks) and how to organise teams (personnel, tasks, and coordination). In doing so, the staff decided to limit the individual autonomy of their work, but the new locally developed standards also provided the opportunity to focus on the core treatment and care tasks without being encumbered by organisational uncertainties. It was furthermore clear that these standards only received support from the staff because they were developed locally, and that the top-down attempt to, for instance, provide SOP for morning meetings had already failed.

A global overview of one's work, especially regarding potential possibilities for change gets lost in the daily work pressure. In this respect, problems tend to be explained either in terms of lack of resources, poor management decisions, and/or individual attitudes. The VSM workshops revealed to the employees that their potential span of control and change was larger than they initially expected. In the third workshop, the head nurse in ward E asked the consultant: "Couldn't we extend the project to a 3-year project? It makes far more improvements, than I could dream of. Additionally, it involves the staff in a fantastic way in developing the ward."

13.7 Conclusion

Lean has a reputation for putting more pressure on the employees to work faster, but the results from the experiments with P-lean indicate that lean can also be used to benefit the employees. P-lean is a potentially powerful tool for workplace innovation. It provides, through simple means, possibilities for the employees to be involved in local innovative activities where they improve not only their own work environment but also the quality and productivity of the work to benefit the patients.

Although, in this project, we tested P-lean in a hospital setting, we believe that, with some careful consideration of context and employee involvement, it can also be utilised in other sectors.

Acknowledgements This project is supported by a grant from the Danish Work Environment Research Fund.

References

Adler, P. S., & Borys, B. (1996). Two types of bureaucracy: Enabling and coercive. *Administrative Science Quarterly, 41*(1), 61–89.

Adler, P. S., Borys, B., Bennis, W., Lawler, E. D., Scott, B. S., Bailyn, L., et al. (1996). Two types of bureaucracy: Enabling and coercive we owe thanks too for helpful comments from. *Administrative Science Quarterly, 41,* 61–89.

Appelbaum, E., Bailey, E., Berg, P., & Kalleberg, A. L. (2000). *Manufacturing advantage: Why high-performance work systems pay off.* Ithaca & London: Cornell University Press.

Bolton, S. C. (2004). A simple matter of control? NHS hospital nurses and new management. *Journal of Management Studies, 41*(2), 317–334.

Brännmark, M., & Håkansson, M. (2012). Lean production and work-related musculoskeletal disorders: Overviews of international and Swedish studies. *Work-A Journal of Prevention Assessment & Rehabilitation, 41,* 2321–2328.

D'Andreamatteo, A., Ianni, L., Lega, F., & Sargiacomo, M. (2015). Lean in healthcare: A comprehensive review. *Health Policy, 119*(9), 1–13.

Dellve, L., Williamsson, A., Strömgren, M., Holden, R. J., & Eriksson, A. (2015). Lean implementation at different levels in Swedish hospitals: The importance for working conditions and stress. *International Journal of Human Factors and Ergonomics, 3*(3–4), 235–253.

Glouberman, S., & Mintzberg, H. (2001a). Managing the care of health and the cure of disease—Part I: Differentiation. *Health Care Management Review, 26*(1), 56–71.

Glouberman, S., & Mintzberg, H. (2001b). Managing the care of health and the cure of disease—Part II: Integration. *Health Care Management Review, 26*(1), 70–84.

Harrison, B. (1994). *Lean and mean. The changing landscape of corporate power in the age of flexibility.* New York: The Guildford Press.

Hasle, P. (2010). Professional assessment and collective standards—Lean in a cancer department. In P. Vink & J. Kantola (Eds.), *Advances in occupational, social, and organizational ergonomics* (pp. 345–354). London: CRC Press/Taylor & Francis Ltd.

Hasle, P. (2014). Lean production—An evaluation of the possibilities for an employee supportive lean practice. *Human Factors and Ergonomics in Manufacturing, 24*(1), 40–53.

Hasle, P., Bojesen, A., Jensen, P. L., & Bramming, P. (2012). Lean and the working environment: A review of the literature. *International Journal of Operations & Production Management, 32* (7), 829–849.

Hasle, P., Nielsen, A. P., & Edwards, K. (2016a). Application of lean manufacturing in hospitals—The need to consider maturity, complexity, and the value concept. *Human Factors and Ergonomics in Manufacturing, 26*(4), 430–442.

Hasle, P., Starheim, L., Jensen, P. L., & Diekmann, B. J. (2016). Value stream mapping as a tool for systematic employee based improvement of the psychosocial work environment in hospitals. In *23rd EUROMA Conference—Interactions* (pp. 1–6). Trondheim: NTNU.

Henrique, D. B., Rentes, A. F., Filho, M. G., & Esposto, K. F. (2015). A new value stream mapping approach for healthcare environments. *Production Planning & Control, 7287,* 1–25.

Holden, R. J., Eriksson, A., Andreasson, J., Williamsson, A., & Dellve, L. (2015). Healthcare workers' perceptions of lean: A context-sensitive, mixed methods study in three Swedish hospitals. *Applied Ergonomics, 47,* 181–192.

Hood, C. (1995). The "new public management" in the 1980s—Variations on a theme. *Accounting Organizations and Society, 20*(2–3), 93–109. Retrieved from WOS: A1995QQ37800001.

Hood, C., & Dixon, R. (2015). *A government that worked better and cost less? Evaluating three decades of reform and change in UK central government.* Oxford: Oxford University Press.

Jarebrant, C., Winkel, J., Hanse, J. J., Mathiassen, S. E., & Öjmertz, B. (2016). ErgoVSM: A tool for integrating value stream mapping and ergonomics in manufacturing. *Human Factors and Ergonomics in Manufacturing & Service Industries, 26*(2), 191–204.

Karasek, R. A., & Theorell, T. (1990). *Healthy work, stress, productivity, and the reconstruction of working life.* New York: Basic Books.

Landsbergis, P. A., Cahill, J., & Schnall, P. L. (1999). The impact of lean production and related new systems of work organization on worker health. *Journal of Occupational Health Psychology, 4*(2), 108–130.

Leka, S., Cox, T., & Zwetsloot, G. (2008). The european framework for psychosocial risk management (PRIMA-EF). In S. Leka & T. Cox (Eds.), *Policy* (pp. 1–16). Nottingham: Institute of Work, Health and Organizations, University of Nottingham.

O'Flynn, J. (2007). From new public management to public value: Paradigmatic change and managerial implications. *The Australian Journal of Public Administration, 66*(3), 353–366.

Parsons, T. (1969). Research with human subjects and the "professional complex". *Daedalus, 98* (2), 325–360.

Pawson, R., & Tilley, N. (1997). *Realistic evaluation.* Los Angeles, London, New Delhi, Singapore, Washington DC: Sage.

Prætorius, T., Hasle, P., Edwards, K., & Nielsen, A. P. (2015). Towards the collaborative hospital—harnessing the potential of enabling care processes and structures. In M. Elg, P.-E. Ellström, M. Klofsten, & M. Tillmar (Eds.), *Sustainable development in organizations* (pp. 57–76). Cheltenham UK; Northhampton, MA, USA: Edgar Elgar Publishing Ltd.

Regioner, D. (2014). *Løbende offentliggørelse af produktivitet i sygehussektoren—Udviklingen fra 2013–2014.*

Rother, M., & Shook, J. (2009). *Learning to see: Value stream mapping to add value and eliminate MUDA (version 1.4).* Cambridge, MA: Lean Enterprise Institute.

Siegrist, J. (1996). Adverse health effects of high-effort/low-reward conditions. *Journal of Occupational Health Psychology, 1*(1), 27–41.

Simonet, D. (2013). The new public management theory in the british health care system: A critical review. *Administration & Society, 47*(7), 802–826.

Stanton, P., Gough, R., Ballardie, R., Bartram, T., Bamber, G. J., & Sohal, A. (2014). Implementing lean management/six sigma in hospitals: Beyond empowerment or work intensification? *The International Journal of Human Resource Management, 25*(21), 2926–2940.

Starheim, L., Hasle, P., Jensen, P. L., & Juul Diekmann, B. (2016). Integrating Lean end employee involvement in developing the work process and well-being in a hospital setting. *European Work and Organizational Psychology in Practice, In press.*

Stewart, P., Richardson, M., Danford, A., Murphy, K., Richardson, T., & Wass, V. (2009). *WE sell our time no more—Workers' struggle against lean production in the British car industry.* London: Pluto Press.

Trägårdh, B., & Linberg, K. (2004). Curing a meagre health care system by lean methods—Translating "chains of care" in the Swedish health care sector. *International Journal of Health Planning and Management, 19,* 383–398.

Ulhassan, W., Westerlund, H., Thor, J., Sandahl, C., & von Schwarz, U. T. (2014). Does lean implementation interact with group functioning? *Journal of Health Organization and Management, 28*(2), 196–213.

Van Eerd, D. L., Cole, D., Irvin, E., Mahood, Q., Keown, K., Theberge, N., et al. (2010). Process and implementation of participatory ergonomic interventions: A systematic review. *Ergonomics, 53*(10), 1153–1166.

Westgaard, R. H., & Winkel, J. (2011). Occupational musculoskeletal and mental health: Significance of rationalization and opportunities to create sustainable production systems—A systematic review. *Applied Ergonomics, 42*(2), 261–296.

Womack, J. P. (1996). The psychology of lean production. *Applied Psychology-An International Review-Psychologie Appliquee-Revue Internationale, 45*(2), 119–122.

Worrall, L., Mather, K., & Seifert, R. (2010). Solving the labour problem among professional workers in the UK public sector: Organisation change and performance management. *Public Organization Review, 10*(2), 117–137.

Young, T. P., & McClean, S. I. (2008). A critical look at lean thinking in healthcare. *Quality and Safety in Health Care, 17*(5), 382–386.

Author Biographies

Liv Starheim Ph.D. is a researcher at the Implementation and Performance Management group at the Management Engineering Department at the Technical University in Denmark. Her main research area is in regulating and handling the challenges in the psycho-social working environment.

Peter Hasle is professor in sustainable work processes at the Centre for Industrial Production, Aalborg University Copenhagen with former positions at public and private research institutions and internationally with the ILO. His research interests cover among others lean, organisational social capital, OHS management and healthcare.

Chapter 14
Workplace Innovation Context in Poland: Between Structure and Agency

Marta Strumińska-Kutra, Boleslaw Rok and Zofia Mockałło

14.1 Introduction

Our analysis explores the cultural and institutional context of the process of creating and implementing significant changes and innovations, as well as the effects of their implementation. We concentrate in this chapter specifically on workplace innovation (WPI) defined as a social and participatory process which shapes work organisation and working life, combining their human, organisational and technological dimensions (Howaldt et al. 2016: 3). We depart from the emphasis of the social need as a trigger for innovation and assume that the ability to fulfil the employees' needs at the workplace shapes the quality of working life.

Therefore, in this chapter we perceive WPI as a form of social innovation, understood as 'new ideas (products, services and models) that simultaneously meet social needs (more effectively than alternatives) and create new social relationships

Research presented in this chapter is part of a project financed in the field of research and development by the Ministry of Science and Higher Education in Poland/National Centre of Research and Development under the III phase of the national programme "Improvement of safety and working conditions" (2014–2016), coordinated by the Central Institute for Labour Protection—National Research Institute.

M. Strumińska-Kutra (✉)
VID Specialized University, Oslo, Norway
e-mail: martastr@kozminski.edu.pl

M. Strumińska-Kutra · B. Rok
Kozminski University, Warsaw, Poland
e-mail: brok@kozminski.edu.pl

Z. Mockałło
Ergonomics Department, Laboratory of Occupational Psychology and Sociology, Central Institute for Labour Protection—National Research Institute, Warsaw, Poland
e-mail: zomoc@ciop.pl

P.R.A. Oeij et al. (eds.), *Workplace Innovation*, Aligning Perspectives on Health, Safety and Well-Being, DOI 10.1007/978-3-319-56333-6_14

or collaborations' (Murray et al. 2010). We show that workplace innovative measures are implemented under the existing social, institutional and cultural circumstances that, to varying degrees, can be conducive to the diffusion and implementation of innovations in organisations, and influence their effectiveness as tools that positively affect the quality of life and company performance.

We start with the outline of the cultural, institutional and economic context of innovation in Poland. Cultural and institutional patterns present in the Polish context contradict patterns supportive for WPI. It may be related to the post-communist background. Research suggests that employees in Central and Eastern Europe (including Poland) display external types of motivation typical for traditional bureaucracies, requiring close monitoring from the supervisors and do not feel responsible neither for problems solving nor for the pursuit of corporate mission (Marody and Lewicki 2010; Gadomska-Lila 2011).

In the empirical part of this chapter, we reconstruct workers' perceptions of the implementation of innovation in their organisations. We focus on specific relations between (1) attitudes and aspirations of employees at different levels, and (2) numerous aspects of formal and informal organisational structures, such as relations of dependency, hierarchy and leadership styles, conventional communication methods, reward and discipline systems.

14.2 Specificity of Institutional and Cultural Conditions of WPI in Poland

Polish enterprises are among the least innovative in Europe. The percentage of innovative enterprises engaged in technological (product and process) innovation in Poland stands at 16%, while those implementing non-technological (marketing and organisational) innovation account for 15.5% of all enterprises (Fig. 14.1). Poland occupies the penultimate position among European countries (Niec 2015). It may be related to the Polish post-communist background, e.g. communist legacy leads Polish employees to maintain external rather than internal motivation to work and to expect rules and regulations rather than experiment and make own decisions (Marody and Lewicki 2010). It also concerns Polish managers who are not likely to trust their employees enough to let them take risk (Gadomska-Lila 2011).

Experience from this communist economic system can contribute to lack of willingness and know-how to work in teams and in a context of cooperation. Another example of a consequence of the post-communist regime is that Polish employees seem to be not interested in interaction and building a workplace community (Marody et al. 2010) which is crucial for innovation development (Kohtamäki et al. 2004).

Low levels of innovation come with low labour productivity. The latter systematically increases through the years, nevertheless the productivity is still three times lower than the average of the European Union. Polish workers typically spend

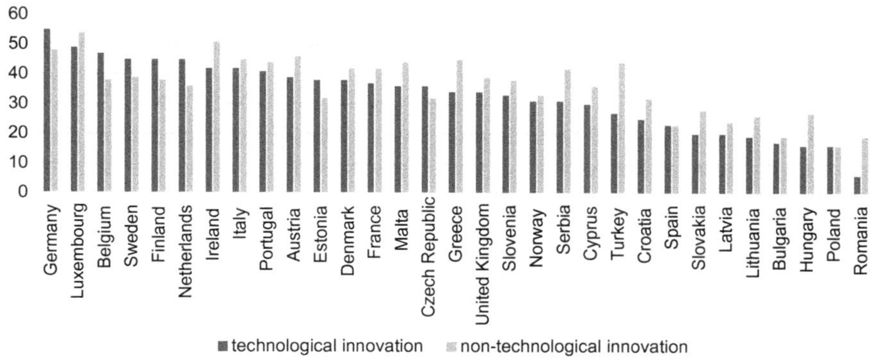

Fig. 14.1 Share of innovative enterprises in EU (technological and non-technological innovation combined). *Source* study based on Eurostat data (Niec 2015, p. 16)

39.9 h per week at work; in terms of working hours, they rank seventh among OECD countries and first among the Member States of the EU with the average 36.8 h per week at work; the lowest rate is found in the Netherlands—28.9 h (OECD 2014). This means that although Polish employees work long, their labour productivity is among the lowest in Europe.

When juxtaposing the information about low labour productivity with data about low levels of employee participation in decision making processes WPI appears as a great opportunity. Harnessing workers' knowledge and skills to streamline organisational processes and raise the quality of working life seems to be a significant win-win solution. But the temptation to use 'ready to wear' models of WPI, especially those developed in Western European countries, should be resisted. On the one hand the barriers to the introduction of workplace innovation do not significantly differ from those observed elsewhere in the CEE (see e.g. for Hungary, France, Netherlands, Spain, UK: Makó and Illéssy 2015), namely: narrow and anachronistic understanding of innovation as a technological phenomenon. This is evident at different levels of policies supporting innovation in CEE Member States like Poland and Hungary; conservative and formal nature of public aid criteria; unfavourable cultural patterns, such as low propensity for risk taking among investors, excessive faith in the free market and competition as the sole sources of innovation.

On the other hand, this standard list should be supplemented with some information about barriers connected with the specificity of Polish socio-economic cultural and institutional context, which we consider especially unfavourable for implementation of cooperative, empowering and participatory initiatives at the workplaces, especially those aiming at improving the quality of working life.

First, in terms of socioeconomic context, Poland, along with other post-communist countries of Central and Eastern Europe (Hungary, Czech Republic, Slovakia) is implementing a distinctive model of capitalism called *dependent market economy* (Nölke and Vliegenthart 2009) or *liberal dependent post-communist economy* (King 2007). Dependent economies have comparative advantages in the assembly and production of relatively complex and durable consumer goods. These comparative advantages are based on institutional complementarities between skilled, but cheap, labour; the transfer of technological innovations within transnational enterprises; and the provision of capital via foreign direct investment. Combination of these factors heavily influences innovation patterns within Polish economy.

Local innovativeness in general is not incentivised—multinational corporations often import innovation into the region because they consider CEE economies as production places for semi standardised goods and services and not for research & development (Högselius 2003; Bendyk 2015). Additionally, since the comparative advantage of these economies relays upon low wages, the space for the quality of work improvement is limited. Indeed, Polish wages are one of the lowest in the EU. Median gross hourly earnings in Poland is 4,3 EUR and 7,7 PPS (Purchasing Power Standard) which ranks Poland 21st and 17th respectively, among EU countries with the average 12 EUR and 12 PPS, according to Eurostat.[1]

An expected, negative influence of above mentioned factors on the possibilities to create, develop and implement WPI is potentially strengthened by a specific cultural feature of Polish enterprises marked by low levels of employee initiative and participation. Research interprets these phenomena as consequences of high power distance, propensity to adopt and accept autocratic leadership (Dunn 2004; Stocki et al. 2008), as well as generally low levels of social trust (Czapinski and Panek 2015). Interestingly the passive attitudes persist even in the cases when employer expects the initiative, cooperation among employees and autonomy in problem solving (Kochanowicz et al. 2007).

Research suggests that employees in CEE display external types of motivation typical for traditional bureaucracies, requiring close monitoring from the side of supervisors and exact adaptation to existing procedures. They do not feel responsible neither for problem solving nor for the pursuit of corporate goals (Mockałło 2016). Additionally, Polish workers find it difficult to reconcile cooperative attitudes with competitive and antagonistic perceptions of relationships among workers, especially those occupying different levels of the organisational ladder (Kochanowicz et al. 2007).

Hence cultural and institutional patterns present in the Polish context (and to the large extend in other countries of CEE region) contradict patterns supportive for WPI. Further we can expect that lack of employees' participation and the general pressure on the maintenance of low labour costs (typical for dependent market economies) results in highlighting organisational performance as a major goal of innovation and neglecting the goal of quality of working life improvement. Indeed, goals of implementing innovations in Polish SME are mainly profit increase, company development, market share increase or clients' expectations (Sieradzka 2014). These expectations are confirmed by some quantitative research conducted

[1]http://appsso.eurostat.ec.europa.eu/nui/submitViewTableAction.do Accessed April 12, 2017

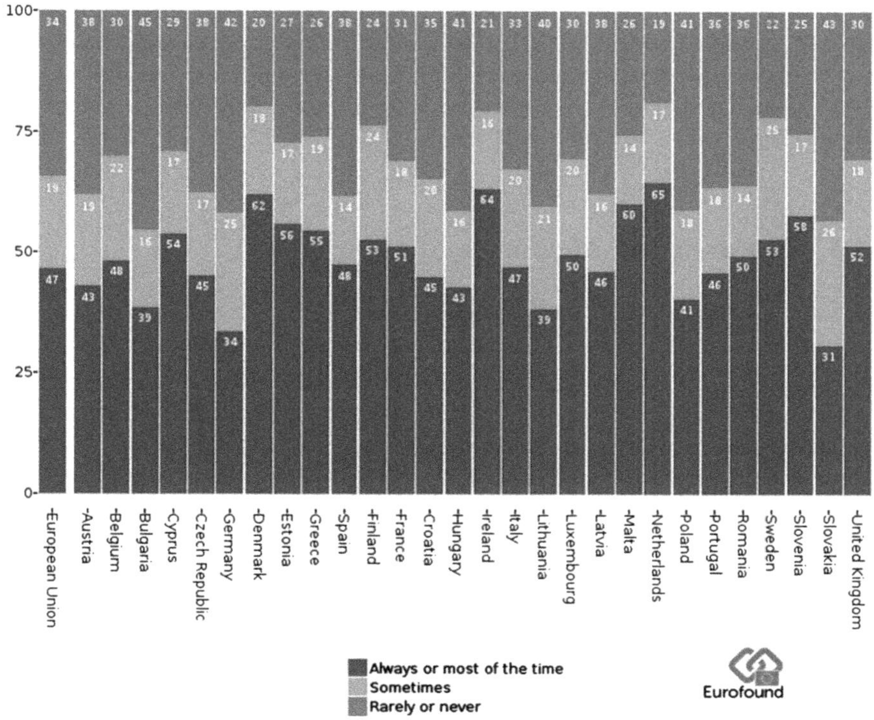

Fig. 14.2 Participation in improving the work organisation or processes, by country (%). *Source* https://www.eurofound.europa.eu/pl/surveys/data-visualisation/european-working-conditions-survey-2010?locale=EN&dataSource=EWCS2010&media=png&width=740&question=y10_q51d_3&plot=euBars&countryGroup=linear&subset=bd_age&subsetValue=All. Accessed April 12, 2017

on WPI in Poland. European Working Conditions Study reveals that 41% of Polish employees has rarely or never participated in improving work organisation or processes (see Fig. 14.2) comparing to 34% EU-level average.

Considering the historically rooted passivity of workers, it is not surprising that 73% of Polish employees are satisfied with the way their opinions were considered when decisions were made about their work (European Comission 2014). This result ranks Poland even above the average EU-level (70% total satisfied employees).

Moreover, quantitative research conducted among Polish workers confirms that there is a significant relationship between social trust, attempts to introduce WPI and the outcomes of these attempts. A research conducted among 500 Polish service sector employees showed the relationship between social capital at work and WPI. Social capital at work was measured by levels of trust, solidarity and reciprocity that are shared among members of the same community, as well as networks, collective action and mutual responsibility (Kouvonen et al. 2006).

Lower levels of social capital at work were related to lower levels of all WPI factors analysed: strategic orientation on the environment, organising smarter, product—market improvement and flexible work. Simultaneously it was also found that lower WPI was related to lower employees work engagement, work ability and higher intent to leave work (Mockałło 2016).

14.3 Research and Methodology

14.3.1 Purpose of This Research

Our goal is to further explore the role of social trust and employees' participation in the processes of innovation creation, development and implementation. We assume that unfavourable features of Polish cultural and institutional context for WPI deliver a good opportunity to reflect on the role of these factors. Research on WPI related organisational processes and causal relationships between different organisational factors 'producing' innovation was focused on best practices and innovation leaders (Hogan and Coote 2014). But the incongruity between the original cultural, institutional and organisational context and the participatory and empowering nature of WPI can impede the introduction of changes prescribed by various WPI models and hence influence the possibilities to achieve the goals like the rise in organisational performance and the quality of working life.

Even authors developing WPI models based on best practices warn to use them as a blueprint or recipe book and indicate that "even within more flexible structures, mistrust and disempowerment can be embedded in the systems and processes that shape decision-making, resource allocation, standard operating procedures and performance management. They can reflect a culture of centralised control and micro-management which requires careful dismantling (Totterdill et al. 2016).

Yet, the issue of cultural, institutional and organizational context unfavourable for WPI, as well as issue of its dismantling remains under researched. Our research problem is exploring failed and mixed experiences of WPI introduction which aims at contributing to both theorising and the practical implementation of WPI.

14.3.2 Research Questions

To investigate how the innovating process aiming at the improvement of the quality of working life and of organisational performance (i.e., WPI) does enfold in an average Polish enterprise on the level of local perceptions and practices we choose a qualitative, interpretive approach facilitating in-depth understanding of social processes (Silverman 2013). In qualitative research the understanding is acquired through reconstruction of worldviews and practices displayed by the participants of a given social situation. We regard expressions of regulative, normative, and cognitive structures as institutions; institutions are standards as to what is to be held to be normal, what must be expected, what can be relied upon, which rights and duties are attached to which positions, and what makes sense in the community or social domain for which an institution is valid (Offe 1998, p. 200). We choose an institutional perspective as a departure point since it addresses the core of our considerations: question of how specific features of organisational environments like values, norms, rules and beliefs influence perceptions and practices within organisations itself (Hatch and Cunliffe 2006).

While observing innovation process we focus on the issues of employee participation. It is crucial since WPI "enables employees to participate in organisational change and renewal and hence improve the quality of working life and organisational performance" (Oeij et al. 2015). That is also why we treat employees' needs and problems as a departure point of innovation process in our research. We assume that if the quality of working life is to be improved, the needs and problems employees are implicitly addressed as well. That is why employees' participation is needed from the very beginning of the innovation process.

In line with our institutional approach we assume that employees' ability to participate may be influenced by the individual's attitudes to innovation (e.g. importance of innovation and innovation processes, trust they may or may not have towards other co-workers and managers) and by the formal and informal organisational structures (e.g. norms, procedures, routines connected to performance assessment, incentives systems, organisational hierarchies). Hence, we ask the following research questions: What is the role of employees in the process of designing and implementing workplace innovation? What is the impact of formal and informal organisational structures in this process? While our research is on innovation, it is not exactly on workplace innovation, but on innovation and change in which we studied the role of employees. We nonetheless speak of workplace innovation or workplace related innovation, because one of the central aspects of WPI is the engagement and involvement of employees in the innovation process (Totterdil et al. 2016), and that is what is exactly the case in our Polish study.

14.3.3 Data Sources and Methods

There were two phases in the research. As a start, we conducted ten semi-structured interviews (I) with managers and four focus group interviews (FGI) with 32 employees from diverse companies that were engaged in workplace related innovation and organisational change to reconstruct a general narrative about innovation and change within organisations. In line with the conceptualisation of WPI as a form of social innovation and the intention to reconstruct local understandings and practices, research tools were prepared to explore the process of innovating with the focus on:

(a) needs and problems as triggers of innovation (what are the most pressing problems? do workers try to address them, if yes how, if not why? what were the most significant innovations influencing their workplace?);
(b) participatory aspects of innovation process (how does the process look like? who initiates innovation? how does the employee involvement looks like on different stages of the process? who cooperates and communicates with whom with what result?);
(c) outcomes of innovations (what were the most important outcomes of the process? were the problems solved/needs addressed? examples of successful and unsuccessful innovations).

Selection of interviewees followed a two-stage procedure. First selecting organisations experiencing innovation/ organisational change in their latest year, and then selecting an employee who was engaged in the innovation process. Since our purpose was to observe typical rather than extreme patterns (Patton 1990) we did not seek for neither success nor failure stories. To account for variables typically influencing organisational processes, practices and perceptions (Hammerslay and Atkinson 2007) we diversified the sample according to the size of the organisation (beneath and above 50 employees), location (big and small and big city—above and below 100 thousand inhabitants), and industry (production and services). When selecting the interviewees within organisations we used their position within the organisational hierarchy (line employees, middle and senior management) as a criterion. We have conducted four FGI with eight participants (per interview) being front line employees and representing different organisations (in sum 32 employees from 32 different organisations) and ten semi-structured interviews with middle management representatives (10 participants from different organisations).

In the second phase, to deepen our knowledge with a context related information we have conducted an exploratory field research. Using the same selection criteria, we have chosen four organisations which were observed during a one-day field visit, in each of them two interviews were conducted (with a top-level manager and a line worker). Further we have conducted participatory observation with a shadowing technique (Czarniawska 2007) in eight organisations. Eight top-management representatives were accompanied ('shadowed') and interviewed by a researcher during their working day. Protocols were organised around the process of innovation and implementation with a special focus on the role of individuals and formal and informal organisational structures. Table 14.1 summarizes our data selection and methods.

Interviews and observation notes were transcribed. Predetermined codes, based on theoretical categories describing the process of innovation, were used to order and analyse the empirical data (Miles and Huberman 1994; Creswell 2014). Chunks of data were bracketed according to a. expressions of needs/problems at the workplace (as the starting point for innovation); b. descriptions of ideas, their generation and communication; c. perceptions of implementation method and d. implementation effects. Through the analysis we used data triangulation and methodological triangulation (Denzin 1978). First one was used in order to compare

Table 14.1 Basic information on the methodology of the qualitative research

Method	Sampling criteria	Number of managers/employees
Focus group interview (FGI)	Industry (production and services) Position (front line employees) Location (big cities)	4 groups × 8 participants = 32 participants (employees)
Individual in depth interview (IDI)	Industry (production and services) Position (middle level management representatives) Location (small and big cities)	10 interviewees (managers)
Participant observation (shadowing)	Industry (production and services) Position (top management rep.) Location (small and big cities)	8 locations × 1 participant = 8 participants (employees and top managers)
Mini-case studies (interview and a field visit)	Industry (production and services) Position (top management and front line employees) Size of the company (beneath and above 50 employees)	4 organisations × 2 interviewees in each = 8 participants (top managers)

perspective of line employees, middle and top management representatives, as well as to juxtapose descriptions coming from those employed in services versus production and small versus big enterprises. Methodological triangulation enabled us to confront individual accounts of innovation process with data coming from observation of organizational practices.

14.4 Results of the Empirical Analysis: From Needs Through Implementing Innovations

In line with the social innovation approach (Murray et al. 2010) we began by analysing interviewees' statements relating to the need for change in terms of behaviour, processes or the material dimension of the workplace. The problems spontaneously discussed by the employees and managers were emerging around three major areas:

(a) Work organisation:

- excessive bureaucracy that hinders and drags out the process of fulfilling the simplest needs;
- low level of autonomy, preventing e.g. the introduction of flexible working hours;
- inadequate work organisation by the management (failure to respect working hours and the right to breaks, mismatch between tasks, working time and capabilities of an employee; incorrect settlement of overtime; improper delegation of work in the situation of redundancies and/or absences);

(b) Management procedures:

- communication: not listening to the opinions and ideas of employees, lack of a rapid and more effective change introduced as a response to the needs of employees;
- incentives systems: lack of salary adjustments, unfair remuneration for overtime work;

(c) Changes regarding the physical work environment:

- technology changes, equipment—its replacement and/or more effective maintenance;
- general working conditions—failure to adapt the workspace and offices to the needs of employees, e.g. replacing open spaces with smaller offices, comfortable temperature, parking spaces.

It is worth noting that these topics prevailed in FGI with front line employees regardless of the industry (production, services) or location (large cities, small towns). The attempts to innovate within all these three areas become point of relevance in subsequent discussions and narratives lead by interviewees.

14.4.1 Generating and Implementing Innovation: Top Down versus Bottom up Initiatives

The studied organisations lack an effective system encouraging two-way communication that would allow employees to talk openly about their needs, and—according to research participants—if such systems have been put in place, they do not work properly. Even if employees have expressed a need for change, their voice is not heard, which discourages them from becoming actively involved. Employees under research feel that they have no influence on changes and actions taken by their firm and that their opinions are not considered. Following quotes relating to the phase of ideas development illustrate this problem: "*Nobody pays any attention to employees' ideas. We have no influence on anything. For the management, we are anonymous*". (FGI 4, BC, S)[2]. "*We used to share [problems and ideas], but… there was never any response*". (FGI 2, BC, S).

In addition, according to the FGI participants trade unions do not act as intermediaries in the bottom-up communication of the needs and problems experienced by the employees. The employees attitude towards unions is therefore ambivalent. Among interviewees trade unions have an image of organisations acting only on behalf of their members. Despite having considerable bargaining power, trade unions do not represent employees who are non-members. They do not offer support or provide information that employees would expect.

The relationship between regular employees and middle management is perceived by those employees as highly conflictual, also when the processes of innovating are concerned. One of the crassest illustrations of this antagonistic vision is a narrative about managers "stealing" the ideas of employees depicted by the interviewees as a quite common phenomenon. "*We tend to keep our ideas to ourselves, because the manager snitches them and presents as his own to company owners from Denmark. So, let's say, if he gets four questionnaires, he's got a whole bunch of achievements or successes ready to use*" (IDI 8, SC, P).

Furthermore, employees tend to avoid taking initiatives for change with a risk of being loaded with additional responsibilities and in case of initiatives failure, with a risk of being held entirely accountable for the negative consequences. Involvement does not bring tangible benefits, e.g. it is not taken into account in the employee performance review. Neither does it translate into promotion or a pay raise. "*There is the so-called annual assessment. (…) I am required to [complete a questionnaire and] demonstrate what I have achieved in a given year, what I have changed, and so on (…) Our questionnaires are then only forwarded to the HR Director, and eventually end up in the trash bin.*" (IDI 8, SC, P). The same interviewee continues: "*As our ideas do not affect our salary, it makes no sense to go the extra mile.*"

Given the negative experiences of those who have attempted to make changes, the strategy of not suggest improvements in the workplace has become the predominant attitude among employees under research. They are afraid to act, react, and report problems and needs. They fear that their active attitude may be interpreted as demands and eventually lead to their dismissal. Employees know from experience that taking initiative may turn against them. It is hard for them to unite and take joint action. When directly asked about the reasons, they refer to the Polish mentality, reluctance to participate in collective action, negative experiences of those who have taken initiative. "*Everybody wants changes, but nobody wants to speak to the boss about it.*" (FGI 2, BC, P).

Consequently, much of the innovation processes observed during the field visits and innovations that are implemented, have been initiated by senior staff and firm

[2]The coding (FGI 4-BC-cost control specialist-rail transport) indicates the research method (FGI—Focus Group Interview, IDI—In-Depth-Interview, Sh - Shadowing, Mini Case—MC), observation number (e.g. FGI number 4), location (BC—big city, SC—small city), industry (S-services, P production), size of the company (used only for MC—SE as small enterprises: beneath 50 employees and BE: above 50 employees).

owners. Even more; they have consulted shop-floor employees. Interviewed employees complain that 'usually' information about changes to be introduced reaches them as rumours. It is only later that they receive an official notice by e-mail or in writing. Ideas have rarely been presented to them, for instance by their superiors, let alone discussed. In organisations under research it is common practice to oblige employees to sign declarations asserting that they have become familiar with information about changes and their implementation.

From these examples, we draw the following tentative conclusion. Given the lack of procedures facilitating cooperative idea generation and development and the lack of proper communication channels, or their misuse, employees' ideas either go unnoticed or are not appreciated in the process of change and innovation.

14.4.2 Goals and Outcomes of Innovation Processes: Between Declared Objectives and Instrumentalisation

Interviewees' narratives about innovations are rather cynical. In their opinion innovations are not implemented by the management with a view to improve an organisational performance. The way they describe context of innovation processes suggest that innovation and change is instrumentally applied as a disciplining tool and/or legitimizing the position of newly employed managers. The latter aspect can be illustrated with the following quotes from line-workers: *"'Innovations' are implemented all the time [sarcastically], particularly when a new manager is appointed."* (FGI 3, BC, S). *"From my perspective, it seems that, at the moment, the director's son is trying to show off, to make his mark, secure his position within the company and make sure he won't be marginalized."* (MC, SC, LE).

The use of innovation by managers for the purposes of raising employees control results in a tendency to associate innovation with downsizing or increased responsibilities. One of the interviewees says: "*The introduction of a new system will improve working time organisation and we have been told that it will be implemented and that one person will be kicked out (...), unless he or she will be needed in another department.*" (FGI 3, BC, S). It is particularly interesting that the instrumental use of innovation to discipline employees is also referred to by middle management representatives. One of them explains the practice of innovation implementation in a following way: "*[when an innovation is introduced], employees are not spoken to directly and reassured that their position is not at risk. This kind of uncertainty is good: it is a psychological trick used to make employees aware that someone watches over them at all times, that their mistakes can be frowned upon and that they always need to do their best and remain on guard.*" (IDI 9, SC, S).

The use of innovation as a discipline tool generates stress when used in conjunction with an authoritarian management style, i.e. the mobilisation of workers through fear. From the view point of FGI participants this tendency is evident in the case of those who have been promoted internally: they are depicted as particularly cruel and strict. Interestingly, workers emphasize that managers usually have good intentions and are trying to be supportive. This apparent contradiction is exemplified with the following quote: "*in a large group, he puts on a show and humiliates the staff; however, he is accommodating and kind when dealing with individual employees; (...) he is a bit of a despot who has introduced strict discipline, but does it in good faith and with the company's growth and development in mind.*" (FGI 3, BC, S). One can hypothesize that this contradiction is rooted in the lack of appropriate training or mentoring in human resource management directed to those who have been promoted to managerial positions.

14.5 Conclusions: Trust and Participation Between Structural and Individual Influences

Our considerations were organised around an issue of transplanting solutions like WPI models developed in Western European countries into environments that lack cultural infrastructure on which the "original" can rely (Offe 1998). The existing body of research on Polish socio-economic environment shows that there is a significant shortage of this cultural infrastructure in Polish enterprises. We have expected that Polish companies do not apply workers' participation when they innovate. From our research, we can tentatively conclude that this indeed seems to be the case. Among major reasons we could list a lack of participatory approaches and a lack of cooperative experiences. The results clearly indicate the lack of social trust on the employees' side and on the managers' side regards innovations.

Further, we have expected that the lack of employees' participation and the general pressure on the maintenance of low labour costs (typical for dependent market economies) results in highlighting organisational performance as a major goal of innovation and neglecting the goal of quality of working life improvement. If this is true – how are WPI processes enfolding in these unfavourable conditions? In answering this general question, we have focused on the role of employees and formal and informal organizational structures in the process of designing and implementing workplace innovation.

In line with the institutional approach we assumed that employees' ability to participate may be influenced by the individual perception of innovation and innovation processes, trust towards other co-workers and managers and by the formal and informal organizational structures (e.g. norms, procedures, routines connected to performance assessment, incentives systems, organisational hierarchies). Through focusing on mediocrity—as opposed to focus on good practices and innovation leaders—there was a possibility to grasp the impact of structural and individual elements on shaping innovative activities. Consequently, we were able to specify which structures and individuals jointly produce undesirable, or even adverse effects. We have intentionally focused on these negative patterns since our aim was to explore under researched area of cultural, institutional and organizational contexts unfavourable for WPI.

Individual employees' responses, even if initially inconsistent with the roles imposed by the structure (see attempts at influencing change) eventually succumb to expectations contained within structures, as the latter "reward" behaviour that is in line with expectations and punish behaviour that deviates from the accepted formal and informal norms which presuppose the passivity of employees. Examples of this mechanism include situations in which ideas shared by employees are never responded to (obstructed bottom-up communication channels), employees discover that their line managers have appropriated their ideas (no direct channels of communication with senior staff), new ideas generate new obligations (inadequate work organisation management processes within a specific post and failing coordination between posts), or the risk of defeat, while the initiative itself is rarely rewarded. Lack of positive incentives and the predominance of negative stimuli further deepen the passivity of employees, characterized by their reluctance to cooperate and an antagonistic attitude towards line managers and the employer.

In these circumstances, the very notion of workplace innovation has negative connotations among lower-level employees in Poland, because the management uses it instrumentally. Innovations are not introduced to improve processes or working conditions, but rather to strengthen the manager's position within the organisation, e.g. when a new manager enacts changes in the department to legitimize their appointment for the position and to discipline employees, or when failure to adapt to discretionary changes introduced and controlled by managers is punished with low assessment scores, or even dismissal. Within this mechanism attempts to introduce WPI yield results very different from these assumed by a model developed based on best practices, usually performed in Western European countries.

14.6 Discussion and Further Research

Qualitative research is used here to provide contextual understanding of social processes in which participants of these processes shape their worldviews and make sense of surrounding reality. Therefore, our findings do not reflect an overall picture of innovation processes in the population of Polish enterprises nor they provide conclusive information about prevailing attitudes and behaviours among the employees of different levels. Hence our ability to draw conclusions about frequency of phenomena under research is limited. We can only infer about their relative commonality at the background of quantitative and qualitative research presenting specificity of Polish context when processes of innovation and employee participation are concerned.

Nevertheless, an analysis of qualitative data enabled us to notice significance of certain structural and individual factors in the process of WPI development and implementation which can be explored in further research to draw theoretical generalizations about the nature of processes under research. Themes addressed in the study can be interpreted as a feedback mechanism between formal and informal rules of operation of the organisation and individual actions taken by its members. At the forefront, there is negative feedback from the point of view of innovation processes: interviews, observations and case studies unveil how specific types of organisational structures, the system of assigning and executing tasks in the workplace shape individual attitudes—among leaders and their subordinates alike—that are unfavourable to the development of workplace innovation.

Qualitative research focused on everyday practice related to change and innovation in a variety of organisations has helped us to understand innovation processes in terms of structure and agency. The first category captured formal and informal rules, procedures, cognitive and behavioural patterns intertwined in the process of innovating. The second category encompassed individual responses and practices to these rules. We contribute to the existing literature on workplace innovation by emphasizing interaction between individual and organizational levels of analysis, thus setting the stage for thinking about workplace innovation in terms of a phenomenon produced by the interaction of structural and individual factors. Some conclusions on the role of employees and formal and informal organizational structures in the process of workplace innovation design, implementation and institutionalization can be drawn when we compare the results of quantitative and qualitative research, in which we are involved now as a next step of our research project.

Further research verifying the validity of the 'structure-agency WPI model' could also include multiple case analysis with a case selection based on a maximum variation logic (Patton 1990). The variation should refer to the structural (e.g. organic and mechanistic systems) and individual aspects (e.g. passive versus active employee attitudes, different types of leadership) as well as to the outcomes of WPI (successful versus unsuccessful). Another possible venue of exploration is a comparison of WPI initiatives displaying in organisations functioning within different institutional and cultural environments, since both structures and individuals within an organisation are influenced by the characteristics of these environments (Hatch and Cunliffe 2006).

In theoretical terms our model contributes to deeper understanding of WPI because instead of focusing on different aspects separately or through investigating correlations between these aspects and WPI, what is more common in quantitative research (e.g. Oeij et al. 2012), as it focuses on their interaction displaying in time. In other words, we propose a framework to understand not whether certain factors are connected and with what significance, but how and why—through what kind of processes—they become connected.

References

Bendyk, E. (2015). Białe plamy na mapie, czyli o ukrytych potencjałach innowacyjności. In P. Zadura- Lichota (Ed.), *Innowacyjna przedsiębiorczość w Polsce. Odryty i ukryty potencjał polskiej przedsiębiorczości* (pp. 91–108). Warszawa: Polska Agencja Rozwoju Przedsiębiorczości.

Creswell, J. W. (2014). *Research design: Qualitative, quantitative, and mixed methods approaches* (4th ed.). London: Sage.
Czapiński, J., & Panek, T. (2015). *Diagnoza społeczna 2015. Warunki i jakość życia Polaków*. Warszawa: Rada Monitoringu Społecznego.
Czarniawska, B. (2007). *Shadowing and other techniques for doing fieldwork in modern societies*. Malmo: Liber.
Denzin, N. (1978). *Sociological methods: A source book* (2nd ed.). New York: McGraw-Hill.
Dunn, E. (2004). *Privatizing poland. Baby food, big business, and the remaking of labor*. Ithaca, NY: Cornell University Press.
European Commission. (2014). Flash Eurobarometer 398 working conditions report. Retrieved October 8, 2016, from http://ec.europa.eu/public_opinion/flash/fl_398_en.pdf
Gadomska-Lila, K. (2011). Budowanie kultury innowacyjności w świetle badań empirycznych. *Współczesne Zarządzanie, 1,* 124–133.
Hammersley, M., & Atkinson, P. (2007). *Ethnography: Principals in practice* (3rd ed.). London: Routledge.
Hatch, M. J., & Cunliffe, A. L. (2006). *Organization theory: Modern, symbolic, and postmodern perspectives*. New York, NY: Oxford University Press.
Hogan, S. J., & Coote, L. V. (2014). Organizational culture, innovation, and performance: A test of Schein's model. *Journal of Business Research, 67*(8), 1609–1621.
Högselius, P. (2003, January 16–18). *Can East European Countries Innovate?* Paper presented at the Danish Research Unit for Industrial Dynamics Winter Conference, Aalborg, Denmark. Retrieved October 16, 2016, from http://www.druid.dk/uploads/tx_picturedb/dw2003-708.pdf
Howaldt, J., Oeij, P., Dhondt, S., & Fruytier, B. (2016). Workplace innovation and social innovation: An introduction. *World Review of Entrepreneurship, Management and Sustainable Development, 12*(1), 1–12.
King, L (2007). Central european capitalism in comparative perspective, w: B. Hancké, M. Rhodes, M. Thatcher (Eds.), *Beyond varieties of capitalism: Conflict, contradictions, and complementarities in the european economy* (pp. 307–327). Oxford: Oxford University Press.
Kochanowicz, J., Mandes, M., & Marody, M. (2007). *Kulturowe aspekty transformacji ekonomicznej*. Warszawa: Instytut Spraw Publicznych.
Kohtamäki, M., Kekäle, T., & Viitala, R. (2004). Trust and innovation: From spin-off idea to stock exchange. *Creativity and Innovation Management, 13,* 75–88.
Kouvonen, A., Kivimaki, M., Vahtera, J., Oksanen, T., Elovainio, M., Cox, T., et al. (2006). Psychometric evaluation of a short measure of social capital at work. *BMC Public Health, 6,* 251.
Makó, C., & Illéssy, M. (2015). *Innovation policy review. National and European experiences*. QuInnE Working Paper 1. Retrieved September 19, 2016 from http://bryder.nu/quinne1/sites/default/files/Working%20Paper%201%20revised%2020160425.pdf
Marody, M., & Lewicki, M. (2010). Przemiany ideologii pracy. In M. Marody, & J. Kochanowicz (Eds.), *Kultura i gospodarka* (pp. 85–125). Warsaw: Wydawnictwo Naukowe Scholar.
Miles, M. B., & Huberman, A. M. (1994). *Qualitative data analysis: An expanded sourcebook*. London: Sage.
Mockałło, Z. (2016). Innowacje w miejscu pracy i ich związek z jakością życia zawodowego—kapitałem społecznym, zaangażowaniem w pracę i samopoczuciem pracowników. In M. Strumińska-Kutra, & B. Rok (Eds.), *Innowacje w miejscu pracy. Pomiędzy efektywnością a jakością życia zawodowego,* Warsaw: Poltext.
Murray, R., Caulier-Grice, J., & Mulgan, G. (2010). *The open book of social innovation*. London: The Young Foundation and Nesta. Retrieved September. 19, 2016 from http://youngfoundation.org/wp-content/uploads/2012/10/The-Open-Book-of-Social-Innovationg.pdf
Nieć, M. (2015). Działalność innowacyjna przedsiębiorstw w Polsce na tle krajów Europy. In P. Zadura-Lichota (Ed.), *Innowacyjna przedsiębiorczość w Polsce. Odryty i ukryty potencjał polskiej przedsiębiorczości* (pp. 11–46). Warszawa: Polska Agencja Rozwoju Przedsiębiorczości.
Nölke, A., & Vliegenthart, A. (2009). Enlarging the varieties of capitalism: The emergence of dependent market economies in East Central Europe. *World Politics, 61,* 670–702.

OECD (2014) Employment database—Employment indicators. Retrieved September, 19, 2016 from http://www.oecd.org/employment/emp/employmentdatabase-employment.htm
Oeij, P. R. A., Dhondt, S., Kraan, K., Vergeer, R., & Pot, F. D. (2012). Workplace innovation and its relation with organisational performance and employee commitment. *LLINE Lifelong Learning in Europe, 4*. Retrieved September 19, 2016 from http://www.elmmagazine.eu/articles/workplace-innovation-and-its-relations-with-organisational-performance-and-employee-commitment/
Oeij, P. R. A., Žiauberytė-Jakštienė, R., Dhondt, S., Corral, A., Totterdill, P., & Preenen, P. (2015). *Workplace innovation in European companies.*, Eurofound Study Luxemburg: Office for Official Publications of the European Communities.
Offe, C. (1998). Designing institutions in East European transitions. In R. E. Goodin (Ed.), *The theory of institutional design* (pp. 199–226). Cambridge: Cambridge University Press.
Patton, M. (1990). *Qualitative evaluation and research methods*. Beverly Hills, CA: Sage.
Sieradzka, K. (2014). Barriers to innovation activity of Polish small and medium-sized enterprises. *Przedsiębiorstwo i region, 6,* 88–96.
Silverman, S (2013). Doing qualitative research, 4th Edition, London: Sage.
Stocki, R., Prokopowicz, R., & Żmuda, G. (2008). *Pełna partycypacja w zarządzaniu. Tajemnica sukcesu największych eksperymentów menedżerskich świata*. Warszawa: Wolters Kluwer.
Totterdill, P., Pot, F. D., & Dhondt, S. (2016). Defining workplace innovation. In M. Strumińska-Kutra, & B. Rok (Eds.), *Innowacje w miejscu pracy. Pomiędzy efektywnością a jakością życia zawodowego*, forthcoming. Warsaw: Poltext.

Author Biographies

Marta Strumińska-Kutra, Ph.D. is a Associate professor in social innovation, Organization and leadership, VID Specialized University in Oslo, Norway and Assistant professor at Social Sciences Department, Kozminski University in Warsaw, Poland. Visiting Fellow (2014) and Associate Fellow (2014–2016) at the Institute for Science, Innovation and Society (Oxford University, Oxford, UK). She conducts research on: Corporate Social Responsibility, social and sustainable innovation, public governance.

Boleslaw Rok, Ph.D. is a Professor of Management at Kozminski University, Department for Entrepreneurship and Business Ethics. Director of Positive Entrepreneurship Research Lab, Academic Director of Corporate Responsibility Executive Programme, consultant to many professional associations, governmental and commercial organizations. His research interests lie in the intersection of social innovation, P2P economy, corporate responsibility, sharing economy, sustainability and strategic management.

Zofia Mockałło, MA is a researcher, Department of Ergonomics, Laboratory of Occupational Psychology and Sociology, Central Institute for Labour Protection—National Research Institute, Warsaw, Poland, member of a PEROSH Wellbeing at Work Initiative. Her research area is focused on work & organizational psychology, sources and effects of employees' wellbeing with special interest in social capital and innovation.

Chapter 15
Workplace Innovation and the Quality of Working Life in an Age of Uberisation

Chris Warhurst, Chris Mathieu and Sally Wright

15.1 Introduction

That innovation can both create and destroy jobs has long been accepted (e.g. Schumpeter 1939). Recent debate on this issue has focused on technological innovation and the purported effects of a mix of automation and robotisation—mass job losses (Frey and Osborne 2013). The basic and highly influential argument is that some jobs are more susceptible to replacement than others, with Frey and Osborne estimating almost half of US jobs are at risk—a figure echoed by a Bank of America Merrill Lynch report (cited in Stewart 2015). In the UK, leading parliamentarians, Bank of England officials and respected consultancies have picked up this theme (Spence 2016). The message of each in their own way is one of threat. Deloitte suggests that between 20 and 25% of jobs in the UK will go but expect the 'pace of automation to increase exponentially' so that more jobs safe today will become unsafe tomorrow (cited in Spence 2016).

What has been too readily overlooked in this debate about the rise of the robots is that Frey and Osborne focused exclusively on potential job destruction. As Osborne (2015) later pointed out, the original study offered no commentary on potential job creation opportunities. A range of possible impacts of technology on jobs exist he stated: job loss, job change and job creation. Indeed each time technology anxiety has been stoked, for example in the 1960 s when firms first started

C. Warhurst (✉) · S. Wright
Institute for Employment Research, University of Warwick, Coventry, United Kingdom
e-mail: C.Warhurst@warwick.ac.uk

S. Wright
e-mail: S.A.Wright@warwick.ac.uk

C. Mathieu
Department of Sociology, Lund University, Lund, Sweden
e-mail: Christopher.mathieu@soc.lu.se

P.R.A. Oeij et al. (eds.), *Workplace Innovation*, Aligning Perspectives on Health, Safety and Well-Being, DOI 10.1007/978-3-319-56333-6_15

installing computers, more jobs have been created than destroyed the *Economist* (2016) points out. Hole-in-the-wall cash dispensers (or automated teller machines, ATMs) might have replaced bank tellers, for example, but call centre banking jobs emerged. Nevertheless the latest claim for the *death of jobs*, recast as the death of (human) work or labour, triggered by Frey and Osborne, has gained much traction, shaping public and policy debate.

Whilst debate rages, another digital disruption is quietly taking place; one that seems less worrying to the public given that so many are already willingly using it —the platform economy. This economy builds on the sharing economy albeit with transactions migrating to websites and apps, performed by strangers and in which goods and services are monetised (Katz 2015). As such, what was once an informal economy—friends, families and neighbours lending tools and time to each other, for example—has become formalised and commercialised. Examples are numerous but the most high profile globally are TaskRabbit (disrupting the handyman industry), Airbnb (disrupting the hotel industry) and Uber (disrupting the taxi industry), in which, nominally, people do tasks for others, hosts share an unused room in their home with guests and those with a car offer rides to those who need a lift somewhere respectively. Many of these companies demonstrate how different forms of innovation, technological and non-technological, can exist simultaneously and synergistically.

What is significant about this form of digital disruption is that workers are no longer employed by a company but are paid by individual buyers to perform a single task: assemble flat-pack furniture, lend a bed or offer a lift in the case of TaskRabbit, Airbnb and Uber respectively. As Christopher Degryse of the European Trade Union Institute points out, technology is creating 'Digital Taylorism' whereby production processes are being disaggregated into 'tiny simple repetitive tasks … offered to the "community" of crowdworkers' (quoted in *Labour Research* August 2016, p. 14). The latter observation, about crowdworkers, is also significant, for the key innovation of this business model may be the *death of employment* as workers become freelancers and micro-entrepreneurs (Singer 2014), no longer employees paid a wage by employers in exchange for labour, governed by a contract that typically allows the employer a managerial prerogative to direct, monitor, evaluate, discipline and reward 'their' workers (Warhurst 1997). Instead 'a vast … pool of freelancers who are available to work on demand … on a mind-boggling array of … tasks', some specialised, some mundane (O'Connor 2015, p. 17).

Although framed in popular parlance as 'freelancers' and 'entrepreneurs', the operationalisation of this new business model is making it increasingly difficult to determine whether the legal status of a worker is that of an employee, an independent contractor or self-employed (Tomassetti 2016). Moreover, despite claims of liberating workers from wage-slave drudgery, the remaining task-based work has been criticised for being of poor quality, detrimental to workers' well-being. As such, with its aspiration to improve the quality of working life (Pot et al. 2016), this development throws down a challenge to advocates of workplace innovation. It also signals the need to examine the relationship between innovation and job quality, not just job creation/destruction (cf. Mako et al. 2015).

This chapter examines this development, focusing specifically on Uber and the effects of Uberisation on job quality. This focus on Uber is deliberate; despite the plethora of companies that claim to specialise in 'disruptive innovation' (cf. Naughton 2016). Uber is the one held up as the 'leading case' (Eisenbrey and Mishel 2016) and the 'poster child' (Smith 2016) for the platform economy. The chapter first locates Uber and Uberisation amidst the various types of innovation and outlines its challenge to advocates of workplace innovation. It then outlines the model of work offered by Uber and its impact on the quality of working life of Uber driver-partners. The chapter ends by highlighting how Uberisation resonates more widely in the platform economy, flagging some of the current and possible responses to it and calling for this issue to have greater salience in public and policy debate.

15.2 Innovation and Uber

There are different types of innovation and debates about their efficacy (see Fagerberg et al. 2005). A commonly accepted articulation of these types is that provided by the Oslo Manual (OECD 2005), which defines four types of innovation. These four types are: Product innovation, Process innovation, Marketing innovation and Organisational innovation. The first two types can be grouped as technological innovations; the second two types non-technological innovations.

Evidence suggests that technological innovation represents the exception rather than the rule and that non-technological innovations are more prevalent. Importantly, evidence also suggests that non-technological innovation, specifically organisational innovation, can be at least as impactful economically as technological innovations (Sanidas 2005). Other studies show that there are correlations amongst the different types of innovation and, furthermore, that there are synergies amongst the different types, with technological and organisational innovations complementing rather than substituting for each other (Battisti and Stoneman 2010). Policy-makers are increasingly not just recognising but trying to encourage non-technological innovations and establish a more balanced emphasis in policy (e.g. OECD 2010; European Commission 2015).

The fourth type of innovation, Organisational, whilst prominent within the Oslo Manual and promoted by the European Commission (2015), is vaguely defined and poorly measured (Mako et al. 2015). However attempts to better frame it suggest that it envelops 'people management', itself broadly referring to the practices pertaining to 'renewals' in work and employment within organisations and which can include the use of variable pay as a change to reward systems or atypical employment contracts (Eurofound 2012; see also Ramstad 2009).

Focused on the organisational design and operations management that comprise production systems (Eurofound 2015), workplace innovation similarly focuses on work organisation and human resource management combined with technology, broadly defined. It is defined by EU-OSHA (2012, p. 4) as 'an organisation's practice of managing, organising and deploying its human and non-human resources'. The difference between Organisational andWorkplace innovation however lies in the explicit normative underpinnings of the latter (Kesselring et al. 2014). In terms of inputs, its emphasis is on employee participation and involvement in a stakeholder dialogue about innovation. It also argues that outcomes should include an improvement in the quality of working life not just high performing and sustainable organisations, and as such it should offer win-win benefits for employees and employer (Pot et al. 2016).

California-based Uber was founded in 2009, launched in 2010 and by 2015 had expanded to more than 400 cities worldwide (Knight 2015). It is a good example of multiple, synergistic innovations that have created a distinct business model that, as Tomassetti (2016) points out, seems to epitomise the current defining of the firm as 'hollowed out'. Most attention has focused on Uber's technological and marketing innovations. A smartphone app connects drivers, on demand, with service users. Uber claims not to be a taxi service. Rather it styles itself as a digital platform that matches task-doers with task-seekers; in other words simply a 'connecting service' (Adam et al. 2016). More specifically, Uber insists that it is a technology provider—an app—with drivers paying to access the app to connect them to users. These users pay Uber, which then gives part of the service fee to drivers. Uber sets the fares, with price not based on ownership of the means of production but usage in a spot market and for the type of service offered e.g. UberX, UberXL, UberSelect, UberPOOL or BLACK. Drivers are not employed by Uber; they are 'partners' who use their own vehicles. Uber provides its drivers with a personal identification number, its app and a smartphone if the driver does not have one. In addition to no employment being offered to these drivers, no work is guaranteed. Uber does not set maximum or minimum hours that a driver might work; instead drivers determine their own working time and intensity (Smith 2016; Tomassetti 2016). The premise, Rogers (2015, p. 99) suggests, is that drivers have the flexibility and freedom to 'work whatever hours they please': it is self-management by 'independent entrepreneurs'.

Until recently, when debate has occurred about the Uber business model, it has tended to focus on its disruptive impact on the state and other businesses. In this regard, there are two dominant accounts of Uberisation: one benign, the other malign. The benign account rests on the platform economy being a version of neighbours helping neighbours in a sharing economy. One form of this help is non-profit car-pooling and Uber explicitly draws on this imagery, stating that it offers 'ride sharing'. This sharing, it is argued, helps reduce the non- or under-use of resources. It is a 'new age' in which 'underutilised assets become peer-to-peer services for hire, enabled by the Internet and smartphones' (Cusumano 2015, p. 33).

Even better, it offers a pur*er* market for consumers, proving an on-demand service overcoming spatial clustering of service providers; that is taxis that stay close to where more pick-ups might occur, for example around airports or city centres. This potential to be beneficial for consumers is lauded. Providers are 'everyday entrepreneurs … seeking to shake up the market by solving other people's problems' according to the UK Government (HM Government 2015, p. 1). The malign account criticises the 'permissionless innovation' (Thierer 2014) of the platform economy and companies such as Uber for operating outwith regulatory frameworks. In the context of European governments having to make spending cuts (Streeck 2015), the primary concern is tax evasion and Uberisation stimulating and folding into the informal economy. In this respect, that Uber drivers do not offer receipts to passengers and may not be paying tax on their earnings is a problem. In addition, Uber itself is not making 'meaningful tax payments' in Hungary and the UK (Adam et al. 2016). Furthermore there are concerns about the creation of unfair competition for existing businesses (e.g. Eurofound 2016a). Jurisdictions grant licences to registered taxi drivers providing them limited entry to operate in the industry in exchange for regulation adherence. By contrast, Uber drivers are exempt the regulation usual for taxis and do not have taxi licences. Moreover they may also lack insurance. For example, in the US, Uber carries an insurance policy for all drivers while they are logged into their app. (http://www.idrivewithUber.com). In the UK, drivers bear the cost of their insurance (Knight 2015). At the very least therefore insurance coverage is patchy. Furthermore there is no employer to ensure compliance with consumer equality rights legislation. On the basis of this alleged unfair competition there have been legal challenges to Uber in Europe and the US centred on: in Finland and the Netherlands, illegally charging passengers without a professional qualification; in Sweden, from illegally profiting from car-pooling; in Italy and Hungary, being comparable to taxi services and so needing to adhere to existing regulations; and in the US ensuring compliance with the Disabilities Act (Adam et al. 2016, Rogers 2015). Employer organisations in Norway are clear: digital platform companies such as Uber need to accept regulation, pay their taxes and have insurance (Eurofound 2016b).

These accounts need to be taken seriously but what is striking about them is their narrow focus: benefits for consumers, problems for governments and existing businesses. For example, the UK's HM Government (2015) review of the sharing economy is framed exclusively in terms of the benefits for consumers arising from technological innovation. If there are concerns, it is to ensure consumer protection and that calculating tax liability is made easier for providers (Wosskow 2014). What is missing is any consideration of its impact—beneficial or otherwise, on producers—that is, those who work within the Uberisation model. This myopia needs to be addressed. As *Labour Research* (August 2016, p. 13) succinctly summarises, 'Those working through these platforms do so outside of traditional employment relationships and without the rights and protections that come with normal employment contracts.' The business model shifts the risk from the platform provider (Uber) to task-providers (Uber drivers). The potential therefore for Uber drivers to have a poor quality of working life cannot be ignored. The next section of this chapter turns to focus on this issue.

15.3 Uber and the Quality of Working Life

The Uberisation model can be understood in two ways. The first is as a business model; the second is as an operational model. As a form of disruptive innovation, the first model was outlined in the previous section. In this section we examine the second—the operational model underpinning the relations between the platform company (Uber), the service provider (the Uber driver-partner, hereafter the driver) and the customer. More specifically, within this operational model, we examine the quality of working life of the driver. Defining and measuring job quality is difficult, with little disciplinary or methodological consensus (Findlay et al. 2013; Muñoz de Bustillo et al. 2011; Wright 2015). In this analysis the focus is provided by the lens of organisation and workplace innovation: the work and 'employment' of drivers mediated by technology, and, whether this human resources management offers win-win benefits for the company and worker, including whether the latter's quality of working life benefits. Without longitudinal analysis, it is difficult to assess if drivers' quality of working life is improving or deteriorating as Uber drivers per se. However it is possible to measure their quality of working life against the benchmark provided by the claim of self-management and being an independent contractor as well as comment on the implications generally of Uberisation for the quality of working life.

The general relationship between Uber and its drivers is a core issue in this analysis. Uber makes it quite clear that it facilitates a client flow via the app to drivers but that the company owes no obligation to drivers. The relationship is based on mutual interest, rather than a set of mutual rights and obligations, not least to a minimum income or amount of work. It is a rudimentary pecuniary win-win offered by Uber—the company makes money from its app, the drivers make money from the app. Nevertheless despite the use of the term 'independent contractor', the drivers are both dependent upon a technology, the app, and the technology company that controls access to the app. In other words, there is a difference between, for example, a ditch-digger who can buy a backhoe (from one of several manufacturers) and then autonomously disposes over this piece of technology as he or she sees fit, and the driver who is wholly reliant on having access to Uber's app, which is granted by the technology company and without which this particular work cannot be carried out. Thus, the technology can never be autonomously controlled by the driver—its use is conditionally permitted and a relation of dependence on it remains central.

Another aspect of technology dependence is the requirement that drivers acquire, maintain and absorb all costs associated with the other means of production—the vehicle used to provide passenger transportation. This demand is both a requisite burden but also allows a degree of autonomy and discretion over the immediate work environment of the driver. However this autonomy and discretion can again be curtailed by Uber. Uber keeps comprehensive data on its drivers' trips, fares and time and uses this data to measure their waiting location and time, time-to-task and working time as well as monitor, manage and reward driver performance with productivity-based incentive schemes. Furthermore it can, and does, deactivate drivers for poor productivity (Eisenbrey and Mishel 2016).

Probably the most important factor that differentiates Uber drivers from independent contractors is that they are subject to standardized price-setting—that is to say, they cannot set their own rates for their services or negotiate them collectively. Not only does Uber set the price of rides, it unilaterally alters and varies the price of rides. Likewise, drivers largely cannot choose their customers. In other words, Uber drivers, unlike independent contractors, do not build their own client base; instead, the client base comes through the app.

A final relational factor is that the driver can choose when to activate the app, and thereby choose their own working hours, as well as what geographic locations to work in. However, when a driver is logged onto the app, there is an explicit obligation to accept trips. The acceptance rate has been labelled by Eisenbrey and Mishel (2016) as being 'engaged to wait'—that once a driver turns on the app he or she is expected to accept at least 80% of the ride requests given through the app. This obligation is phrased as 'acceptance rates' and if Uber finds a driver is not consistently accepting trip requests, they may be temporarily logged out of the app (see Uber's Driver Deactivation Policy).

Looking at the reward or pay system, Uber drivers do not receive wages but their opportunities for earning income are in part controlled by UberUber and in part controlled by themselves. The latter refers to the time drivers decide that they are available to work. It is the former that is of most interest. Uber uses a cashless system where the customer (or 'ride sharer') pays via a debit or credit card before the ride. This system removes cash transfers between the driver and the ride sharer. Uber takes a share of the driver's fare (or in Uber's parlance, a driver pays a fee to license the technology). Significantly, Uber has the power to unilaterally change (i.e. increase) the fee it charges the drivers. For example, in the US the fee started at 20%, however Uber increased it to 25% in some US markets and in 2015 began testing the impact of increasing the fee to 30% (Huet 2015). Along similar lines, in London Uber raised the fee it takes from new drivers from 20 to 25% in December 2015 (Hayward 2017).

As Uber sets pricing, it is beyond the control of the driver. As drivers are compensated by tasks carried out rather than time in total service (i.e. time spent with the app on, which would include time spent waiting for as well as time spent carrying out tasks), driver income is dependent on an inflow of tasks. As the flow of customers is controlled by the app, the driver has little opportunity to control or impact the flow of clients. Increasing income thus is primarily a function of working more hours—work extendification, which can be more or less efficient in terms of returns for hours invested in time with the app on.

Nevertheless, Uber says that its drivers are well-paid, claiming, for example, that New York drivers receive a median of $90,000 per annum. However, as Sherman points out, to make this amount at $25 per h, a driver would need to be working a 70 h week, 52 weeks per year. In France, one driver who was working a 60–-70 h week claimed to be earning only €2000 per month because Uber had lowered its prices for users and increased the number of Uber drivers. As such, he said, it was becoming increasing more difficult to earn a living (Eurofound 2016c). Indeed in the UK, the GMB union released the income details of one driver who, working 234 h in one month, had received £5.04 per h and so considerably less than the UK's National Minimum wage at the time of £6.50 per h (*Labour Research,* August 2016). As Singer (2014) suggests, these micro-entrepreneurs of the platform economy are 'struggling to piece together a living wage from small, individual events'. Drivers, moreover, are not provided with entitlements such as sick pay and holiday pay. The upshot for drivers is a fixed cost envelop, over which they have little control and variable, often low, income, over which they only have partial control.

Work duration, scheduling, flexibility, regularity and clear boundaries between work and non-work are all a matter of two app buttons controlled by the driver—on and off. The second set of variables about intensity of pace are largely beyond the control of the driver, as once the app is on, the driver is expected to take the customers who are directed to the driver, and the driver, again because of acceptance rate quotas, cannot realistically engage in other productive activities that as mentioned above cannot be dropped in a matter of minutes, such as driving for a competitor such as Lyft in order to minimize economically unproductive waiting time. In other words, pace and intensification is variable, beyond the control of the driver and can be a matter of too much or too little. In London, for example, drivers have 15 s to decide whether or not to accept a job. If they refuse three jobs in a row, they are logged out of the system for 10 min (Knight 2015).

With regard to physical and psychosocial health, Uber is reliant upon providing a safe and trusting environment for both drivers and riders, otherwise drivers would not drive and customers would not ride. Uber does background checks on drivers and uses a mutual rating systems. Aside from securing a sufficient degree of mutual trust between riders and drivers, Uber relinquishes responsibility for health related issues (such as for health insurance in the US or workers' compensation cover in Australia) to the driver as an 'independent contractor'. Arguably, Uber's cashless payment system reduces the chance of drivers being robbed and/or attacked. However there are other health and safety issues linked to Uber's operational model. For example, Knight (2015) highlights an example of a London Uber driver who injured his back in a car accident after dropping off a customer. It seems that insurance policies in some jurisdictions only cover drivers when they are logged on to the app. While attempts to launch women-only ridesharing services in the US have been faced with anti-discrimination legislation, concerns about safety and sexual harassment for both drivers and passengers have been the subject of media attention (Zalkind 2016), though how different this harassment is from that experienced by taxi drivers is an issue for empirical research.

Dialogue between Uber and its drivers on these matters is limited. Uber is categorical on the point that there exists an arm's length distance between it and its drivers. While Uber encourages communication and dialogue with drivers and riders, it unilaterally sets the conditions for use of its app, and uses economic means (such as surge pricing) and behavioural rules and standards for controlling the actions of the drivers without formal mutual consultation. In fact, the relationship between Uber and its drivers is characterized by a very high degree of behavioural regulation. This behavioural regulation entails warnings, temporary suspension, permanent disassociation, as well as behavioural training in terms of tips from other drivers found on the Uber website, general behavioural guidelines and recommendations. Uber gives detailed instructions about how work is to be carried out, as in which routes to drive for example.

There is thus nominal personal autonomy and strict company control, with an absence of any structured dialogue to bridge the two. If participation is a matter of dialogue and influence over the practice of work with colleagues and proximate management, the basic conditions for workplace participation are largely absent due to the structural conditions of the Uber driver role. The degree of autonomy enjoyed by the Uber driver is impacted by structural conditions—working alone and the opportunity to turn off the app are conditions related to high autonomy but working under very proximate economic and behavioural regimes, monitored both by Uber and the ratings that riders give drivers introduces a very strong, and in the latter dimension, potentially capricious, restriction on autonomy, ultimately based on the threat of deactivation.

Despite the existence of significant structural barriers, there have been recent instances of Uber drivers 'organising' or 'collectivising'. The capacity for organisation is enhanced, according to Rogers (2015, p. 99) by Uber drivers being 'naturally tech-savvy' and able to mobilise social media campaigns. One example of organising has seen the International Association of Machinists and Aerospace Workers form the Independent Drivers' Guild—an association for New York City's drivers (estimated to total around 35,000 drivers in May 2016). The guild was formed to represent the drivers in meetings with Uber, including when drivers appeal against Uber decisions to deactivate them.

In addition to governments, trade unions and other business having concerns about Uber's business model, its operational model too is not without complaint. Researchers Rosenblat and Stark spent ten months monitoring online forums of Uber drivers in the US and after analysis of driver views, studying their experience of surges, nudges from the platform, threats of deactivation for their ratings or refusals of incoming jobs, they described the way that Uber controls its drivers through a set of mostly automated commands as 'algorithmic management' which narrows the choices that drivers have to accept (cited in Knight 2015). This research suggests that the claim by Uber that its drivers are independent contractors is not supported by the evidence: any possibility of self-management is usurped by this algorithmic management. In sum, work quality for Uber drivers can literally be described as ‘digital’ in the sense of it being a matter of two positions—off or on. When it is off, drivers are not subject to the economic and behavioural regime that Uber dictates; when on, they are, and their quality of working life is seemingly poor.

15.4 Conclusion

Uber is highly innovative, causing digital disruption to governments, existing businesses and employment. This innovation extends synergistically beyond technological to non-technological innovation. Uber, however, is not alone in this disruption; there are others with similarly provided services. If Uber really is the avant garde exemplar company of the platform economy (Smith 2016; Eisenbrey and Mishel 2016) then it is not inappropriate to talk of Uberisation as a business model that is and will continue to be adopted so that it has the capacity to expand well beyond one company—see for example the points made about TaskRabbit by Kuttner (2013) and Biddle (2014), and, more recently, about Hermes couriers (Booth and Osborne 2016).

Given this possibility, it is important to appreciate that Uber represents more than a technological and, indeed marketing, innovation, it is a clear organisational innovation but which, for advocates of allied workplace innovation, offers a questionable prospect of a decent quality of working life for its drivers. Operationally, once Uber drivers are logged on, their work is tightly controlled, monitored and assessed—as much as for any employee. Driver earnings can be low despite long working hours. Significantly drivers are not in a position to charge their own rates to so-called ride sharers: prices too are strictly controlled by Uber. However because of the arm’s length relationship between the company and the drivers, there is no stakeholder dialogue about this human resource management, let alone the business model. Beyond the switching on and off of the app, self-management by independent entrepreneurs is illusory in the case of Uber. Instead it is, as Rosenblat and Stark suggest, management by algorithm (cited in Knight 2015); and which, to be clear, is a management imposed on the drivers. As a consequence, Uberisation fails to meet the criteria set out by advocates of workplace innovation; indeed it looks set to undermine the normative aspirations of workplace innovation (see Pot et al. 2016).

The implications are manifold. At the very least there is a need to stop using the euphemism of the ‘sharing economy’ in relation to Uberisation. Companies on the platform economy, such as Uber, need to make profits through the monetisation of exchange—a point made by Lizette Risgaard, President of the LO union in Denmark (Eurofound 2016a). As with any labour process converting materials into goods and services through the application of technology and human labour and intended to generate surplus value, which is then appropriated by a non-producer (typically an employer), that non-producer will seek to exercise a managerial prerogative to direct, evaluate, discipline and reward that labour (Warhurst 1997). Even without the issue of employment contracts, Uber is no exception.

If the technology anxiety in current public and policy debate is being over-played, with claims of the death of jobs needing to be tempered (Stewart 2015), it might be that a different type of anxiety is warranted: innovation anxiety, and focused on the death of employment and, with it, the development of poor quality of working life. Indeed, if digitalism can create new forms of work that diminish the quality of working life, 'vigilance' is needed, according to Adam et al. (2016), to prevent a 'race to the bottom in employment (work) standards'. What constitutes employment and with it what constitutes decent, let alone good, job quality will need to start featuring more prominently in public and policy debate about digital disruption.

The hope for proponents of workplace innovation is that change comes with challenge. Certainly there are a number of responses currently to Uber. These responses centre on the classic 'market versus regulation versus organised labour' options. With the first, market, the assumption is that any deficiencies in the model —for example that it offers poor quality of working life—will, with sufficient public awareness, be ironed out by self-correcting markets with consumers of the service —ride sharers—migrating away from Uber as it becomes a lightning rod for discontent due to its prominence (Rogers 2015). The second, regulation, is one that is now beginning to be debated—though has two variants: one, led by Uber and its political lobbying, centres on the creation of new laws and regulations, needed, it is argued, because the platform economy is different (e.g. Harris and Krueger 2015)[1]; the other, championed by the social partners, centres on Uber and other platform companies complying with existing laws and regulations (e.g. Adam et al. 2016). The third, a challenge from organised labour, is gaining traction. While Uber does not formally recognise the Independent Drivers' Guild for collective bargaining purposes union (Alba 2016), as we noted above (e.g. Alba, 2016; *Labour Research,* August 2016), drivers are organising and, in some cases, collectively withdrawing their labour over pricing.

Whichever response is more efficacious, movement for change is already underway: in 2016 the UK union GMB brought a test case in a London employment tribunal on behalf of two drivers who claimed Uber had acted unlawfully by not offering holiday pay and sick pay. This was the first time that Uber faced legal action in the UK over whether their drivers are workers or self-employed. The employment tribunal ruled that Uber must pay drivers the UK national minimum wage and holiday pay, leaving Uber open to claims from all of its UK drivers. Uber has said that it will appeal the decision however if the decision is upheld it will have widespread implications for Uber UK as well as other UK companies that use apps and the internet to match customers with workers (Osborne 2016). Similarly, in

[1]For an excellent review and critique of the arguments underpinning this claim of difference, see Tomassetti (2016).

France, a labour court was asked by one driver to change his 'partnership contract' to an employment contract, with the decision pending (Eurofound 2016c)

In the longer term, the other hope is that there is (again; see Mokyr et al. 2015) a tendency to over-estimate the long-term difference made by some innovations. The platform economy is embryonic; as it matures there are a number of different possible trajectories. The two most obvious ones are 'normalisation' and entrenchment. The first, normalisation, centres on platform companies such as Uber becoming like other companies, mainstreamed as they eventually comply with existing laws and regulations such as the Disability Act in the US; as Rogers (2015) notes, the US Justice Department affirmed that while Uber may contract out its service, it cannot contract away is responsibilities. In Denmark, trade union leader Lizette Risgaard has not only called for platform companies to comply with existing Danish law, including labour law, but, in a clear indication of attempts to normalise Uber, also called for the company to join a Danish employers organisation and, with that membership, cover drivers with a collective agreement (Eurofound 2016a). It might be, however, that the 'labour problem'—that is, the usual directing, monitoring, evaluating and, disciplining and rewarding of employees who have their own interests (Warhurst 1997)—will solve itself. In this regard one other scenario is that Uber dispenses with drivers altogether and starts to use driverless cars. Driving, while not necessarily a routine task, can be automated and robotised. In this scenario it is not the death of employment that Uber portends but, in terms, of providing transportation—for people—and delivery—for products—the death of the jobs (or, again, more accurately, death of human work or labour) of not only taxi drivers but also truck/lorry/van drivers (cf. Frey and Osborne 2013). Whether Uber drivers are employees would thereby become irrelevant, if not meaningless, as technology substitutes for labour. However any new technology is not determinant: how it is used and for what purpose by and within organisations is shaped by human agency; by government, social partners, users and providers, and the public (e.g. Clark et al. 1988). Whether it is the job or employment that is under threat, in an age of Uberisation, with its normative underpinnings, the debate about which jobs remain and how residual human labour is to be managed and organised is one that advocates of workplace innovation need to enter and try to influence.

Acknowledgements The authors would like to thank the Horizon 2020 Framework Programme of the European Commission and award project number 649497 'Quality of jobs and Innovation generated Employment outcomes', without which this chapter would not have happened.

References

Adam, D., Bremerman, M., Durna, J., Fontanarosa, F., Kraemer, B., Westphal, H., et al. (2016). *Digitalisation and working life: Lessons from the Uber cases around Europe*. Dublin: Eurofound.

Alba, D. (2016, May 10). NYC Uber drivers are organizing—Just don't call it a union. *Wired*. Retrieved November 3, 2016, from http://www.wired.com/2016/05/nyc-Uber-drivers-organizing-just-dont-call-union/

Battisti, G., & Stoneman, P. (2010). How innovative are UK firms? Evidence from the fourth community innovation survey on synergies between technological and organisational innovations. *British Journal of Management, 21*(1), 187–206.

Biddle, S. (2014, July 23). If TaskRabbit is the future of employment, the employed are fucked. *Valleywag*. Retrieved November 3, 2016, from http://valleywag.gawker.com/if-taskrabbit-is-the-future-of-employment-theemplyed-1609221541

Booth, R., & Osborne, H. (2016, July 18). How Hermes couriers shoulder insecurity of internet shopping boom. *The Guardian*. Retrieved November 3, 2016, from http://www.theguardian.com/business/2016/jul/18/how-hermes-couriers-shoulder-insecurity-of-internet-shopping-boom

Clark, J., McLoughlin, I., Rose, H., & King, R. (1988). *The process of technological change: New technology and social change in the workplace*. Cambridge: Cambridge University Press.

Cusumano, M. A. (2015). How traditional firms must compete in the sharing economy. *Communications of the ACM, 58*(1), 32–34.

Economist. (2016, June 25). *March of the machines: What history tells us about the future of artificial intelligence and how society should respond* (vol. 12). Retrieved November 3, 2016, from http://www.economist.com/news/leaders/21701119-what-history-tells-us-about-future-artificial-intelligenceand-how-society-should.

Eisenbrey, R., & Mishel, L. (2016, March 17). *Uber business model does not justify a new independent worker category*. Report. Economic Policy Institute. Retrieved November 3, 2016, from http://www.epi.org/publication/Uber-business-model-does-not-justify-a-new-independent-worker-category/

EU-OSHA. (2012). *Review of workplace innovation and its relation with occupational safety and health*. Luxembourg: Office of the European Union.

Eurofound. (2012). *Work Organisation and innovation* (Cox, A., Rickard, C., Tamkin, P.). Luxembourg: Publications Office of the European Union.

Eurofound. (2015). *Workplace innovation in European countries* (Oeij, P. R. A., Žiauberytė-Jakštienė, R., Dhondt, S., Corral, A., Totterdill, P., & Preenen, P.). Luxembourg: Office of the European Union.

Eurofound. (2016a, January 30). *European observatory of working life database*. Denmark: Trade union LO open to dialogue with Uber. Retrieved November 3, 2016, from http://www.eurofound.europa.eu/observatories/eurwork/articles/working-conditions-industrial-relations/denmark-trade-union-lo-open-to-dialogue-with-uber

Eurofound. (2016b, March 8). *European observatory of working life database*. Norway: Change of attitudes to the "sharing economy". Retrieved November 3, 2016, from http://www.eurofound.europa.eu/observatories/eurwork/articles/industrial-relations/norway-change-of-attitudes-to-the-sharing-economy

Eurofound. (2016c, March 2). *European observatory of working life database*. France: New union covers Uber drivers.. Accessed November 3, 2016, from http://www.eurofound.europa.eu/observatories/eurwork/articles/industrial-relations/france-new-union-covers-uber-drivers

Commission, E. (2015). *State of the innovation union 2015*. Brussels: European Commission.

Fagerberg, J., Mowery, D. C., & Nelson, R. R. (Eds.). (2005). *The Oxford handbook of innovation*. Oxford: Oxford University Press.

Findlay, P., Kalleberg, A. L., & Warhurst, C. (2013). The challenge of job quality. *Human Relations, 66*(4), 441–452.

Frey, C. B., & Osborne, M. A. (2013, September 17). *The future of employment: How susceptible are jobs to computerisation?* Retrieved November 3, 2016, from http://www.oxfordmartin.ox.ac.uk/downloads/academic/The_Future_of_Employment.pdf

Harris, S. D., & Krueger, A. B. (2015). Proposal for modernising labour laws for twenty-first-century work: The "independent worker". *Discussion paper 2015-10*, December 2015, The Hamilton Project, Washington, DC: Brookings Institution. Retrieved November 3,

2016, from http://www.hamiltonproject.org/assets/files/modernizing_labor_laws_for_twenty_first_century_work_krueger_harris.pdf
Hayward, S. (2016, November 1). *Become an Uber driver UK: All you need to know about pay, tax and your driving license*. Retrieved November 3, 2016, from https://www.lovemoney.com/news/51136/uber-driver-uk-job-pay-requirements-license-tax
Hayward, S. (2017). *Become an Uber driver UK: All you need to know about pay, tax and your driving license*. Retrieved November 3, 2016, from https://www.lovemoney.com/news/51136/uber-driver-uk-job-pay-requirements-license-tax
HM Government.(2015). *Independent review of the sharing economy: Government response*. London: Department of Business, Innovation and Skills. Retrieved November 3, 2016, from http://www.idrivewithUber.com/how-much-do-Uber-drivers-make.
Huet, E. (2015, September 9). Uber raises UberX commission to 25% in five more markets. *Forbes online*. Retrieved November 3, 2016, from http://www.forbes.com/sites/ellenhuet/2015/09/11/Uber-raises-Uberx-commission-to-25-percent-in-five-more-markets/#65f0401464b5.
Katz, V. (2015). Regulating the sharing economy. *Berkeley Technology Law Journal, 30*(385), 1067–1126.
Kesselring, A., Blasy, C., & Scoppetta, A. (2014). *Workplace innovation: Concepts and indicators*. Brussels: European Commission.
Knight, S. (2015, April 27). How Uber conquered London. *The Guardian*. Retrieved November 3, 2016, from https://www.theguardian.com/technology/2016/apr/27/how-Uber-conquered-london
Kuttner, R. (2013, September/October). The task rabbit economy. *Prospect*, 46–55.
Labour Research. (2016). *Are workers getting a fair share of this gig?* Labour research, August, 13–15. Retrieved November 3, 2016, from http://www.lrdpublications.org.uk/publications.php?pub=LR&iss=1835&id=idp8202736
Makó, C., Illéssy, M., & Warhurst, C. (2015). The evolution of EU innovation policy relevant to job quality and employment. *QuInnE Working Paper*. Retrieved November 3, 2016, from http://bryder.nu/quinne1/sites/default/files/Working%20paper%202.pdf
Mokyr, J., Vickers, C., & Ziebarth, N. L. (2015). The history of technological anxiety and the future of economic growth: is this time different? *Journal of Economic Perspectives, 29*(3), 31–50.
Muñoz de Bustillo, R., Fernández-Macías, E., Antón, J., & Esteve, F. (2011). *Measuring more than money: The social economics of job quality*. Cheltenham UK: Edward Elgar.
Naughton, J. (2016, November 22). Uber is certainly slick, but it's not the future, The new review. *The Observer*, 21. Retrieved November 3, 2016, from https://www.theguardian.com/commentisfree/2015/nov/22/uber-slick-but-not-disruptive.
O'Connor, J. (2015, November 29). The human cloud. The new review. *The Observer*, 17.
Osborne, H. (2016, November 28). Uber loses right to classify UK drivers as self-employed. *The Guardian*. Retrieved November 3, 2016 from https://www.theguardian.com/technology/2016/oct/28/uber-uk-tribunal-self-employed-status
Organisation for Economic Co-operation and Development. (OECD). (2005). *Guidelines for collecting and interpreting innovation*. Paris: OECD.
Organisation for Economic Co-operation and Development (OECD). (2010). *Innovative workplaces: Making better use of skills within organisations*. Paris: OECD.
Osborne, M. A. (2015). What impact does the digital impact have on work and inequality? *Social Europe Journal, 9*(1), 23–28.
Pot, F., Totterdill, P., & Dhondt, S. (2016). Workplace innovation: European policy and theoretical foundation. *World Review of Entrepreneurship, Management and Sustainable Development, 12*(1), 13–32.
Ramstad, E. (2009). Expanding innovation system and policy—An organisational perspective. *Policy Studies, 30*(5), 533–553.
Rogers, B. (2015). The social costs of Uber. *The University of Chicago Law Review Dialogue, 82* (85), 85–102.

Sanidas, E. (2005). *Organisational innovation and economic growth*. Cheltenham: Edward Elgar.
Schumpeter, J. A. (1939). *Business cycles*. New York: McGraw-Hill.
Singer, N. (2014, August 16). In the sharing economy, workers find both freedom and uncertainty. *New York Times*. Retrieved November 3, 2016 from http://www.nytimes.com/2014/08/17/technology/in-the-sharing-economy-workers-find-both-freedom-and-uncertainty.html?_r=0.
Smith, J. W. (2016). The uber-all economy of the future. *Independent Review, 20*(3), 383–390.
Spence, P. (2016, July 12). Robots will replace a quarter of business services workers by 2035, says Deloitte. *The Telegraph*. Retrieved November 3, 2016 from http://www.telegraph.co.uk/business/2016/07/11/robots-will-replace-a-quarter-of-business-services-workers-by-20/
Stewart, H. (2015, November 5). Resistance is useless: Rise of the robots will transform the human workplace by 2035. *The Guardian*. Retrieved November 3, 2016 from https://www.theguardian.com/technology/2015/nov/05/robot-revolution-rise-machines-could-displace-third-of-uk-jobs
Streek, W. (2015). The rise of the European consolidation state. In *MPlfG Discussion Paper 15/1*. Cologne: Max Planck Institute for the Study of Societies.
Thierer, A. (2014). *Permissionless innovation: The continuing case for comprehensive technological freedom*. Arlington, Virginia: Mercatus Center at George Mason University.
Tomassetti, J. (2016, June 24–26). Does Uber redefine the firm? The post-industrial corporation and the guy with a car as entrepreneur. In *Paper to the 28th annual conference of the society for the advancement of socio-economics*. California: University of Berkeley.
Uber US website. (2016). *Driver Deactivation Policy*. Retrieved November 3, 2016 from https://www.uber.com/legal/deactivation-policy/us-en/
Warhurst, C. (1997). Political economy and the social organisation of economic activity: A synthesis of neo-institutional and labour process analyses. *Competition and Change, 2*(2), 213–246.
Wosskow, D. (2014). *Unlocking the sharing economy: An independent review*. London: Department of Business, Innovation and Skills.
Wright, S. (2015). Challenges in researching job quality. In A. Knox, & C. Warhurst (Eds.), *Job quality in Australia: Perspectives, problems and proposals* (pp. 15–36). Annandale, NSW: Federation Press.
Zalkind, S. (2016, April 21). Confessions of a female Uber driver: Women-only rideshare has many pluses. *The Guardian*. Retrieved November 3, 2016 from https://www.theguardian.com/technology/2016/apr/21/chariot-for-women-female-only-rideshare-Uber

Author Biographies

Chris Warhurst is a Professor and Director, Institute for Employment Research, University of Warwick. Chair of the Editorial Management Committee of Human Relations. Fellow of the Royal Society for the Arts. Current research: job quality; workplace innovation; graduate labour; vocational education and training; aesthetic labour

Chris Mathieu is a Director of Studies and organisational sociologist at the Department of Sociology, Lund University, and Project Leader for the Horizon 2020 project *QuInnE*: Quality of Jobs and Innovation Generated Employment Outcomes (quinne.eu). Current research focuses on job quality, innovation, assessing competence among surgeons, cultural production and photography.

Sally Wright is a Senior Research Fellow, Institute for Employment Research, University of Warwick. Current research: job quality; workplace innovation; decent work; restructuring and change; pay and working conditions.

Chapter 16
Towards the High Road of Workplace Innovation in Europe? An Illustration of the Usefulness of the Dataset of the European Working Conditions Survey

Agnès Parent-Thirion, Greet Vermeylen, Mathijn Wilkens, Isabella Biletta and Frank D. Pot

16.1 Introduction

The European Union is committed to the improvement of working conditions, the promotion of job quality and productivity at work, increasing its competitiveness and recovering from the crisis through the promotion of smart, inclusive and sustainable growth.

Work, alongside 'family' plays a fundamental role in our societies and is a key ingredient for their cohesion and capacity to integrate all citizens. Expectations that

---

The views in this paper are those of the authors and do not necessarily reflect the official position of Eurofound.

A. Parent-Thirion (✉) · G. Vermeylen · M. Wilkens · I. Biletta
Eurofound, Dublin, Ireland
e-mail: apt@eurofound.europa.eu

G. Vermeylen
e-mail: gve@eurofound.europa.eu

M. Wilkens
e-mail: mwi@eurofound.europa.eu

I. Biletta
e-mail: ibi@eurofound.europa.eu

F.D. Pot
Radboud University Nijmegen, Nijmegen, The Netherlands
e-mail: frank.pot@ardan.demon.nl

P.R.A. Oeij et al. (eds.), *Workplace Innovation*, Aligning Perspectives on Health, Safety and Well-Being, DOI 10.1007/978-3-319-56333-6_16

people have of their own work are high and diverse; they are looking for income and security, high quality interpersonal relationships and opportunities for personal development, fulfilment and self-expression at work (Méda and Vendramin 2017).

Europe avails of an increasingly highly skilled workforce. Companies and workers alike may benefit from being able to deploy and develop the skills of this workforce, by, for instance, providing them opportunities to build on their tacit knowledge. The work of a highly skilled workforce may be more difficult to monitor and control, therefore, forms of work organisation that include employee involvement may be best suited for this type of work. In addition, companies may want to ensure retention of this highly skilled workforce.

Investing in initial education is important but the development of skills over the life course is key in a context of an ageing and declining population in the EU. Understanding how companies contribute to the development of competences of their workforce matters.

Benefits of such an approach would even go beyond worker and company benefits alone as they would also support welfare state gains and the competitiveness of individual countries. Therefore, the purpose of this chapter is to focus on the relevance of job autonomy and organisational participation for job quality across Europe, as drivers of workplace innovation.

16.2 Methodology

16.2.1 The European Working Conditions Survey

Eurofound is implementing "the EWCS" which is a survey among workers and employees. Six editions of the survey have taken place so far, the last one in 2015. The 2015 edition (Eurofound 2016) covers 35 European countries: EU28, the five EU candidate countries (Albania, the former Yugoslav Republic of Macedonia, Montenegro, Serbia, and Turkey), Switzerland and Norway. The minimum sample size per country is 1000 responding employees (exceptions were made for Poland 1200, Italy 1400; France 1500; United Kingdom and Slovenia 1600; Germany and Turkey 2000; Belgium 2500 and Spain 3300).

The sample of the EWCS is representative for the working population in each country, of those persons aged 15 and over (16 and over in Bulgaria, Norway, Spain and the United Kingdom), living in private households and being in active employment (i.e., those who performed at least one hour of work for pay or profit during the week preceding the interview). In each country, a multistage stratified random sampling design was used. Individual sampling frames have been used in 5 countries (Denmark, Finland, Norway, Portugal and Switzerland) and address registers in 11 countries (Belgium, Bulgaria, Croatia, Greece, Ireland, Lithuania, Luxembourg, Montenegro, The Netherlands, Turkey, and United Kingdom). For the other countries where no high-quality registers could be accessed, addresses were enumerated following a random procedure that was separated from the interviewing stage. The survey interviews were carried out face to face using computer assisted personal interviewing (CAPI). The average duration of interviews was 45 min. The overall response rate is 42.5%. The response rate has decreased by on average 1.7% for the group of countries that were included in the last edition of the survey but increased in 23 of the 33 comparable countries.

The EWCS collects information on 'real work' as experienced by the worker rather than prescribed and described by other workplace parties (such as their managers). Real work refers to the concrete achievement of the tasks at hand, the work by a specific operator. Real work will depend on respondents' skills and characteristics, their interpretation of the tasks, their operating mode as well as how they adapt to unforeseen circumstances and the variants that will present themselves in tasks. Indeed, the performance of work as an activity depends on the components of the situation of work (the working environment and job design) as well as on the individual worker.

EWCS builds on workers' answers during face to face interviews in their home regarding their experience of and behaviour in the workplace. Because the information is collected face to face in people's homes, this means that there is no burden placed on the companies which employ them, and also allows for a more trusting interview. No proxy interview is allowed as a household member would not be in a position to provide valid answers on the working conditions of another household member.

Since 1991, when the first EWCS took place, the questionnaire has expanded and been revised in line with users' feedback, expert advice and Eurofound's stakeholders' guidance. The interview lasts forty-five minutes on average and covers job and worker characteristics, employment conditions, working time duration, scheduling and organisation, exposure to physical risks, the intensity of work and pace determinants, skill use and autonomy, worker participation and employment representation, the social environment at work, remuneration and job (in)security, work life balance, engagement and motivation, health and well-being.

16.2.2 Job Quality Indices

The 6th EWCS overview building on Trends in Job quality (Eurofound 2012b) has constructed seven job quality indices: physical environment, skills and discretion, social environment, work intensity, working time quality, earnings, and prospects. Information gathered from the EWCS respondents who report on their own circumstances, such as their work life balance, their own employability as well as their company's workplace practices such as social climate, the practices of team work and task rotation, is not included in the seven indices. A full description of the indices is included in Eurofound (2016).

The job quality indices share a number of features:

- They are developed at the level of the job and gather information pertinent to that level only.
- The indices gather together those characteristics of work and employment which a number of theories (job demand-control, effort reward imbalance, job demand-resource model) have identified as key to meeting people's needs at work and for which high quality epidemiological studies empirically support the existence of a causal relationship with health and well-being.
- The indices include both positive and negative job features. For example, the social environment includes workers' experience of adverse social behaviour incidents (negative features) together with social support and management quality such as 'good leadership' (positive features).
- Except for earnings, which is the least elaborated index, the indices are built on a number of indicators and try to capture the phenomenon under investigation with the support of a number of individual questions.

The seven job quality indices are not combined together into one single index but rather are presented separately to allow for a more detailed description of job quality. This approach is not only transparent but also highlights the multidimensionality of job quality.

16.2.3 Types of Work Organisation

In this chapter, we focus on a specific part of the 'skills and discretion index', namely, decision latitude at the individual job level (job content) and participation in shop floor consultancy at the department level. We do so, because these are core elements of workplace innovation (Eurofound 2012a, pp. 62–64). An analysis carried out in 2013 by Gallie and Zhou (Eurofound 2013) of the 2010 5th European Working Conditions Survey data has confirmed that employees working in high involvement work systems can support the emergence of both company performance as well as employee well-being. Dhondt et al. (2014) used the data of the 5th EWCS as well and found that the interaction of job autonomy and organisational level decision latitude (organisational participation) provided higher levels of subjective well-being.

Only Employees of the EU28 are included in this analysis. The EU28 analysis is based on 28,551 interviews that took place between February 2015 and September 2015.

In our analyses, we constructed four types of work organisation which build on workers' experience of discretion at work (also called job autonomy or decision latitude) and organisational participation, assessed as their concrete experience of involvement related to work organisation practices.

- High involvement organisation (high discretion at work and high organisational participation)
- Discretionary organisation (high discretion at work but low organisational participation)
- Consultative organisation (high organisational participation but low discretion at work)
- Low involvement organisation (low on both dimensions)

Discretion at work and organisational participation are components of the skills and discretion index.

Descriptive results are reported. Figures have been rounded up. Separate regression models **controlling for country, economic sector of activity, company size, gender and age have been carried** out (but not reported) to assess whether differences in dependant variables are significant. Because the survey is cross sectional all our conclusions speak to levels of association between variables and not to causal relationships.

16.3 Analysis

16.3.1 Employee Involvement Related to Types of Work Organisation

Organisational participation refers to the departmental or organisational level. The majority of employees in the EU28 reports having some influence in how their work is organised: 66% of employees are at least 'sometimes' involved in improving the work organisation or work processes of their department or organisation. Similarly, 69% of employees feel that they can influence decisions that are important for their work. These questions are combined into an index 'organisational participation' (Cronbach alpha 0.78).

Discretion at work refers to job level and the ability to change the order of tasks (65% of employees), to select methods of work (66% of employees) and to change the speed or rate of work (68% of employees). They are combined into the index task discretion (Cronbach alpha 0.72).

Organisational participation and discretion at work were considered in combination to determine the extent to which employees are involved in their organisations. Each index was dichotomized into high and low. High discretion at work was considered to be present when respondents reported discretion in the methods of work, the order of tasks and in changing the speed of work, whereas low discretion at work was considered to be present when they reported discretion at work in only two of the three dimensions of autonomy at work. High organisational participation is a score above the median, whereas low organisational participation is a score below the median.

Approximately a third of the respondents work in either low involvement organisations (34%) or high-involvement organisations (31%); 18% of employees work in discretionary organisations whereas 17% of employees work in consultative organisations.

16.3.2 *Strong Structural Differences in Forms of Employee Involvement*

Empirical analysis of EWCS 2015 data shows that the types work organisation vary significantly across countries, economic sectors and occupational categories, whereas the size of the company, age and gender do not seem to matter as much. Although there are substantial differences across occupations (Fig. 16.1), sectors (Fig. 16.2) and countries (Fig. 16.3), each type of work organisation is

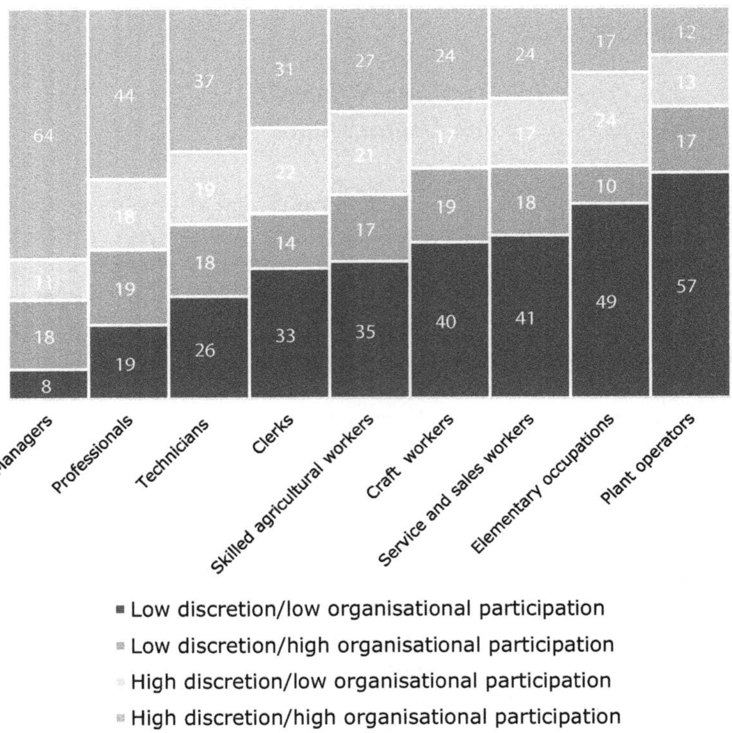

Fig. 16.1 Four employee involvement types of work organisation by occupation (main group of ISCO 1—names shortened), employees only, EU28 in %. *Source* Eurofound EWCS data, 2015, own calculation

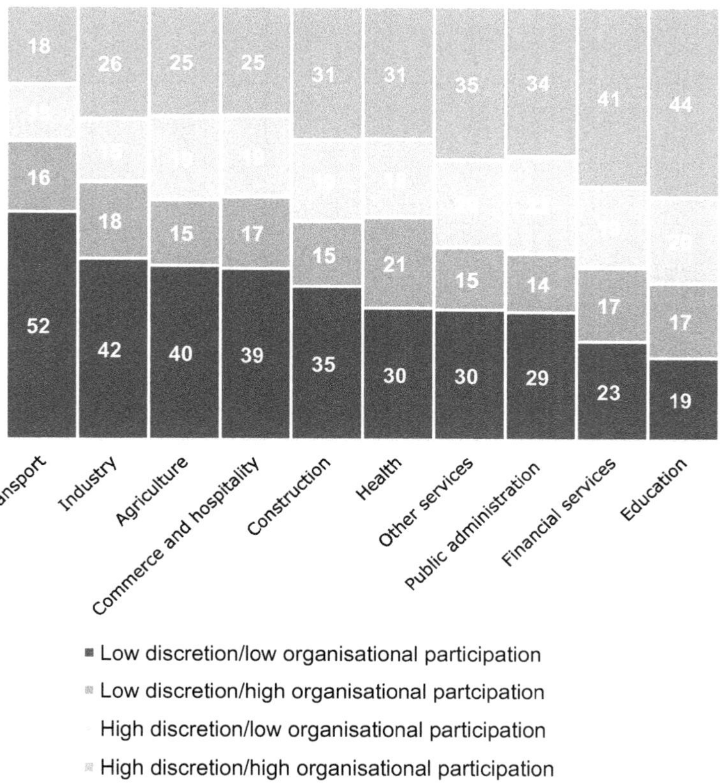

Fig. 16.2 Four employee involvement types of work organisation by economic activity (main groups, nace rev. 2), employees, EU28 (%). *Source* Eurofound EWCS 2015 data, own calculation

present in every sector and occupation, confirming that it is possible to organise work in different ways.

There are important differences in the distribution of various occupational groups across types of work organisation: 64% of managers and 44% of professionals work in high involvement organisations; 49% of elementary occupation and 41% of service workers work in low involvement organisation types. Occupational differences regarding the consultative and discretionary type of organisation are smaller. Each type of work organisation, however, can be applied to all occupations.

As Fig. 16.2 shows, low-involvement organisations are more prevalent in transport (52%), industry (42%), agriculture (40%) as well as commerce and hospitality (39%). High-involvement organisations are more common in service-oriented sectors such as education (44%) and financial services (41%).

There are important country differences regarding these four organisational types (Fig. 16.3):

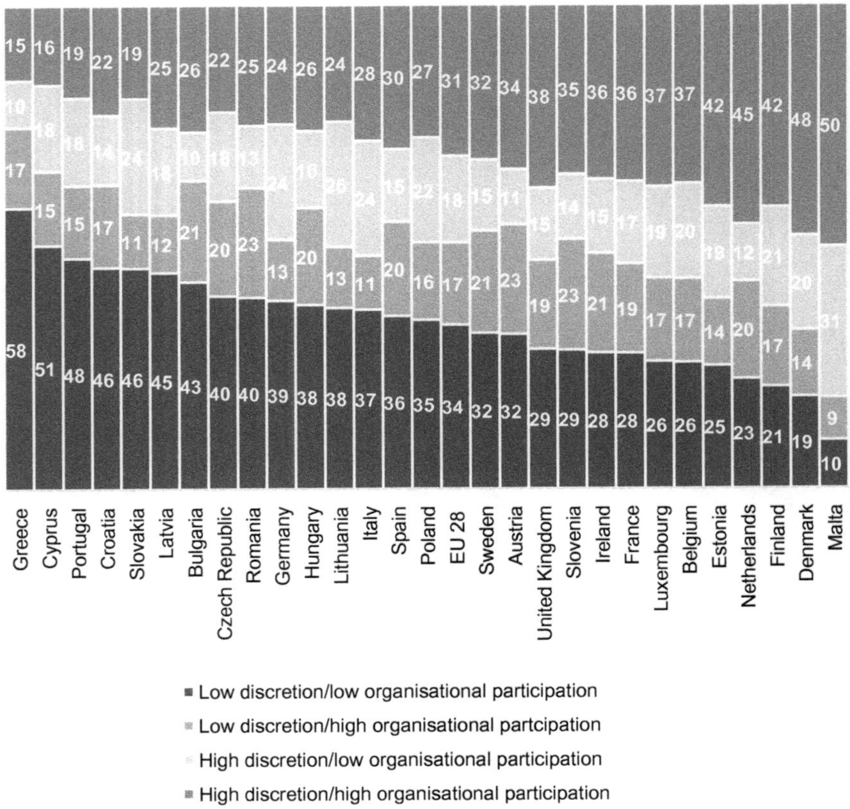

Fig. 16.3 Four employee involvement types of work organisation by country, employees, EU28 (%). *Source* Eurofound EWCS data 2015, own calculation

- Country differences regarding the incidence of low involvement organisations amongst European countries are very high. The majority of employees in Greece (58%) and Cyprus (51%) work in low involvement organisations; these organisational types are also quite frequent in Portugal (48%), Slovakia, Croatia (both 46%), and Latvia (45%), whereas only 10% of Maltese employees report working in organisations of this type.
- High involvement organisation types are more frequent in Malta (50%), Denmark (48), the Netherlands (45), Finland and Estonia (both 42%), United Kingdom (38%) as well as Belgium and Luxembourg (both 37%).
- 17% of employees in Europe work in consultative types of organisations. A higher proportion of employees in Romania, Slovenia and Austria (each three 23%) work in this organisational type.
- The discretionary organisation type is more frequent in Malta (31%), Lithuania, Slovakia and Germany (all three, 24%).

There are very few differences when company or establishment size are taken into consideration.

The proportion of men and women working in four types of work organisations is relatively equal. When taking age into account, older workers (over 50 years of age) report more frequently than younger workers to be working in high involvement organisations (34% of 50 years plus versus 31% of middle aged workers and 27% of workers under 35 years of age). Conversely, 38% of the workers under 35 years of age report working in a low involvement organisation, as do 34% of middle aged workers and 31% of older workers.

16.3.3 Forms of Employee Involvement Work Organisation and Job Quality

In this section we describe how the four types of employee involvement and work organisation fare in relation to the job quality indices developed by Eurofound (2016).

All job quality indices are measured on a scale from 0 to 100, except for monthly earnings which is measured in euros. The other 6 indices are compiled from answers to 10 or more questions and focus on the characteristics of work and employment which have been shown in a number of epidemiological studies to have a positive or negative impact on health and well-being. Each index is subject-matter specific. As a consequence, each index includes different combinations of positive and negative features. A higher value except for work intensity, means a higher quality.

As discretion at work and organisational participation are elements of the skills and discretion index, we can only analyse the remaining 6 indices in relation to types of work organisation (Fig. 16.4).

Higher scores imply a better job quality, except for work intensity.

The *physical environment* job quality index measures exposure to physical risks in the workplace. The index comprises exposure to vibrations, noise, high and low temperatures; breathing in smoke, vapours, or tobacco smoke; handling or being in contact with chemical products, infectious materials, as well as tiring or painful positions, lifting or moving people, carrying or moving heavy loads, and repetitive hand or arm movements.

Employees working in high discretion/high organisational participation types of work organisations as well as those working in a type of work organisation with high task discretion and low organisational participation report a slightly higher (better) level of physical environment job quality. The pattern is statistically significant when controlling for country, economic sector, occupation, company size, age and gender. Interestingly, after controlling for these factors, the effect is only evident for male employees.

Job quality index	Low discretion /low organisational participation 'low involvement'	Low discretion / high organisational participation 'consultative'	High discretion / low organisational participation 'discretionary'	High discretion / high organisational participation 'high involvement'
Physical environment	81	83	85	86
Working time quality	70	70	73	72
Earnings (in euros)	1,207	1,503	1,377	1,769
Prospects	60	68	62	70
Work intensity[a]	36	36	31	34
Social environment	71	81	73	82

Fig. 16.4 Six job quality indices related to four employee involvement types of work organisation, employees, EU28 (%). *Source* Eurofound EWCS data 2015, own calculation ([a]For all job quality indices except work intensity, a higher value means a higher job quality)

Working time quality comprises four dimensions: duration, discretion, atypical working time and working time arrangements organised around discretion and predictability, as well as flexibility. Moreover, it contains positive and negative elements. There are very few but significant differences in working time quality amongst employees in different forms of work organisation which is higher (better) for employees in high discretion and low organisational participation (73) and in high discretion and high organisational participation forms of work organisation (72). These effects remain significant when all factors mentioned above are controlled for.

The item non-response for work related *earnings* was quite high (20% of the full dataset for all 35 European countries). Further analyses suggest that non-response was systematically higher for certain groups of respondents. Based on the answers provided by respondents we believe that, as a result of this, the level of earnings in the EWCS is underestimated. Work related earnings are highest in high discretion and high organisational participation organisations. The differences are significant and remain so when all other factors are controlled for.

The job quality index *prospects* measures the continuity of employment, as measured through a person's employment type of contract, job security and career prospects as well as recent downsizing in the company. Prospects are higher (better) in high discretion and high organisational participation type of organisations (70) as well as low discretion and high organisational participation (68) types of work organisation. These effects are high and significant when the other factors are controlled for.

The *work intensity* index measures the level of exposure to quantitative and emotional demands as well as pace of work and employee interdependency. Contrary to the other job quality indices, a higher level of work intensity denotes a lower level of job quality. Work intensity is lowest in high discretion and low organisational participation (31) and high discretion and high organisational participation (34) types of organisations. These effects hold when all other factors are controlled for.

Finally, the *social environment* index measures management quality, social support from managers and colleagues (all three positive) as well as the experience of adverse social behaviour (negative). The highest levels of social environment job quality are experienced by workers in high discretion and high organisational participation (82) and employees in low discretion and high organisational participation (81) types of organisations.

Overall, employees working in types of work organisation characterised by high discretion at work and high organisational participation report the highest levels of job quality in terms of physical environment, earnings, social environment and prospects. They also score second best in working time quality and work intensity. This is consistent with previous empirical research and the theoretical foundations of workplace innovation (as described in Chaps. 6–14 in Part III Research).

Employees with a high level of discretion at work and low levels of organisational participation experience the highest levels of working time quality and work intensity and score second in their experience of the physical environment. Interestingly, they score second to last on measures of social environment, earnings and prospects.

Employees in low discretion and high organisational participation types of organisations report second best levels in earnings, prospects and social environment, are third in physical environment and among the worst two regarding working time quality and work intensity.

Employees working in low discretion and low organisational participation report the lowest job quality on all dimensions.

16.3.4 Types of Work Organisation and Other Human Resource Practices

In this section we describe how other human resources and work organisation practices support employee involvement: role clarity, training and consultation in times of organisational restructuring. Multivariate analysis has confirmed the correlations reported below.

16.3.4.1 Role Clarity

Employees working in high involvement (97%) and consultative types of work organisation (94%) report higher role clarity than employees in the other two types of work organisations (91% in discretionary organisation and 90% in low involvement organisations).

16.3.4.2 Training and Learning

Employees working in high involvement and consultative types of work organisation (low task but high organisational participation) report a higher incidence of training in the last 12 months (51 and 50% respectively) than employees working in discretionary types of work organisation (36%) and low involvement organisations (32%). A similar picture emerges on learning new things on the job. Furthermore, employees in high involvement and consultative types of work organisation who had participated in some training, more strongly agree with the statement that the 'training has helped them improve the way they work'.

16.3.4.3 Management of Change

Employees in consultative and discretionary work organisations report a slightly higher incidence of restructuring (26 and 30%) than employees in consultative (24%) and low involvement types of work organisation (21%). They also report more frequently that they had been informed of the forthcoming changes and were asked to give their opinion. This is more the case for employees in high involvement organisations (84% vs. 82% in consultative types of work organisations, 78% in discretionary types of work organisation and 70% in low involvement types of work organisation). Half of the employees (52%) in high involvement organisations and 39% of those in consultative organisations, who have experienced restructuring, also report that, prior to the reorganisation, they had been asked to give their opinion.

From this we may conclude that the high involvement and consultative types of organisation provide the best role clarity, training and learning opportunities and consultation in case of restructuring. This is consistent with expectations regarding the higher reported incidence of organisational participation.

16.3.5 Meaningful Work, Engagement and Well-Being

So far we have discussed more or less objective characteristics of the work situation. In this section we focus on subjective aspects. How do employees assess their work from a point of view of meaningfulness? How do the more or less objective

	Low discretion /low organisational participation 'low involvement'	Low discretion / high organisational participation 'consultative'	High discretion / low organisational participation 'discretionary'	High discretion / high organisational participation 'high involvement'
Your job gives you the feeling of work well done - always (always and most of the time)	28 (70)	44 (86)	30 (78)	47(90)
You have the feeling of doing useful work - always (always and most of the time)	39 (74)	57 (89)	43 (80)	61(94)
The organisation I work motivates me to give my best job performance - strongly agree (strongly agree and tend to agree)	13 (49)	27 (70)	14 (53)	33 (75)
Satisfied with working conditions - very (very satisfied and satisfied)	16 (78)	27(91)	21(84)	37(93)
Engagement	63	68	66	71
Well-being	64	70	67	72
Proportion of workers at risk of mental health problems	8	5	6	3

Fig. 16.5 Meaningful work, motivation, satisfaction with working conditions, engagement and well-being related to four employee involvement types of work organisation, employees, EU28 (%). *Source* Eurofound EWCS data 2015, own calculation

work characteristics influence their subjective well-being? The answers are important for employee motivation and level of engagement.

Much research indicates that employee involvement through job autonomy and organisational participation can improve performance as it has a positive impact on employee engagement and motivation. Especially, in the context of an increasingly skilled workforce, meaningful work takes on a new significance.

Figure 16.5 shows Meaningful work, motivation, satisfaction with working conditions, engagement and well-being related to four employee involvement types of work organisation.

Meaningful work is measured by 2 items: (1) "Your job gives you the feeling of work well done", (2) "You have the feeling of doing useful work".

Although many questions are related to motivation there is one direct question connected to motivation: "The organisation I work for motivates me to give my best job performance".

Whether employees are satisfied with working conditions in general is measured by one question: "How satisfied are you with working conditions in your main paid job?"

Work engagement is a positive, fulfilling, work related state of mind that is characterized by vigour, dedication and absorption and it has been associated with a number of positive outcomes for both workers and companies. It is distinct but overlaps to a certain extent with work motivation (Schaufeli and Salanova 2007). Our engagement index is based on answers to the following questions: (1) "At work I feel full of energy", (2) "I am enthusiastic about my job", (3) "Time flies when I am working", (4) "I feel exhausted at the end of the day (R)", (5) "I doubt the importance of my work (R)". Based on the answers to these questions (the mean score of all dimensions), an engagement index ranging from 0 to 100 was constructed (Eurofound 2016, p. 103).

Subjective well-being was measured through the World Health Organization's well-being index—WHO-5. In a series of five questions, interviewees were asked to indicate if over the previous two weeks they: (1) "felt cheerful and in good spirits", (2) "felt calm and relaxed", (3) "felt active and vigorous", (4) "woke up feeling fresh and rested", and (5) "their daily life was filled with things that interest them". Through statistical operation a scale was constructed reaching from 0 to a potential maximum score of 100. A high score is associated with a good level of psychological well-being while a low score (below 28) indicates that the person is at risk of mental health problems, including depression. A recent systematic review of the literature on the WHO-5 index shows that the questionnaire assessing subjective psychological well-being has adequate validity as a screening tool for depression, as an outcome measure in clinical trials and can be applied successfully as a generic scale for well-being across a wide range of study fields (Topp et al. 2015).

As Fig. 16.5 shows, employees in high involvement organisations have the best scores on all issues: meaningful work, motivation, satisfaction with working conditions, engagement and subjective well-being. Employees in consultative work organisations are second best and employees in discretionary organisations third. Employees in low involvement organisations have the lowest scores on all issues. Results remain significant when controlling for country, economic sectors, company sizes, age and gender.

16.4 Conclusions and Discussion

Approximately a third of the employees in our sample work in either low involvement organisations (34%; low levels of both discretion at work and organisational participation) or high-involvement organisations (31%; high levels of both discretion at work and organisational participation). Moreover, 18% of the employees in our sample work in discretionary organisations where they experience a high level of discretion at work but less influence on organisational decisions whereas 17% of the employees work in consultative organisations (low levels of discretion at work but high levels of organisational participation).

There are important differences in the distribution of various occupational groups across the different types of work organisation: 64% of managers and 44%

of professionals work in high involvement organisations; 49% of elementary occupation and 41% of service workers work in low involvement organisation types. Occupational differences regarding the consultative and discretionary type of organisation are smaller. Each type of work organisation, however, can be applied to all occupations.

Low-involvement organisations are more prevalent in transport (53%), industry (42%), agriculture (40%) as well as commerce and hospitality (39%). High-involvement organisations are more common in service-oriented sectors such as education (44%) and financial services (41%).

Country differences on the incidence of low involvement organisations amongst European countries are very high. The majority of employees in Greece (58%) and Cyprus (51%) work in low involvement organisations. These organisational types are also quite frequent in Portugal (48%), Slovakia, Croatia (both 46%) and Latvia (45%) whereas only 10% of Maltese employees report working in organisations of this type.

The best job quality is reported in high involvement organisations followed by discretionary organisations. Employees in high involvement organisations score best on physical environment, prospects and social environment. Employees in discretionary organisations have the highest scores on working time quality and the lowest (best) on work intensity. Employees in low involvement organisations have the lowest scores in general (and highest for intensity in a tie with the consultative type of work organisation).

The high involvement and consultative types of organisation provide the best role clarity, training and learning opportunities and consultation in case of restructuring. This is consistent with expectations regarding the higher reported incidence of organisational participation.

Employees in high involvement organisations have the best scores on: meaningful work, motivation, satisfaction with working conditions, engagement and subjective well-being. Employees in consultative work organisations are second best and employees in discretionary organisations third. Employees in low involvement organisations have the lowest scores on all issues.

The picture is quite clear. Evidently, organisational participation and discretion at work both make a difference. Employees in high involvement organisations are far better off than employees in low involvement organisations. This is in line with the results of Gallie and Zhou based on the 5th EWCS data (Eurofound 2013). Employees in consultative organisations score better than discretionary organisations on a number of dimensions. The question emerges whether organisational participation (high in consultative organisations and low in discretionary organisations) is even more important than discretion at work (job autonomy or decision latitude on the task level). This question has also been raised by the results of the analysis of EWCS 2010 data by Dhondt, Pot and Kraan (2014), who used the concept 'organisational level decision latitude'. Interestingly the new Job Content Questionnaire (JCQ2) of Karasek et al. (see Foreword) has also added organisational level decision latitude to the task level autonomy. More research to clarify this question is more than welcome. Regarding implications for organisational design, human resource management and leadership style, our analysis shows again that the combination of job autonomy and participation in decision making at the organisational level results in high quality jobs and higher levels of engagement and well-being. These benefits reported by employees are also likely to lead to better organisational performance. Taking into account that the percentage of high involvement organisations is only 31% there is most certainly room for improvement.

The EWCS data provide important input for the discussion about workplace innovation in Europe. Job autonomy and organisational participation are crucial elements for workplace innovation. The wealth of data about employees across Europe can be used to answer research questions that explore elements of workplace innovation in relation to good quality of work and working life, and organisational characteristics. Apart from employee data, Eurofound also collects data about companies and organisations through its European Company Survey (ECS). These ECS data also represent a wealth of information about organisational level indices that can be connected to workplace innovation (as has been done already in the recent past, see Chaps. 8–10). Our contribution can be understood as an invitation to fully utilise these data sources for research related to workplace innovation.

Acknowledgements The authors would like to thank their Eurofound colleagues Jorge Cabrita (research manager) and Oscar Vargas Llave (research officer).

References

Dhondt, S., Pot, F. D., & Kraan, K. O. (2014). The importance of organizational level decision latitude for wellbeing and organizational commitment. *Team Performance Management: An International Journal, 20*(7/8), 307–327.

Eurofound. (2012a). *Overview report 5th european working conditions survey*. Luxembourg: Publications Office of the European Union.

Eurofound. (2012b). *Trends in Job quality*. Luxembourg: Publications Office of the European Union.

Eurofound. (2013). *Work organisation and employee involvement in Europe*. Luxembourg: Publications Office of the European Union.

Eurofound. (2016). *6th EWCS overview report*. Luxembourg: Publications Office of the European Union.

Méda, D., & Vendramin, P. (2017). *Reinventing work in Europe, dynamics of virtual work*. London/New York: Palgrave-Macmillan.

Schaufeli, W. B. & Salanova, M. (2007). Work engagement: An emerging psychological concept and its implications for organizations. In S. W. Gilliland, D. D. Steiner, & D. P. Skarlicki (Eds.), *Research in social issues in management* (Vol. 5, pp. 135–177). Managing social and ethical issues in organizations. Greenwich, CT: Information Age Publishers.

Topp, C. W., Ostergaard, S. D., Sondergaard, S., & Bech, P. (2015). The WHO-5 well-being index: A systematic review of the literature. *Psychotherapy and Psychosomatics, 84*(3), 167–176.

Author Biographies

Agnès Parent-Thirion is a Senior researcher at Eurofound, the European Foundation for the improvement of working and living conditions.

Greet Vermeylen is a Senior researcher at Eurofound, the European Foundation for the improvement of working and living conditions.

Mathijn Wilkens is a Research officer at Eurofound, the European Foundation for the improvement of working and living conditions.

Isabella Biletta is a Research manager at Eurofound, the European Foundation for the improvement of working and living conditions.

Frank D. Pot is an Emeritus professor of Social Innovation of Work and Employment, Radboud University Nijmegen (NL) and former director of TNO Work and Employment.

Part IV
Practice

Chapter 17
Towards a Total Workplace Innovation Concept Based on Sociotechnical Systems Design

Pierre van Amelsvoort and Geert Van Hootegem

17.1 Introduction

Given growing global competition and the predicted shortages in the labour market, organisations, nowadays, face the dual challenge of creating workplaces that are, on the one hand, more productive, flexible, and innovative, and on the other hand, healthy places to work. There seems to be a need for workplace innovation (WPI) to transform traditionally monolithic bureaucratic organisations into modern organisations that meet these challenges. A workplace innovation (WPI) is a developed and implemented practice or combination of practices that either structurally (through division of labour) or culturally (in terms of empowerment of staff) enable employees to participate in organisational change and renewal and, hence, improve the quality of working life and organisational performance (Oeij et al. 2015). WPI is based on flexible instead of bureaucratic ways of organising, and, therefore, WPI-practices could help to transform bureaucratic, inflexible, and, consequently, organisations with limited innovative capability. Bureaucratic organisations, however, are defined by and embedded in their structures, support systems, decision making systems, facilities and IT systems. Bureaucracies are, due to their focus on maximising the division of labour and central control of the work processes, designed for stable environments and mass production. Hence they are not well-suited to respond to the need to be agile in a dynamic environment with ever changing customer demands. Therefore, to realise new ways of organising through workplace innovation, an integrated approach to systemic change in the organisation is needed. We call this Total Workplace Innovation, which we define as a

P. van Amelsvoort (✉) · G. Van Hootegem
Catholic University Leuven, KU Leuven, Leuven, Belgium
e-mail: pierrevanamelsvoort@ziggo.nl

G. Van Hootegem
e-mail: geert.vanhootegem@kuleuven.be

P.R.A. Oeij et al. (eds.), *Workplace Innovation*, Aligning Perspectives on Health, Safety and Well-Being, DOI 10.1007/978-3-319-56333-6_17

renewal of the organisation of work with an integrated view on the division of labour, working relations and the supporting systems, with the dual aim of improving both performance and quality of working life. This definition is in line with the main starting points of STS-D theory (sociotechnical systems design), as will be shown later.

In this contribution, we suggest the concept of Total Workplace Innovation (TWIN) as a potential way to create common ground among researchers and practitioners interested in understanding and implementing workplace innovation (WPI). We hope to change the discussion around workplace innovation from a scattered focus on identifying one best approach to developing an integrated framework that takes the whole organisation into account. Indeed, we aim at developing a more comprehensive theory for stakeholders who are involved in design processes for workplace innovation, by starting from sociotechnical design (STS-D) and by exploring how we can broaden this perspective with other approaches (see Table 17.1), to also cover issues such as IT-design and HR-design. We need a combination of core systems and support systems to develop a complete alternative for the bureaucratic monolithic organisation.

Although STS-D theory and WPI (see Chap. 5) have much affinity, they are not the same and could mutually reinforce each other. In this chapter, we regard STS-D as a design approach that focuses mainly on the design of the core work process. If we add elements from other theories to this design, then we slowly but surely broaden the STS-D design perspective in a manner that advances the goals of WPI. Therefore, we selected a number of different theoretical approaches and practices that complement STS-D in addressing the design of various support systems. Notably, this selection is based on the fact that all of these approaches take the core

Table 17.1 Additional theoretical concepts to complete TWIN

Lean thinking	Provides guidance and tools for creating quick response (Just in time) logistics systems for the different order streams Provides guidance and tools for continuous improvement to increase job control aimed at coping with interference and waste
Total productive maintenance	Provides guidance and tools for autonomous maintenance in whole task groups (teams) Provides guidance and tools for improving the collaboration between operators and maintenance staff
Relational coordination theory	Provides guidance and tools for horizontal coordination in the control structure and job control
New world of work	Provides practices and tools for designing the infrastructure of facilities regarding job demand
HRM	Provides guidance and tools for recruitment, rewarding and developing employees for humane and productive organisations
Archipelago ICT	Provides guidance for designing IT systems based on variety and job control
Sociocracy	Provides guidance for democratic strategic decision making to complete the design of the control structure and to increase job control possibilities

work processes as a starting point but add an important ingredient in terms of support systems. Moreover, they are explicitly focused on improving employee engagement, which is a core feature of WPI approaches (see Chap. 5 in this volume). A focus on employee involvement remains rather implicit in the design steps of STS-D, which sees job quality as logically emerging from choices made in the design of the structure of production and the distribution of management tasks. Therefore, these additional theoretical perspectives could complement STS-D theory (Kuipers et al. 2010).

We will first discuss sociotechnical systems design (STS-D) theory and its control structure as a base for the design of the core work processes, that is, the primary process (this is equivalent to Modern STSD or M-STSD in Chap. 5). STS-D theory and practices emphasize the joint optimisation of the technical and social aspects of the organisation, with a simultaneous focus on achieving improved productivity and improved quality of working life. STS-D also strives for employee participation. STS-D has a long tradition in the design of the division of labour with the aim of creating innovative and humane organisations (Pot and Dhondt 2016). Indeed, although STS-D practices show variation across different regions of the world (Baxter and Sommerville 2011; Mohr and van Amelsvoort 2016; Mumford 2006; van Eijnatten1993), the applied basic theory regarding the division of labour remains the same, namely creating conditions for increasing the speed in the production flow, job control or self-organisation for employees, teams, units and communities of work. The overall aim is to help organisations become more agile with the help of engaged employees. However, STS-D theory only addresses the design of core work processes, whereas a systemic approach to redesigning the organisation for TWIN also requires a focus on the design of support systems. Examples of support systems in organisations are ICT-systems, HR, sales and marketing, purchasing, distribution and dispatch, and maintenance.

Because the starting point for the design of an organisation is its core work process, we start with STS-D. To aid our discussion on the design of the support systems we add a number of other theoretical concepts, such as: Lean Thinking (focuses on logistics and quality management); Total Productive Maintenance (stresses the integration of operations and maintenance); HRM theories (put a to focus on human resources policies); Relational Coordination theory (focuses on improving lateral communication); ICT theories (focus on the design of the information infrastructure); the concept of the New World of Work (underlines the notion of working independent of time and place and the need to create flexible facilities); Sociocracy (focuses on participative strategic decision-making and policy.

The chapter is structured as follows. First, we will describe STS-D theory, followed by a presentation of its design principles and the design sequence as it relates to TWIN. Second, we will discuss the complementary approaches mentioned above. Finally, we will end with some concluding thoughts.

17.2 STS-D Theory as the Base for TWIN

An organisation's core work process is the primary process of an organisation, such as, making goods or providing services. How these goods or services are produced, i.e., how the core work processes are organised, largely determines the extent to which the organisation's products or services create added value for customers. Hence, orchestrating an organisation's shift towards workplace innovation-related goals—performance and quality of work—typically requires a redesign of the core work process. In this respect, STS-D theory provides a valuable framework, given that core work processes are rooted in a dynamic systems-theoretical perspective of work and organisation (Kuipers et al. 2010; de Sitter 1994; de Sitter et al. 1997). The design of the core work processes determines the needed degree of (central) coordination and the possibilities for (shop floor) self-organising capabilities. A maximum division of labour creates the need for central coordination and hierarchical control whereas a minimum division of labour creates conditions for self-organisation and horizontal coordination (i.e., more job autonomy). Given that organisations are complex social systems, a systemic view as offered by STS-D is helpful in redesigning organisations when required by changing economic circumstances. Bureaucracies have difficulties in coping with economic changes, while flexible, flow-based organisations are better equipped to handle change and turbulence (Kuipers et al. 2010).

STS-D theory suggests that, as a result of the division of labour, the organisation is an interacting network of people executing tasks and roles, using (ICT-) technological instrumentation, tools and machines. These tasks and roles are thus allocated to individuals, teams, departments and business units. STS-D makes the distinction between production and management in the following manner:

(a) the structure of executing activities (the *production structure* of the core work processes—PS) and
(b) the structure of control activities to manage the core work processes (the *control structure*—CS).

In STS-D a role or task is the work that needs to be done, which is often related to the work of other people. All these roles and tasks together constitute the whole of the core work process. In other words, all these roles and tasks together complete the whole task of, for example, a team or an organisation. The notion of whole tasks implies, in theory, that there is no division of labour at all, such as for example, when a team is making a complete end-product from start to finish. This is, however, almost never the case, and therefore, roles can be seen as nodes interacting with other interdependent nodes to complete the core work process (see Fig. 17.1). A node is a point where several inputs and outputs from different interaction partners come together to do the work.

In STS-D, as an offspring of systems theory, inputs are transformed into outputs as in the input-throughput-output model. The core work processes function in a similar vein at every level, such as at the level of tasks, jobs, teams, departments

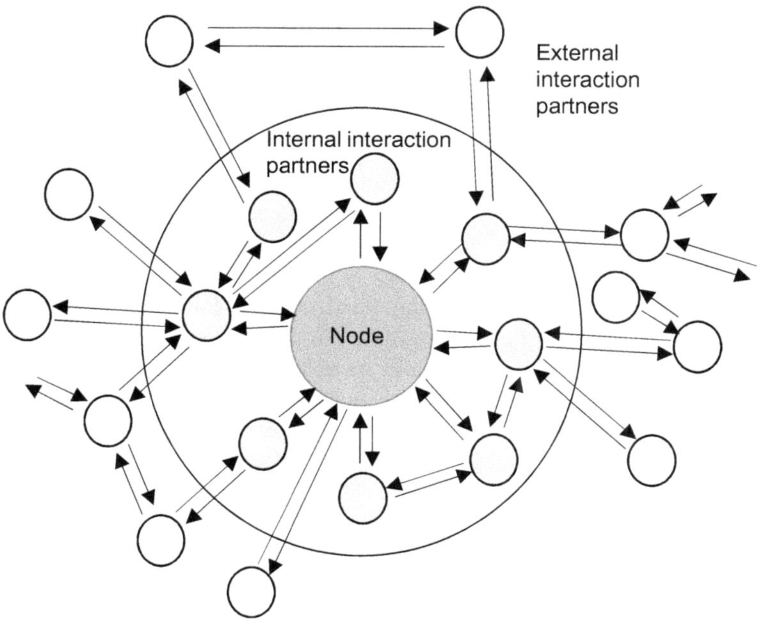

Fig. 17.1 The interaction network with nodes (Kuipers et al. 2010)

and the organisation as a whole. At the nodes, inputs are therefore transformed into outputs or outcomes, meaning that resources are transformed into products or services. Interaction between nodes, for example, the collaboration of individuals in a team, is necessary for a number of reasons, such as, the exchange of information, knowledge creation, planning and/or coordination, and deliberation. Team members are, for example, dependent on each other's task execution. At the nodes, interactions happen with both internal and external interaction partners. In order to ensure productivity either directly or indirectly, these various interactions between nodes need to be established at the right time, between the right jobs, with the right material or information and at the right place. Otherwise, production gets delayed or mistakes become a risk. Figure 17.1 illustrates this point.

However, these planned interactions between nodes can suffer from interference due to variance that is not accounted for in the original planning of the production in the core work processes. For instance, in the building and construction industry, different parties have to collaborate to get the job done as they are connected in specific supply-chain models. If one of the parties withholds information or drops out of the project unexpectedly, this will interfere with the other parties' capability to get the job done. In this sense, a node has to cope with two types of variance:

(a) *external variance*: such as lack of information, communication errors, changing customer demands, incomplete input, conflicting, ambiguous or competing demands;
(b) *internal variance*: human errors, technical disturbance, invalid and inflexible capabilities, shortage of resources.

The key question that arises is how can organisations deal with these types of variance at the nodes in ways that do not disrupt the production process? According to STS-D, to deal with such variance, organisations should on the one hand, redesign the division of labour in such a way that the complexity of the interaction network can be reduced, and on the other hand, increase job control possibilities so that variances can be controlled at the source. In this respect, de Sitter suggested to create simple organisations but make jobs complex, meaning that jobs become rich and varied (de Sitter et al. 1997). In other words, bureaucracies create jobs that are too simple for the complex changes in the environment. TWIN and STS-D create complex jobs so that organisations can deal with that complexity in flexible ways (Mohr and van Amelsvoort 2016).

17.2.1 The Relation Between the Division of Labour and Productivity

The productivity of an organisation is related to its capability to cope with strict external demands, namely, business and customer demands for variety (product mix), and uncertainty about both short- and long-term planning. Therefore, the capability to meet these external demands, is contingent upon the needed internal variety, namely meeting requirements in relation to efficiency, quality, flexibility, and innovation. Only if organisations can internally vary how they operate, are they able to meet the external requisite variety (Ashby 1969).

Bureaucratic organisations are based on the principle of maximum division of labour, which, in turn, leads to complexity and rigidity (Achterbergh and Vriens 2009). This maximum division of labour can be counterproductive for a number of reasons. First, bureaucratic organisations (see Fig. 17.2) tend to be characterized by (1) simple jobs, i.e., the formation of silos between functional departments, each pursuing fragmented goals and interests, and (2) complex interactions, i.e., long hierarchical communication lines, central decision-making, and a large number of rules and meetings. Bureaucracies have many nodes, and are therefore exposed to the risk of much interference in the core work processes when the work cannot be performed as initially planned. Figure 17.2 indicates that the performance of the core work processes requires several dependencies in terms of control (c) and execution.

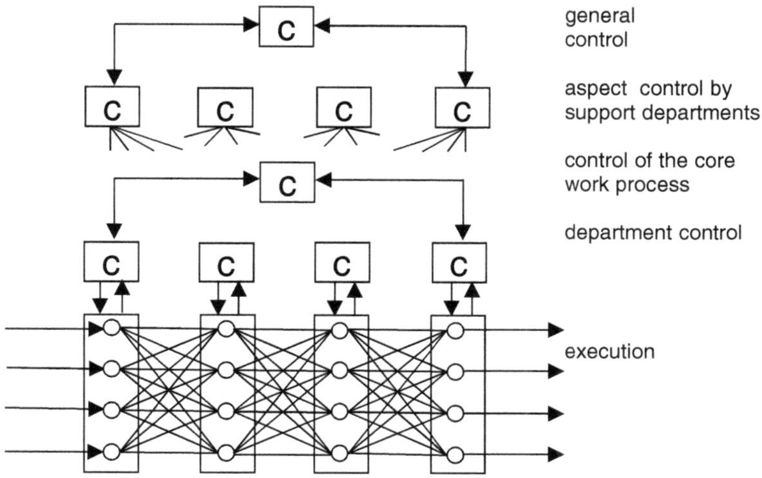

Fig. 17.2 The bureaucratic regime (Kuipers et al. 2010)

Hence, if external pressures on the organisation that threaten the planned process increase, the bureaucratic organisational design will rapidly lead to productivity problems. These problems can be manifested for instance as Kuipers and van Amelsvoort (1990), de Sitter (1994):

- unreliable and long lead times due to poorly harmonized processes;
- slow response times;
- difficulty in quality assurance due to insufficiently managed processes and poor communication;
- poor cost control because actual (hidden) costs cannot be monitored and (too) much interference occurs;
- slow and blind decision-making;
- expensive coordination and control mechanisms;
- lack of employee involvement;
- lack of innovative capability due to poor communication between the business functions, and a lack of initiative.

In general, the traditional, bureaucratic response to these problems is to tighten control (centralisation) and implement more stringent rules and procedures. These measures are counterproductive, because the root cause of these dysfunctions is, in fact, deepened. In contrast, STS-D aims to reduce complexity by minimising the division of labour (see the section on STS-D principles below).

17.2.2 The Relation Between the Division of Labour and Employee Involvement

The division of labour does not only affect productivity but also the quality of working life. For instance, Karasek's *Job Demand-Control* model (Karasek 1979; Karasek and Theorell 1990) (see Fig. 17.3)[1] suggests that work organisation, specifically, high control (autonomy) in performing tasks is crucial in transforming job demands from risks and stress drivers into learning opportunities.

In this model, job demands are seen as stressors such as work overload, unpredictable demands, time pressure, role ambiguity, interference, and emotional and physical demands. Job control is the combination of autonomy, decision latitude, instrumental support from colleagues, constructive performance feedback, craftsmanship, flexible resources, leaders' appreciation and support, accurate information, and communication. In this respect, there is evidence that high job demand and low job control are important predictors of psychological stress and illness. In addition, de Sitter (1994) claims that job control leads to involvement and motivation, which translates into positive effects on indicators such as absenteeism, turnover and stress. Moreover, there is evidence that a combination of high job demand and high job control in the form of active work is a predictor of an innovative organisation (de Sitter 1994).

In sum, job control is an important predictor for employee involvement and, as such, a precursor to workplace innovation. Indeed, STS-D proposes that, by increasing job control, employees are stimulated to learn, better equipped to deal with interference and, thereby, better prepared to respond to challenges arising from job demands. This increased level of job control does not only affect employee involvement but also serves the organisation by affording the possibility to better mobilise the use and development of human talent (de Sitter 1994), and thereby enable the goals of workplace innovation.

17.3 STS-D Design Principles

As previously stated, we take the STS-D perspective as a base for designing the structure of an organisation for TWIN (see Fig. 17.4). To reduce the shortcoming of bureaucracies, de Sitter et al. (1986), de Sitter (1994) developed a three-step design

[1]The Job Demand-Job Control model has affinity with the Job Demands-Resources model (Bakker and Demerouti 2007; Demerouti et al. 2001). However, we see resources as an element of job control, namely in the way that job design should include the possibility to deploy one's resources. For example, the degree of decision latitude determines whether persons can apply their knowledge and talents to solve problems. de Sitter (1994) has pointed out that employees do not get stress from problems in their work, but from the limited autonomy to solve these problems within their designed jobs.

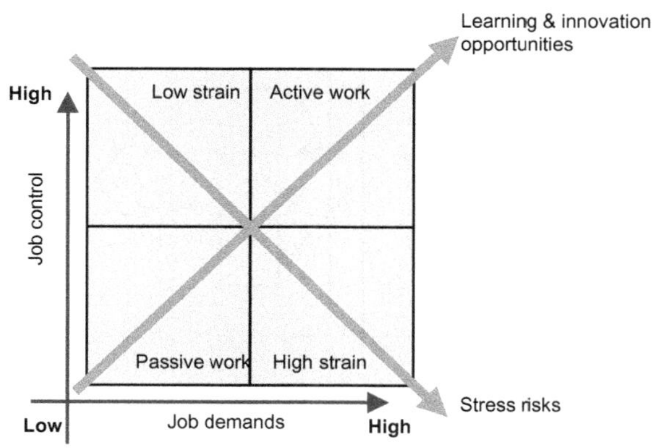

Fig. 17.3 The Job demand job control model of Karasek (1979), Karasek and Theorell (1990)

sequence for reorganising the core work processes. First, one designs the production structure, second the control structure, and third the information structure.

Within STS-D, there is a specific design sequence for the design of organisations (de Sitter et al. 1986).

1. *The design of the production structure, or how an organisation produces its goods or services.* If we assume that strategic positioning, such as the need for flexibility, innovation and healthy work, has been carried out, one needs to first design the core work process. This is done by focusing on the overall picture and then on the details (i.e., first on the whole, then on the parts). Based on the different customer families (see principle 1 below), this means that one starts with creating the different (business) units, then the different departments within these units, and finally, ends with the design of the work teams and jobs.
2. *The design of the control structure, or how the core work process and supporting processes are managed.* The second step is a redistribution of control capabilities through the design of the management structure. This control structure is designed in reverse order, in other words, from the parts to the whole (i.e., bottom-up and not top-down). That is, first one determines what can be controlled at the (lowest organisational) local level (i.e., team and job level), subsequently what can be organized at the level of a larger organisational operating unit (above that level), and finally what needs to be controlled at the (highest) organisational level. Next, the consultation and decision-making structure can be further elaborated in detail. The principle here is that emerging problems require autonomy to solve them at the level where those problems occur. This implies that the task of managing the core work processes should as much as possible migrate to the lowest organisational level.
3. *The design of the information structure (and other support systems), or how information streams support production and management.* Thirdly, the various (technical and support) systems are embedded in the new organisational architecture (see next paragraph). These systems include IT and support

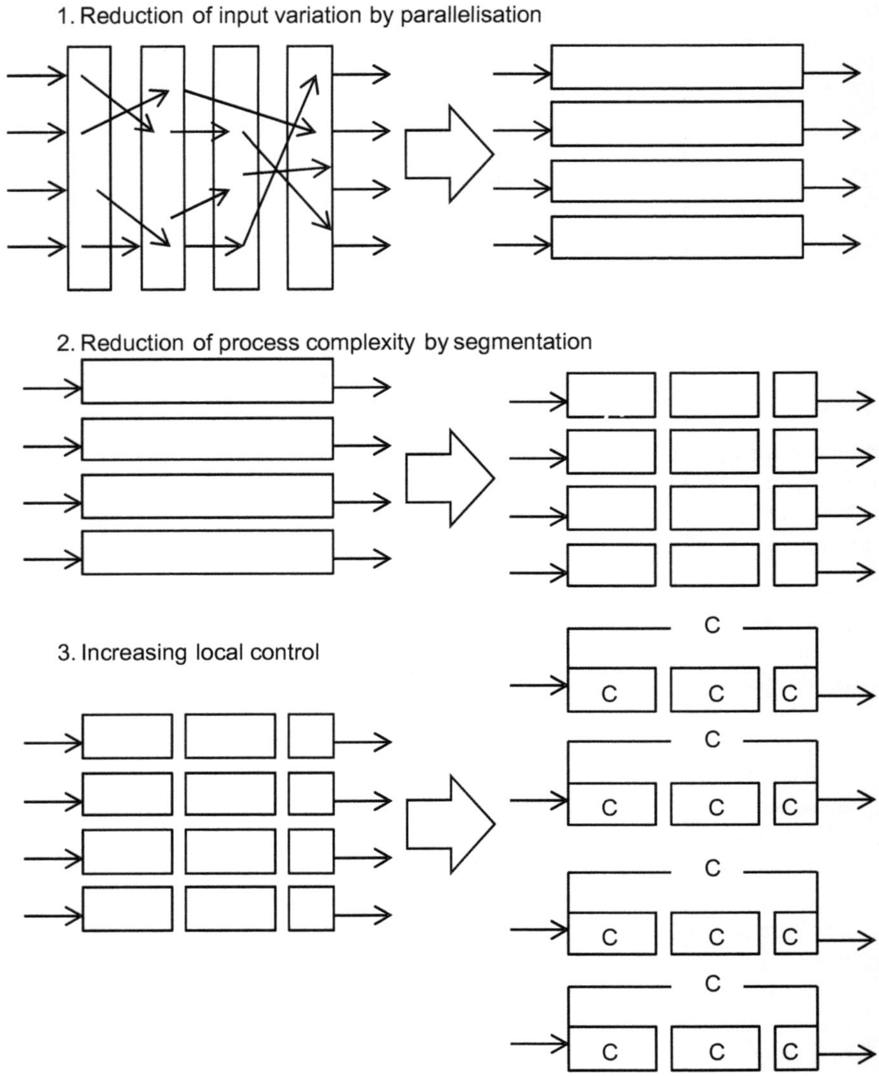

Fig. 17.4 The design sequence of STS-D (Kuipers et al. 2010)

systems. Here the rule is that these systems should support and not control the production and control structure, hence, this information structure should not be designed too soon, and never before step 1 and 2.

We now turn from the sequence of steps to the design rules. Here, again we touch upon the design of the production structure, control structure and information structure, but now in more detail, as the design goes from a crude design to a fine-grained design. This consists of four steps, namely parallelisation, segmentation, local control, and support systems (Fig. 17.4).

The STS-D approach avoids the shortcoming of bureaucracies because it results in a far more flexible design that enables a proper response to change and turbulence. We discuss this design approach in the following from both a strategic and an operational point of view.

17.4 Design as a Strategic Issue

Now that we have explained the general design sequence of STS-D, we address its strategic relevance first. In the next section, we discuss how these strategic choices can result in an operationally robust design. Robust means that interferences in the core work process are minimised. According to the open-system principle, the design of organisations needs to be strategic and should include *all* stakeholder perspectives. This is in stark contrast to the focus on shareholder value alone often witnessed in traditional organisations (Achterbergh and Vriens 2009). From an STS-D perspective, in line with the open-system principle, diagnosing, designing and changing organisations needs to be done by taking into account environmental conditions and strategic business choices. These strategic choices, in turn, impose requirements on the organisation, the "burning platform", and dictate the desired direction (see also Adler and Docherty 1998). Moreover, it is highly recommended that the design is drafted in co-creation with the different stakeholders. Indeed, the best guarantee for success is to *fetch the whole system into the room* (Weisbord 2004). This points to the importance of employee involvement, a hallmark of workplace innovation.

17.5 Robust Organisation Design

Apart from strategic choices, we need robust organisations which can cope with the demands of flexibility and innovation in a dynamic world. Hence, from the STS-D perspective, robust organisation design is based on the following three principles (see Fig. 17.4; van Amelsvoort 2000):

1. *Reduce complexity in the division of labour in the core work processes (PS) by focusing on customer order families*. Reducing complexity can be achieved by the introduction of parallel processing (i.e., factory in a factory). Parallel processes (a) afford a better business focus, and (b) create the conditions for decentralized control (see also principle 2). Parallelisation is defined as creating parallel streams of orders based on different customer families (e.g., markets, type of product). According to this principle, the design of the core work processes is based on the *type* of customers and their orders. This implies

identifying customer families (orders) that show homogeneity in terms of business demands, and, therefore, impose identical constraints on the manner in which the production process must be carried out. Identifying these customer families involves finding criteria to divide customers into relatively homogeneous subsets with different strategic demands. For example, a construction company builds tangible products. However, renovating a house or building a hospital represents completely different core work processes with different strategic demands. Hence, a miniature organisation can be formed around these subsets of customer orders (i.e., one for house renovations and another one for commercial buildings) that each complete the process from a to z for this group of customer orders. In other words, the whole task is performed by a relatively self-organising group (i.e., autonomous work teams). We refer to the process as *parallelisation* (Fig. 17.4). In other words, parallel order streams are created, with each being maximally interdependent within the stream, but minimally dependent across streams. This implies the design of whole tasks and the creation of self-organising groups, units and communities of work which are smaller in scale. *Segmentation* (Fig. 17.4) of the core work processes can help to reduce process complexity and create teams of 8–10 people. Segmentation is defined as cutting the flows of orders into parts, in such a way that a whole task of activities with high interdependency is created (i.e., de Sitter's complex jobs at team level).

2. *Increase the local (job and team) control capability by decentralization: self-organization and a healthy control structure* (Fig. 17.4). In an effective hierarchy designed to deal with turbulence, the different levels of control (i.e. layers of the organisation) have added value in terms of operational and strategic control. That is, flexible and innovative organisations are structured in such a way that they can react fast both at an operational and at a strategic level. To achieve operational control, work teams are self-organised at the operational level. Operational control is the combination of internal control (job autonomy, i.e., decision-making authority, technological variation possibilities, flexible access to means) and external control (coordination, team members' support, recognition, feedback, and influence). According to Ashby's law of requisite variety, control capability at a node (in this case, the self-organised team) is necessary in order to resolve interference at the place where it occurs and to prevent or reduce quality problems, delivery time deviations, or productivity losses (Ashby 1969). To achieve strategic control, different (business) units are set in place. Strategic control is necessary to reduce frequent interference among self-organising units and to explore innovations. Moreover, in dynamic situations, both operational and strategic control imply learning. The preconditions for control and learning are: participation in goal setting and purpose definition as well as effective feedback mechanisms for inspiration and learning, as in the JDJC-model (but now on the level of a team for example).
3. *Congruent infrastructure (technology and facilities) and HR systems: minimum critical specification* (Cherns 1987). Because the units in the organisation have different business demands they will also have different support demands

(Fig. 17.4). A supporting HR system, for example, should differ between teams of technically-skilled employees operating on the shop floor and administrative teams skilled in financial issues working in the office. Therefore, the design of the different support systems and technology should follow the first two principles mentioned above. Moreover, their design should be based on diversity instead of 'one size fits all' and should be focused on providing support instead of controlling. (See next paragraph for a discussion of support systems).

17.6 Complementary Approaches to STS-D for TWIN

So far, the STS-D approach has focussed mainly on the design of the production structure and control structure, i.e., the division of work into tasks and roles for TWIN (i.e., how to produce and how to manage). However, a systemic approach to redesigning the organisation for TWIN also requires the design of support systems (i.e., the information structure). Below, we will discuss each of these different approaches as they relate to TWIN (see also Table 17.1).

17.6.1 Lean Thinking

Lean Thinking and its associated toolbox mainly focus on the reduction of waste. Similar to STS-D, it takes into account the whole core work process, starting with product development and ending with the delivery and subsequent support of a product to the customer (Roos et al. 1991). Lean Thinking's contribution to TWIN is twofold: the principle of Just in Time (JIT), more recently reframed as Quick Time Response (Suri 1998, 2010), and the concept of continuous improvement. Specifically, we propose that the Lean JIT principle can be used to design the logistics systems for TWIN. The continuous improvement element of the Lean toolbox focuses on eliminating all forms of waste. We argue that continuous improvement is an excellent way to enhance job control if applied at the level of individual employees, which in turn, would enable employees to cope with interference, stimulate learning and enhance employee involvement.

17.6.2 Total Productive Maintenance (TPM)

The TPM approach primarily deals with the organization of equipment maintenance in the production industry. Specifically, TPM focuses on the optimal cooperation between the production and maintenance department, which is in line with the aim of TWIN to increase employee job control possibilities. In TPM an important objective is involving all maintenance employees as well as having the operation teams implement practical solutions for further improvement (Nakijama 1988). TPM has added value to STS-D for achieving TWIN, because it increases job enrichment with the focus on the task of autonomous maintenance, i.e., maintenance by operators.

17.6.3 Relational Coordination Theory

In large organisations and networks there is a need for horizontal coordination between teams, unities and communities of work. In this respect, the *Relational Coordination* theory, developed by Gittell (2003), can complement the TWIN model. The theory has been used to examine what sets productive complex organisations, like airlines and hospitals, apart from their less productive competitors. The results suggest that it is the horizontal, informal relationships between employees that made the difference (Gittell 2003). In practice, effective organisations seem to employ a variety of interventions to safeguard the coordination of their internal relationships. However, there are some common characteristics across these interventions. First, these organisations are characterised by a diversity of roles as well as wide-ranging, overlapping roles within order streams. Moreover, they tend to have developed wide-ranging organisation-wide mobility policies where employees can be regularly transferred to another stream or branch. This will facilitate cross-pollination of knowledge regarding customers and operations and will help increase employee job control. Importantly, the time span for organising these coordinated relationships can differ significantly. For example, employees may rotate in their tasks several times within one working day. Security officers in Scandinavian airports (Gittell 2003) are rotating almost constantly, enabling employees to keep an eye on travellers from several different observation points.

17.6.4 The New World of Work

The New World of Work is a combination of practices that focuses on the flexible design of workplace facilities based on time and place independent work (Bijl and Gray 2011). The New World of Work approach complements the STS-D approach to achieve TWIN, by providing tools for designing the infrastructure of the workplace as well as tools for virtual collaboration. Indeed, re-designing an organisation has far-reaching implications for its infrastructural requirements. A flexible organisation, with for instance, project teams working in different time-zones with a dynamic need for deliberations, would need flexible working facilities as well as supporting ICT systems that allow for virtual collaboration.

17.6.5 Human Resources Management (HRM)

STS-D theory informs us on how to design tasks and roles that mobilise human talent (de Sitter 1994), however, it does not provide any answers regarding employee selection, training and development or reward systems. To this end, Human Resources Management (HRM) approaches can be used to complement STS-D in designing a TWIN model. In this respect, we focus on two schools of thought in HRM named after the American Business Schools where they were developed, namely the Michigan (Fombrun et al. 1984) and the Harvard models (Beer et al. 1984). The Michigan approach, generally seen as the "hard" version of HRM, focuses on high performance aligning personnel management with the organisation's strategy. Therefore, the HRM instruments are tailored to ensure that employees add value to the organisation and the focus is on ensuring high employee performance. In contrast, the Harvard model, or the "soft" approach to HRM, focuses on the internal coordination of the expectations and interests of both the business and the employees. Employee involvement is central and the assumption is that it can only be achieved if people are confronted with challenging work. Moreover, it assumes that employee involvement will result in higher productivity, quality and efficiency. Both HRM models assume a balance between management (Michigan) and employee value (Harvard) and they both see HRM as an integral part of the enterprise strategy. STS-D theory claims to balance business demands and employee interests, therefore, for TWIN the combination of both models seem logical.

17.6.6 Archipelago ICT Thinking

Although information technology (ICT) systems play an important role in organisations, ICT has never played a major role in designing organisations from a STS-D perspective. However, ICT systems profoundly determine organisational design choices, as they create the technical context within which many organisations are operating and, hence, they also affect the social work system (Bednar and Welch 2016). In fact, in many cases, ICT is the context within which work takes place. In STS-D, ICT systems are regarded as support systems, hence, in the design sequence, this implies: 'first organise, then automate'. The introduction of traditional enterprise ICT systems, for example, such as enterprise resource planning (ERP), has had negative effects on organisational agility, productivity and organisational and employee health (Govers 2003). This is largely due to the fact that they aim for standardisation and take a 'one size fits all' approach, whereby, all business functions are integrated into one core work process. However, in most organisations, a number of simultaneous processes take place that vary in terms of inputs/outputs and process steps. Moreover, due to this attempt at standardisation, ERPs can lead to a neglect of customer demands as well as decreased job control.

Contrary to this practice, the STS-D principles suggest that archipelago enterprise computerization, when taking into account the necessary variety in work processes, is more suitable for workplaces aiming at TWIN goals (organisational performance and quality of work). Archipelago enterprise computerization is analogous to islands being connected under the water line, while being disconnected above the water line in an archipelago (Govers and Sudmeier 2016). An archipelago enterprise system can consist of various parallel, independent enterprise systems instead of a single organisation-wide one. This would imply that the organisation identifies independent parallel market or production flows (streams) and ideally would provide each stream with its own computerization to deal with the variety and dynamics of that stream. The general information for all streams is the underwater connection. A light version of an archipelago system can be, for instance, a menu card structure. Like in a cafeteria, a menu of an enterprise system is built around clear-cut, varied processes. The archipelago design of IT systems can create the opportunity to provide specific production flow information to the employees and increase job control. This means that you do not have to provide more information than needed, which results in limited complexity for employees.

17.6.7 Sociocracy or the Circular Organisation

Democratic values are important to create innovative and humane organisations. To this end Sociocracy (Lekkerkerk 2016) provides the philosophy and tools to improve strategic decision-making, given that it focuses on participative strategic decision-making. Participative and democratic strategic decision-making enforces strategic job control possibilities, enables employee involvement and serves as an important tool for improving local labour relations.

Sociocracy was developed in The Netherlands in the 1970s (Endenburg 1998; Lekkerkerk 2016), and although the philosophy has spread, it has not been implemented on a large scale. Its more recent US adaptation, that has gained some ground recently, is called Holacracy (Roberson 2015). Sociocracy is a consistent approach to involve employees (including managers) across different hierarchical levels in making non-operational decisions, for instance about strategic choices, including innovation and change (Endenburg 1998). Sociocracy proposes the creation of circles, each consisting of a group of people at the shop or office floor. Depending on the size of the organisation and the operational division of labour, there will be a hierarchy of circles (e.g., operational circle, business unit circle, top circle). The members of a circle elect one of them (the manager excluded) to represent them and their views in a higher-level circle, which is linking related lower-level circles. Thus, a number of layers may be formed until the top-circle is reached at top-management team level. Hence, a 'circle-organization' not only has a normal chain of command hierarchy for operational matters, but also a parallel structure of 'circles' for strategic or policy decision-making, that also serves as a bottom-up feedback channel, increasing the information processing capacity of the organisation.

17.7 Conclusion

STS-D theory and practices have played an important role in designing the structure of humane and innovative workplaces. However, for workplace innovation, simply restructuring units, departments, tasks and roles is not enough. In traditional, bureaucratic organisations, support systems have hidden conservation mechanism to keep the bureaucracy in place. Moreover, for workplace innovation, support and coordination systems as well as democratic strategic decision-making should be included in designing the workplace.

In this chapter, we aimed at developing a more comprehensive design theory for stakeholders who are involved in design processes aimed at workplace innovation, by starting from sociotechnical design and by exploring how we can broaden that perspective with other approaches, to also cover issues such as IT-design and HR-design. Specifically, we focused on the following approaches as potential additional perspectives to STS-D in developing a concept of Total Workplace Innovation:

- The Lean Thinking approach for quick time response (JIT) control systems and for continuous improvement;
- The Relational Coordination theory for supporting horizontal coordination;
- HRM-theories and practices for supporting HR-policies;
- TPM to support the collaboration between maintenance and operation;
- ICT systems design to create effective information support systems, based on variety instead off one size fits all;
- Sociocracy for participative decision-making and democratic involvement on strategic issues.

It is to be noted, that the TWIN concept we presented in this contribution needs to be further elaborated upon. In addition, we admit that we have left several questions unanswered due to limited space, such as, how can these approaches be integrated, how will they affect one another, and what will be the consequences in terms of design steps. Therefore, we suggest a joint journey of practitioners and academic researchers to develop a more profound model and practices for Total Workplace Innovation.

To conclude, the added value of this approach to workplace innovation practices is the understanding of workplace design as a fundamental precondition for the joint optimisation of quality of working life and productivity. The TWIN model starts from a structural design perspective (STS-D), however, to realise TWIN we need to integrate the design perspective with behavioural aspects and specific types of leadership (see Oeij et al. 2015). Realising TWIN in practise, hence, is a simultaneous design and organisational development process.

Acknowledgements The authors and publisher gratefully acknowledge the following permission to use the material in this book: Pierre van Amelsvoort & Geert van Hootegem, Towards a total workplace innovation concept based on sociotechnical systems design. In: European Work and Organizational Psychology in Practice. Special issue on Workplace Innovation, 2017, Volume 1, 31–45.

References

Achterbergh, J., & Vriens, D. (2009). *Organizations: Social systems conducting experiments*. Dordrecht: Springer.

Adler, N., & Docherty, P. (1998). Bringing business into sociotechnical theory and practice. *Human Relations, 51*(3), 319–345.

Amelsvoort, P. (2000). *The design of work and organisation*. Vlijmen: ST-Groep.

Ashby, W. R. (1969). Self-regulation and requisite variety. In F. E. Emery (Ed.), *Systems thinking* (pp. 105–121). London: Penguin Books.

Bakker, A., & Demerouti, E. (2007). The job demands—Resources model: State of the art. *Journal of Managerial Psychology, 22*(3), 309–328.

Baxter, G., & Sommerville, I. (2011). Socio-technical systems: From design methods to systems engineering. *Interacting with Computers, 23*(1), 4–17.

Bednar, P., & Welch, C. (2016). Enid Mumford: The ETICS methodology and its legacy. In B. J. Mohr & P. van Amelsvoort (Eds.), *Co-creating humane and innovative organizations: Evolutions in the practice of socio-technical system design* (pp. 274–288). Portland, Maine, USA: Global STS-D Network Press.

Beer, M. B., Spector, P., Lawrence, D. Q., Mills, R., & Walton, R. E. (1984). *Managing human assets*. New York, NY: Free Press.

Bijl, D., & Gray, M. (2011). *Journey towards the new way of working: Creating sustainable performance and joy at work*. Zeewolde: Par CC.

Cherns, A. (1987). The principles of sociotechnical design revisited. *Human Relations, 40*(3), 153–161.

Demerouti, E., Bakker, A. B., Nachreiner, F., & Schaufeli, W. B. (2001). The job demands-resources model of burnout. *Journal of Applied Psychology, 86*(3), 499–512.

van Eijnatten, F. (1993). *The paradigm that changed the workplace*. Assen/Stockholm: Van Gorcum/Arbetslivscentrum.

Endenburg, G. (1998). *Sociocracy: The organization of decision-making*. Delft: Eburon Academic Publishers.

Fombrun, C. J., Tichy, N. M., & Devanna, M. A. (1984). *Strategic human resource management*. New York: Wiley.

Gittell, J. (2003). *The Southwest Airlines way: Using the power of relationships to achieve high performance*. New York: McGraw-Hill.

Govers, M. (2003). *Met ERP-systemen op weg naar moderne bureaucratieën?* Nijmegen: Radboud Universiteit.

Govers, M., & Sudmeier, P. (2016). Applying information technology from a socio-technical perspective: A socio-technical perspective on computerization. In B. J. Mohr & P. van Amelsvoort (Eds.), *Co-creating humane and innovative organizations: Evolutions in the practice of socio-technical system design* (pp. 289–302). Portland, Maine, USA: Global STS-D Network Press.

Karasek, R. A. (1979). Job demand, job decisions latitude and mental strain; implications for job design. *Administrative Science Quarterly, 24*(2), 285–307.

Karasek, R. A., & Theorell, T. (1990). *Healthy work; stress, productivity and the reconstruction of working life*. New York: Basic Books.

Kuipers, H., & van Amelsvoort, P. (1990). *Slagvaardig organiseren, inleiding in de sociotechniek als integrale ontwerpleer*. Deventer: Kluwer.

Kuipers, H., van Amelsvoort, P., & Kramer, E.-H. (2010). *Het nieuwe organiseren: alternatieven voor de bureaucratie*. Leuven: Acco.

Lekkerkerk, L. J. (2016). Organizing innovation and (strategic) decision making. In B. J. Mohr & P. van Amelsvoort (Eds.), *Co-creating humane and innovative organizations: Evolutions in the practice of socio-technical system design* (pp. 99–122). Portland, Maine, USA: Global STS-D Network Press.

Mumford, E. (2006). The story of socio-technical design: Reflections on its successes, failures and potential. *Information Systems Journal, 16*(4), 317–342.
Mohr, B. J., & van Amelsvoort, P. (Eds.). (2016). *Co-creating humane and innovative organizations: Evolutions in the practice of socio-technical system design Portland*. Maine, USA: Global STS-D Network Press.
Nakijama, S. (1988). *TPM, introduction to TPM*. Portland Oregon: Productivity Press.
Oeij, P., Žiauberytė-Jakštienė, R., Dhondt, S., Corral, A., Totterdill, P., & Preenen, P. (2015). *Third European Company survey—Workplace innovation in European companies*. Luxembourg: Publications Office of the European Union.
Pot, F., & Dhondt, S. (2016). Workplace innovation. In B. J. Mohr & P. van Amelsvoort (Eds.), *Co-creating humane and innovative organizations: Evolutions in the practice of socio-technical system design* (pp. 201–222). Portland, Maine, USA: Global STS-D Network Press.
Roberson, B. J. (2015). *Holacracy, the new management system for a rapid changing world*. New York: Henry Holt & Co.
Roos, D., Womack, J. P., & Jones, D. T. (1991). *The machine that changed the world: The story of lean production*. New York: Harper Perennial.
de Sitter, L. U. (1994). *Synergetisch produceren, Human resources mobilisation in de productie: een inleiding in de structuurbouw*. Assen: Van Gorcum.
de Sitter, L. U., den Hertog, J. F., & Dankbaar, B. (1997). From complex organizations with simple jobs to simple organizations with complex jobs. *Human Relations, 50*(5), 497–534.
de Sitter, L. U., Vermeulen, N., van Amelsvoort, P., van Geffen, L., & Vertroost, P. (1986). *Het flexibele bedrijf*. Deventer: Kluwer.
Suri, R. (1998). *Quick response manufacturing: A companywide approach to reducing lead times*. New York: Productivity Press.
Suri, R. (2010). *It's about time*. New York: Productivity Press.
Weisbord, M. R. (2004). *Productive workplaces revisited: Dignity, meaning and communication in the 21st century*. London: Jossey-Bass/Wiley.

Author Biographies

Pierre van Amelsvoort is an expert in sociotechnical systems design and organisational change. He is a partner of the consulting firm ST-Groep. Between 1994 and 2007 he was professor by special appointment at the Nijmegen School of Management (Netherlands). Since 2010 he is guest professor at the Catholic University Leuven, KU Leuven (Belgium).

Geert Van Hootegem is professor at the Centre for Sociological Research at the Catholic University Leuven, KU Leuven, since 2000, teaching organisational design and change management. He is the founder of Flanders Synergy, a Belgian network of organisations that promote workplace innovation. In 2013 he co-founded Prepared Mind, a consulting firm focussing on Total Workplace Innovation.

Chapter 18
Five Steps to Develop Workplace Innovation

Steven Dhondt, Peter Totterdill, Sylvie Boermans and Rita Žiauberytė-Jakštienė

18.1 Introduction

New technologies change the way we live, consume and meet people. As the Internet has changed our lives, the Industrial Internet is transforming the way we work and produce. The digital revolution is happening. Some sectors experience fast and disruptive changes, others will evolve slowly and steadily. In each case, there is no turning back. Success in the new industrial revolution naturally requires that our industry uses the best available technologies. But technologies alone are not the answer. We should put more focus on human resources and human factors connected to technology. This is our main resource in Europe and we do not make enough use of employees' competences and creativity. Workplace innovation (WPI) not only aims at fostering innovation capacities, it also allows business to remain innovative and adapt to changes more quickly and smoothly.

S. Dhondt (✉) · S. Boermans · R. Žiauberytė-Jakštienė
TNO, The Netherlands Organisation for Applied Scientific Research, Leiden, The Netherlands
e-mail: steven.dhondt@tno.nl

S. Boermans
e-mail: sylvie.boermans@tno.nl

R. Žiauberytė-Jakštienė
e-mail: rita.ziauberytejakstiene@tno.nl

P. Totterdill
Kingston University London, Kingston, United Kingdom
e-mail: peter.totterdill@workplaceinnovation.eu

S. Dhondt
KU Leuven, Leuven, Belgium

P.R.A. Oeij et al. (eds.), *Workplace Innovation*, Aligning Perspectives on Health, Safety and Well-Being, DOI 10.1007/978-3-319-56333-6_18

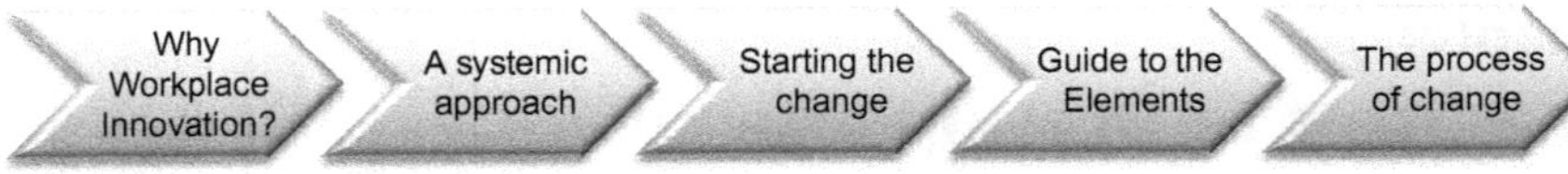

Fig. 18.1 Five steps towards WPI

That is why the European Commission launched the European Workplace Innovation Network, EUWIN, in 2013. In 2016 it connects more than 10,000 companies and other stakeholders, sharing know-how and experience. Its Knowledge Bank[1] is a rich and abundant source of case studies, films, articles and other learning resources.

EUWIN always had a practical approach. Therefore we have also decided to prepare a guide which answers the main question related to workplace innovation: "how can we actually do it?" This short guide will give you practical knowledge, inspire you with great case studies, help you to assess current practice in your organisation, suggest pathways to change and signpost you towards further sources of information and support. Other chapters in this book provide underpinning and references of the statements made in this guide.

It is built around five challenges (also see Fig. 18.1):

1. Why workplace innovation is important for your company?
2. How workplace innovation will transform your organisation?
3. Where to begin?
4. What are the '*Elements*' of workplace innovation?
5. How to achieve commitment from everyone in your organisation?

The end results will deliver what you expect: businesses in every sector and size category throughout Europe report higher performance, better places to work and an enhanced culture of innovation. However, one remark with this guide: we cannot offer a full proof recipe for transforming your organisation because it just doesn't work like that. You have to be prepared to listen, to involve, to experiment, to take risks and to learn.

18.2 Why Workplace Innovation?

The organisation of workplaces plays a vital role in a company's ability to compete and innovate. But what is *workplace innovation*? In short it can be defined as those workplace practices and cultures which enable employees at all levels to use their knowledge, competences and creativity to the full. It builds organisations in which people come to work to do two things: to undertake their functional tasks in the

[1]A link to the Knowledge Bank: http://portal.ukwon.eu/euwin-knowledge-bank-menu-new.

most effective way possible and to improve and innovate the business and organisation.

Such workplaces are likely to include empowering job design; self-organised team working; open and fluid organisational structures; delegated decision-making and simplified administrative procedures; a coaching style of line management; regular opportunities for reflection, learning and improvement; high involvement innovation practices; the encouragement of entrepreneurial behaviour at all levels; and employee representation in strategic decision-making.

Evidence shows that workplace innovation leads to significant and sustainable improvements both in organisational performance and in employee engagement and well-being. In Chap. 6, a list of studies is provided giving evidence of the impact of workplace innovation practices on productivity, manufacturing quality, customer service, financial performance and profitability, and a broad array of other performance outcomes. The chapter shows the complexity of 'proving' the business-case, but also why the positive results should not be dismissed.

18.3 How Workplace Innovation Will Transform Your Organisation

The benefits of workplace innovation are only fully realised when workplace innovation practices run throughout the entire organisation. One of the most significant obstacles to achieving high performance and great places to work is partial change—a failure to recognise that organisations consist of interdependent parts that either nurture or obliterate innovative ways of working.

Transformative changes in performance and working life can be achieved when boards, senior teams, line managers and employee representatives share a common understanding of workplace innovation and a commitment to making it happen.

High performance and great jobs are not the product of a simple initiative or a leadership development programme. They are only found when four basic building blocks (elements) are in place (see Table 18.1).

These four Elements come together to form a system of mutually reinforcing practices which create surprising synergies. *The Fifth Element* represents this synergetic outcome in stronger enterprising behaviour, a culture of innovation, high levels of employee engagement, and organisational and individual resilience. These outcomes are only generated when the four Elements combine (Fig. 18.2).

So if it is that good, why isn't everyone doing it? We know that only a minority of European managers are making full use of these evidence-based practices (Totterdill 2015). Long-established (bureaucratic) ways of doing things are, of course, a powerful force for inertia. Change can be disruptive even when we know that current ways of working are inefficient or prevent us from taking full advantage of new opportunities. Business demands may give little time for considering the

Table 18.1 4 Elements essential for WPI

Element 1	Empowering jobs and self-managed teams
Element 2	Flexible organisational structures, people-centred management practices and streamlined systems and procedures based on trust
Element 3	Systematic opportunities for employee-driven improvement and innovation
Element 4	Co-created and distributed leadership combined with 'employee voice' in strategic decision-making

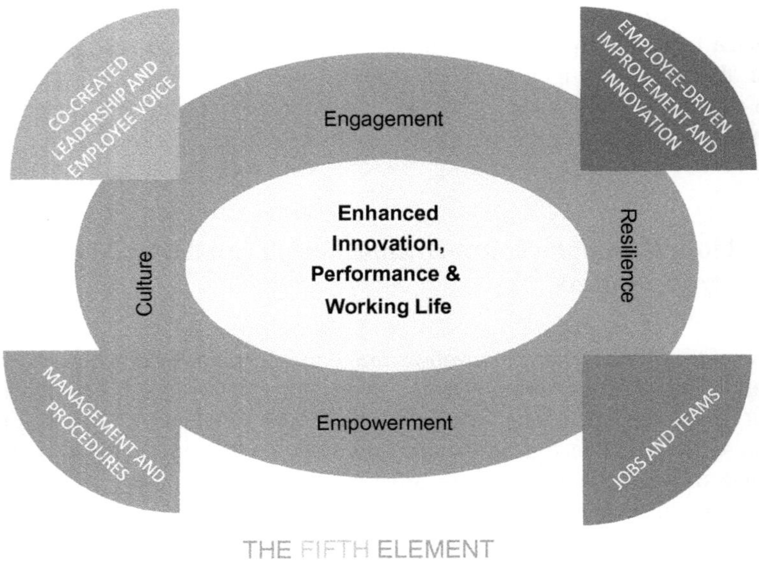

Fig. 18.2 The Fifth Element outcomes

change. From our experience, some managers are simply resistant to the idea that employees should be empowered to play a wider role in decision-making.

18.4 Where to Begin?

The answer to where to start will be different for every organisation, each with its unique history, relationships, challenges and opportunities.

A recent in-depth qualitative investigation into fifty leading companies in Europe identified at least five distinct pathways towards workplace innovation (Oeij et al. 2015). No matter how different the pathways to workplace innovation, the end result was always better workplaces and stronger company performance. The starting point always lies in listening to employees. Employees can help to make understand what business needs and challenges are.

Employee surveys are a powerful means of gaining an overview of what is working and what needs to be fixed, but they do not provide solutions. Yet those being surveyed will have many of the insights and answers necessary to take the next steps. Survey results are best shared with all employees and used to stimulate reflection and dialogue across the organisation. Senior managers demonstrate distrust and feed rumour when they hold on to survey results and worry about sharing them.

Open and inclusive dialogue between managers and employees at every level is indispensable. Be imaginative in the way you stimulate the sharing of ideas and experience. Openness may not come easily, especially where people have not been asked for their ideas previously, and they may need reassurance. Trust and confidence are essential.

Facilitated peer group discussions involving employees at every level can illuminate the causes of deep-seated problems and point the way towards solutions. Managed well, these discussions begin a process of involvement which moves seamlessly from analysis to action.

Other companies' stories in a form of films, case studies, articles and statistics can inspire sceptics and champions alike. Many companies have started on the journey of transformation because they have been stimulated to challenge the status quo by experiencing the effects of workplace innovation in other organisations.

Sometimes the answer is to start with a pressing issue or a challenge such as new technology. How can we find the root causes of recurrent obstacles to high performance or employee engagement? An example? The following inventory of challenges is drawn from our experience of working with companies at the starting point on their journeys towards workplace innovation (see Table 18.2).

How to use your results

Sum up your scores and divide as shown, giving you a figure for each of the four *Elements*. Then ask other colleagues to complete the exercise and compare the results. This will provide you with a strong indication of what is happening in your organisation.

The outcomes provide a great starting point for dialogue and shared reflection involving people throughout the organisation as well as your immediate colleagues:

- **Which is the *Element* with most pressing issues?** Select the highest of the four figures. Do you all agree that this represents the highest priority? If so, where do the root causes for the most pressing issues lie? Is it technology that is blocking change? Or is it leadership?
- **Have you scored over 5 for more than one of the *Elements*?** If you have scored more than 5 for one *Element* it is very likely that you will also have

Table 18.2 An inventory for the identification of challenges

Starting the Journey: which are your most pressing issues?		
Score each issue: 1 = no problem; 10 = severe problem		Score
Element 1: jobs and teams	Frequent delays caused by breakdowns and bottlenecks	
	High levels of employee turnover and/or sickness absence	
	Minor problems are escalated to senior level	
	Poor team cohesion	
	Persistent quality problems	
	Total score for element and divide by 5	
Element 2: structures, management and procedures	Line managers lack team leadership skills	
	Ineffective performance management/appraisal system	
	KPIs/targets drive out opportunities for learning and improvement	
	Departmental/organisational boundaries delay decisions and inhibit innovation	
	Work gets held up by poor co-ordination between departments	
	Total score for element and divide by 5:	
Element 3: employee-driven improvement and innovation	We lack a culture of innovation	
	There are opportunities to improve or innovate but we rarely get around to pursuing them	
	Employees are frustrated that they have no outlet for their ideas	
	We need more effective ways of engaging employees in innovation and improvement	
	Employees are afraid or unwilling to challenge established practices	
	Total score for element and divide by 5	
Element 4: co-created leadership and employee voice	There is a gap between senior management and frontline staff	
	We don't share information with employees unless it is absolutely necessary	
	Senior managers micromanage the work of others rather than empowering them to take decisions	
	Decisions affecting the work of employees are taken without involving them	
	Our culture inhibits change	
	Total score for element and divide by 5	

a high score for some of the others too because different aspects of workplace practice are interdependent. For example, you are unlikely to succeed in stimulating and using employees' ideas for innovation and improvement if performance measures and line management behaviour drive out opportunities for reflection, shared learning and creative thinking.

If all four scores are below 5, you have some assurance that all is going well. It is important to recognise and celebrate what you do well if you want to maintain and develop it in the future. Are good workplace practices and the positive aspects of your organisational culture well enough embedded to survive changes at senior level, new owners or challenging external circumstances?

We have only discussed a small selection of possible issues here and it will be important to provide opportunities for people throughout your organisation to raise their own concerns.

The following sections show how the workplace innovation practices related to each *Element* address the root causes of problems such as these, directly improving both performance and working lives.

18.5 What Are the 'Elements' of Workplace Innovation?

The Fifth Element highlights the interdependence between bundles of workplace practices represented by the four *Elements*. Each *Element* does not exist in isolation but is influenced, for better or worse, by the extent to which the principles that underpin it are aligned with those of the others.

18.5.1 The First Element: Jobs and Teams

Employee participation and the ability to work without close supervision are highly cherished: architects, midwives and refuse collectors perform their jobs well because they make on-the-spot decisions based on background knowledge and experience of 'what works'. They avoid delays caused by unnecessary referral to managers or manuals.

In the best cases they make time to learn and to reflect on what is working well and what should be changed. This generates steady flows of improvement and innovation. Allowing employees discretion in scheduling their own work and in controlling its pace also minimises physical strain and psychological stress (Fig. 18.3).

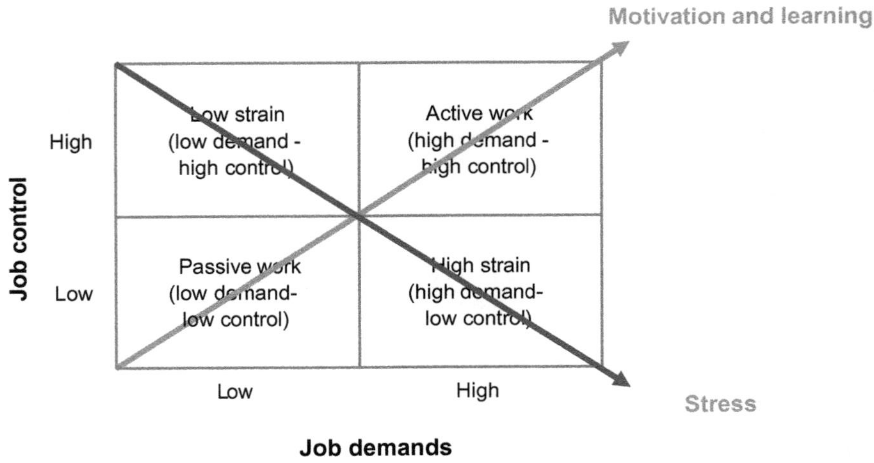

Fig. 18.3 Karasek's demand-control model of occupational stress (Karasek and Theorell 1990)

Karasek's model has had a large influence because it is straightforward and practical. Workplace stress is a function of how demanding a person's job is (factors include interruption rate, time pressures, conflicting demands, reaction time required, pace of work, proportion of work performed under pressure, amount of work, degree of concentration required, and the slowing down of work caused by the need to wait for others) and how much control the person has over their own responsibilities. Control includes the ability to make decisions about their own job, move between a variety of tasks, low levels of repetitiveness, opportunities for creativity, learning and problem solving, and influence in their own work team and in the wider organisation.

Of course individual jobs cannot just be examined in isolation. The ability to share problems and solutions with colleagues, to learn and reflect together, to provide and receive support in challenging times and to celebrate successes plays a vital role in engagement, well-being and performance. Team working lies at the heart of this equation.

Many organisations claim that their people work in teams. But are those teams simply groups of people who sit together and report to the same manager, but rarely co-operate with each other? Some work psychologists call these *pseudo teams* (Katzenbach and Smith 1993).

Real teams consist of people sharing challenges and opportunities in ways that break down barriers and demarcations, and where they generate ideas for improvement, innovation and growth using the insights that day-to-day work experiences give them. Self-managed teams empowered to plan and organise their own work are more productive, whether in factories or offices. They also offer much better places to work.

What are 'good' teams? (West 2012)

- Good teams are clear about their shared tasks, and about precisely who is part of the team. Once teams grow larger than 8–10 members it becomes difficult to maintain cohesion.
- They are clear about the skills the team needs to achieve its purpose.
- The team is empowered to make appropriate choices about recruitment, and recognizes the important of recruiting people who are good at collaboration and sharing.
- Team members need to understand clearly their roles and the roles of other team members, so there is no ambiguity about who is responsible and accountable for each task.
- Good teams set themselves clear, challenging and measureable objectives every year. The aim is not just to get the job done but to achieve significant improvements and innovations. Progress towards achieving these objectives forms an important part of regular team meetings.
- Well-functioning teams assess and seek to improve their effectiveness in working with other teams inside (and sometimes beyond) the organization.
- Teams with a supportive, humorous and appreciative atmosphere deliver better results and their members are significantly less stressed. They are more optimistic, cohesive and have a stronger sense of their efficacy as a team.
- Teams must also meet regularly and have useful discussions, enabling them to reflect on how well they work together and how to improve. Teams that regularly change ways of working are not only more productive but also more innovative than teams that do not. 'We do not have time' is therefore an unacceptable excuse. Such teams are also better able to respond to work pressures and adversity by innovating rather than feeling overwhelmed and helpless.

Resources for Element 1: Job Design and Team Working

- Some of the principles discussed in this section can be explored further in UK WON's resource paper on job design.[2]

[2]The full paper can be found following this link: http://uk.ukwon.eu/pdfs/Job-Design-euwin.pdf.

- Ederfil Becker, a manufacturing plant in the Basque country, created a management model based on self-organisation in which trust, empowerment, communication, transparency and employee voice set the lines of action of its daily operations. A full case study is available.[3]
- At Bombardier in Bruges (Belgium), team working was implemented to shorten communication lines, to increase the sense of ownership at the shop floor level, to avoid disturbances in the production flow, and to enhance the problem-solving capacity on the shop floor. A full case study is available as well.[4]
- Thebe is a Dutch healthcare company that has decided to put professionals at the heart of its strategy and allow them to manage care themselves through self-managed teams. Read a full case study online.[5]

18.5.2 The Second Element: Organisational Structures, Management and Procedures

Truly innovative workplaces recognise the need for a consistent approach to empowerment, learning and development running through every aspect of corporate policy from reward systems and performance appraisal to flexible working and budget devolution.

Many organisations appear to be structured around three assumptions:

1. Hierarchies are just common sense: you need somebody to be in charge.
2. People at the frontline are of lower status and less motivated so they cannot be trusted to make decisions or manage their own work.
3. Other ways of organising may be fine for some companies but they will never work here.

Of course some demarcations may be necessary, reflecting different bodies of expertise and knowledge. Yet management layers inevitably put distance between decision-making and the frontline, disempowering and diminishing the voice of those at the lower levels as well as creating an implementation gap. Hierarchy breeds caution amongst managers, encouraging decisions to be delegated upwards with consequent loss of productivity and responsiveness. Such vertically organised structures create silos and add to the difficulties of building bridges between

---

[3]Read the case study and watch the short film: http://portal.ukwon.eu/EderfilBecker.

[4]Read the full case study: http://uk.ukwon.eu/File-Storage/5114529_7_Bristan-Case-Study.pdf.

[5]Read the full case study: http://portal.ukwon.eu/document-store/putting-professionals-at-the-heart-of-the-organisation-self-organised-teamworking-at-thebe.

functional specialisms. This often causes frustration in resolving day-to-day issues and can have a particularly negative effect on the capacity for innovation.

Conversely a flat structure is the business with either no management layers or a very short chain of command. The aim is often not to erase hierarchies entirely but to allow organisations to form hierarchies naturally. Different groups within organisations can intertwine in ways that help everyone understand other people's jobs, professions, specialisms, priorities, problems and vision. Flat organisations rely on a decentralised approach to management and require a high degree of employee involvement in decision-making. Control in flat companies lies in mutual agreements between self-managing, self-organising and self-designing teams and employees who take personal responsibility for satisfactory outcomes. This in turn empowers employees, facilitates information sharing, breaks down divisions between roles, shares competencies, and uses team or organisation-wide reward systems. These team and organisation-wide reward systems do not focus on the individual task or performance: the incentive should be at the group level performance and at the level of supply of good ideas.

Getting rid of annoying processes

- How much control is actually necessary? Are the risks of loosening control greater than the (real and opportunity) costs of spending senior management time on relatively trivial issues?
- Is the formal appraisal interview a 'tick box'- exercise? Or is it a milestone in a continuous coaching conversation with a line manager who understands how to help each individual develop and deliver their best performance?
- Are there established procedures for shared learning from mistakes and failures in an open and blame-free way?
- Is the importance of shared learning, reflection, improvement and innovation recognised in the organisation's system of KPIs, targets and incentives?

Yet even within more flexible structures, mistrust and disempowerment can be embedded in the systems and processes that shape decision-making, resource allocation and standard operating procedures. They can reflect a culture of centralised control and micro-management. Truly innovative workplaces recognise the need for a consistent approach to empowerment, learning and development running through every aspect of corporate policy from reward systems and performance appraisal to flexible working and budget devolution.

One special remark is that in redesigning your organisation, you need to take into account legal and financial constraints. For example, public institutions cannot easily make the change to workplace innovation if they need to take into account the transparency and other accountability requirements put upon them by national laws. But even so, there are great examples that show how such public institutions can mix workplace innovation with their political embedding. Financial constraints may hinder workplace innovation. Workplace innovation requires managers to not be penny-wise-pound-foolish. Workplace innovation requires investments to provide payback. The many examples from other companies show how one can balance these financial constraints with workplace innovation.

Resources for Element 2: Organisational Structures, Management and Procedures

- In Bristan (UK), high levels of employee engagement are built on a culture of trust and leadership reflected in day-to-day management behaviour. The full Bristan case study is available online.[6]
- Cougar Automation (UK) is a powerful example of strengths-based management in a medium-sized advanced manufacturing company.[7]
- By removing functional boundaries and designing work spaces that encourage people to collaborate, INET-Logistics (Austria) has empowered its employees with high levels of discretion in their day-to-day work.[8]
- Semco (Brazil) is an example of a "radical form of corporate democracy" that aims to distribute decision-making authority out to everyone.
- Breaking away from a traditional hierarchy is difficult for public sector organisations but it formed an essential element in the goal of making the British Geological Survey (UK) a more innovative and responsive organisation. Its matrix structure has created a culture in which employee-led improvement and innovation are valued at every level.[9]

18.5.3 The Third Element: Employee-Driven Improvement and Innovation

A knowledge economy is one firmly rooted in innovation, popularly associated with R&D, ICT investment and high-profile entrepreneurs. However this association can be misleading. Much of the innovation in products, services and processes that leads to enhanced business competitiveness and performance is generated by interaction, dialogue and exploration within the workplace.

---

[6]Read the full case study: http://uk.ukwon.eu/File%20Storage/5114529_7_Bristan-Case-Study.pdf.

[7]More information can be found here: http://portal.ukwon.eu/document-store/cougar-automation-enlightened-management-gets-resultsa.

[8]More information can be found here: http://portal.ukwon.eu/document-store/inet-logistics-gmbh-breaking-down-walls-and-ceilings.

[9]More information can be found here: http://portal.ukwon.eu/document-store/british-geological-survey-bgs-case-study.

Unsurprisingly this is more common in workplaces where employees have greater control over their own work coupled with discretionary opportunities for learning and problem solving (Karasek and Theorell 1990).

Systematic opportunities for shared learning and reflection are well embedded in that type of workplace. They allow employees at every level to reflect on what has gone well and what can be improved in the future, to share knowledge and skills gained in the course of recent work experience, and to anticipate and reflect on the impacts of future challenges and change. This can be reflected in times and spaces where people at work can discuss ideas with their co-workers or in their team meetings. Buzz boards enable ideas to be shared and dedicated spaces can enable people to think in different ways together. Meetings in cafés can offer a creative and reflective time away from the immediate pressures of the workplace.

Sometimes it also involves imaginative opportunities to 'think out of the box' by bringing people together across different departments and divisions to share knowledge and experience and to think creatively together. It can be as simple as establishing regular forums that enable people at all levels of an organisation to leave job titles and hierarchies behind, and to explore new ideas through open and free-thinking discussion.

A growing number of organisations provide employees with regular opportunities to join cross-functional improvement teams to identify and drive forward product or process changes that would otherwise be lost under the pressure of day-to-day workloads. Time-out sessions, 'down-tools weeks' and hackathons, bringing people together who otherwise would not meet, can also become fountains of constructive dialogue and creativity.

Employee-driven innovation and improvement emphasises the importance of aligning the knowledge and expertise of senior teams with the tacit knowledge and experience of frontline workers while recognising and valuing continuing learning. It must also reflect deeper structural practices within each organisation: sustainable and effective employee engagement in innovation and improvement cannot happen in isolation. It must be driven from the top and reinforced by consistent messages from leaders, aligned with organisational structures and procedures, and underpinned by empowerment and discretion in day-to-day working life. Line management culture and performance measurement invariably play a critical role in enabling, or inhibiting, employee-driven improvement and innovation.

Resources for Element 3: Employee-Driven Improvement and Innovation

- The culture of Red Gate Software (UK) "is permission to fail and expectation of failure, if we are not trying hard enough that's when we don't fail, so let's make sure that we're on the edge and we're trying to push ourselves."[10]

---

[10]More information can be found here: http://portal.ukwon.eu/red-gate-software.

- At Innocent (UK), innovation is an everyday behaviour with employees actively encouraged to suggest ways to improve the business, whether it be new products and business streams or better ways of working.[11]
- In The Met Office (UK) a small core of innovators began a process of 'guerrilla' tactics designed to change the culture of a traditional public sector organisation, making innovation part of 'the day job' for everyone.[12]
- Polpharma (Poland) recognises that enabling improvement and innovation requires freedom to act combined with the ability to design the workplace and work environment in ways that employees want.[13]
- Employee-Driven Innovation from a trade union perspective: the Danish Confederation of Trade Unions argues that employee involvement in innovation leads to win-win outcomes for workers and business performance.[14]

18.5.4 The Fourth Element: Co-created Leadership and Employee Voice

It is unsurprising that enlightened leadership often plays a key role in driving workplace innovation within enterprises. Leadership theory is highly contested (Gronn 2002; Khurana 2002) but leadership development has gained increasing prominence through business schools, professional institutions and consultancy.

Early leadership theories were primarily focused on the distinction between "task focus" and "people orientation" and this remains a useful distinction. More recently theories are less concerned with the central, charismatic individual but focus on leadership as a creative and collective process where leadership is co-created through dialogue with and between employees and where employees are empowered to take initiative and contribute to decision making. "Shared and distributed leadership" is a key element of workplace innovation because it focuses on releasing the full range of employee knowledge, skills, experience and creativity. It means that workplace culture and practice provides everyone with the opportunity

[11]More information can be found here: http://portal.ukwon.eu/innocent-smoothie-makers.

[12]More information can be found here: http://portal.ukwon.eu/met-office.

[13]More information can be found here: http://portal.ukwon.eu/Polpharma-Worldwide.

[14]More information can be found here: http://portal.ukwon.eu/document-store/employee-driven-innovation.

Fig. 18.4 Co-created leadership (*Source* http://www.goodworkplaces.net/The-Fourth-Element)

to take the lead in areas which reflect their own expertise or initiative, whether strategic, innovative or operational, while understanding and aligning their actions with those of others.

Leadership is therefore a collaborative or *co-created* process (see Fig. 18.4). It is not dependent on individual charisma or authority but creates shared direction and purpose through organisation-wide opportunities for strategic thinking, shared reflection and learning, and employee voice in decision-making.

Employee voice describes the alignment of strategic priorities and decision-making at senior levels with the practical knowledge, experience and engagement of employees throughout the organisation. It brings together *direct participation* through, for example, self-managed teams and improvement groups, with *representative participation* in the form of employee or union-management partnership forums. These represent times and spaces where senior managers and trade unions or employee representatives get together to tackle big issues in a climate of openness and trust.

Partnership between management, employees and trade unions can take many forms, but always requires openness, transparency and two-way communication. At the very least it can be an effective tool for positive industrial relations, minimising conflict and resistance to change.

An important body of research has begun to show that representative participation structures on their own may have little direct impact on performance or quality of working life. Rather they can exert a positive influence on the development of activities and practices that may do so (Huselid et al. 1997; Teague 2005). In short, they can become the animator and guardian of employee empowerment and engagement.

When partnership arrangements exist alongside the types of participative workplace practices described in the previous three Elements it creates a system of mutually reinforcing practices leading to improved information sharing, greater levels of trust, reduced resistance to change and heightened performance.

Resources for Element 4: Co-created Leadership and Employee Voice

- Workplace partnership and employee consultation can mean more than just good industrial relations. The British company Drax Power's 'Working Together' agreement with trade unions combined with a comprehensive consultation programme helps staff understand the major technological changes faced by the plant, builds trust and demonstrates transparency, stimulates employee ideas and creates a more entrepreneurial culture.[15]
- Novozymes (Denmark) demonstrates that strong partnership between management and unions can transcend traditional industrial relations, empowering frontline workers in day-to-day decision-making and actively involving them in improvement and innovation. 'Shop Floor Management' is a way of creating a forum where there is dialogue between employees who are working in the daily operations, the manager and the technical support.[16]
- Leadership that articulates clear values and lives by them is recognised as a powerful driver of employee engagement and an enabling culture. Skanska UK has been on a sustained journey of transformation since CEO Mike Putnam's appointment in 2009, distancing itself from traditional industry practices by embracing high ethical principles relating to safety, the environment, transparency and quality.[17]
- Becton Dickinson (Ireland) offers persuasive evidence that workplace partnership means more than good industrial relations. Unions can play a key role in engaging workers at all levels in innovation and improvement, creating win-win outcomes for employees and their employers.[18]

18.6 How to Achieve Commitment from Everyone in Your Organisation?

Change is rarely a straightforward linear exercise. It usually involves experimentation, failure and a willingness to see failure as an opportunity for learning and development. It requires consistency of purpose combined with a willingness to

[15]More information can be found here: http://portal.ukwon.eu/document-store/drax-power-limited-working-together-for-change.

[16]More information can be found here: http://uk.ukwon.eu/Novozymes.

[17]More information can be found here: http://portal.ukwon.eu/document-store/skanska-uk-values-driven-leadership-creates-a-culture-of-respect.

[18]More information can be found here: http://portal.ukwon.eu/becton-dickinson.

rethink the vision and objectives set out at the start of the journey. The more you try to change an organisation, the more you learn about it. Your understanding of the nature and extent of the change required will deepen as the journey progresses.

Above all, it means making change happen with people, not to people. They have the knowledge, experience and potential for engagement that can make change happen and make it stick. Below there is a simple checklist for change.

A checklist for change

- Do the board and senior team understand that change will involve asking difficult questions and challenging established practices? Do you really have their support?
- Have you involved all the relevant stakeholders from the beginning? Does everyone understand how they will contribute to the journey?
- Have you anticipated the potential sources of resistance? What is the best way of dealing with it?
- What are the mechanisms for ensuring good two-way communication throughout the journey? How will you evaluate progress? How will stakeholders be involved in shared learning and adjustments to change processes and goals during the journey?
- How will you know whether you have succeeded? And how will you celebrate success?
- How will change be embedded and sustained?

Be creative: *Little details matter for Workplace Innovation*
Workplace innovation is also about attitudes and values. It requires you, manager and employee, to change some of your deepest beliefs of how organisations need to function. You need to be aware of such little details, bring them to a discussion and get a common solution. Below there are a few examples that can give you some ideas.

- *Walking the talk*: get rid of the reserved parking lot for the CEO. In one of our Belgian company cases, the CEO told us that to make the changes acceptable to all the employees, he needed to stop with having a reserved parking space near the factory building. The CEO is now 'one of the guys/girls' and needs to make the long walk as all the other employees.
- *Let employees decide how to get access to management:* in one French company from Normandy, the boss asked the employees to decide where his office space should be and how it should be organised. The boss works for the employees, not the other way round.

- *Your building structures your behaviour:* it is clear that some office building strengthen hierarchy and division. Rearranging your office according to the important processes is helpful to create a new organisation that supports workplace innovation. Making sure that you have sufficient meeting space for the team meetings is crucial for a good workplace innovation environment.

18.7 Onwards

We hope you will find this *Guide to Workplace Innovation*useful in your own organisation. There is so much to gain in terms of improved economic performance and better ways of working!

As indicated at the beginning of this chapter, we cannot offer a full proof recipe for transforming your organisation because it just doesn't work like that. You have to be prepared to listen, to involve, to experiment, to take risks and to learn.

It also takes discipline to maintain the momentum on top of a busy schedule. But it is surprising how far you can go once you engage the trust and enthusiasm of people throughout the organisation.

If you need further tips and examples make full use of the films, case studies, articles and tools in EUWIN's growing Knowledge Bank.

References

Gronn, P. (2002). Distributed leadership as a unit of analysis. *The Leadership Quarterly, 13*(4), 423–451.

Huselid, M. A., Jackson, S. E., & Schuler, R. S. (1997). Technical and strategic human resource management effectiveness as determinants of firm performance. *Academy of Management Journal, 40*(1), 171–188.

Karasek, R. A., & Theorell, T. (1990). *Healthy work; stress, productivity and the reconstruction of working life*. New York: Basic Books.

Katzenbach, J. R., & Smith, D. K. (1993). *The wisdom of teams: Creating the high-performance organization*. Boston: Harvard Business School Press.

Khurana, R. (2002). The curse of the superstar CEO. *Harvard Business Review*, 3–8.

Oeij, P., Žiauberytė-Jakštienė, R., Dhondt, S., Corral, A., Totterdill, P., & Preenen, P. (2015). *Workplace innovation in European companies. Study commissioned by Eurofound.* Luxemburg: Office for Official Publications of the European Communities.

Teague, P. (2005). What is enterprise partnership? *Organization, 12*(4), 567–589.

Totterdill, P. (2015). Closing the gap: The fifth element and workplace innovation. *European Journal of Workplace Innovation, 1*(1). Retrieved from http://journal.uia.no/index.php/EJWI/article/view/166

West, M. A. (2012). *Effective teamwork: Practical lessons from organizational research.* New Jersey: Wiley-Blackwell.

Author Biographies

Steven Dhondt is a Visiting professor, Chair Social Innovation, KU Leuven and senior research scientist at TNO, The Netherlands Organisation for Applied Scientific Research (Leiden, The Netherlands). Current research: workplace innovation, digitalisation and work

Peter Totterdill is a Director, Workplace Innovation Europe; Visiting Professor, Kingston University London (UK)

Sylvie Boermans is a scientific researcher at TNO, The Netherlands Organisation for Applied Scientific Research (Leiden, The Netherlands). During her post-doc, she investigated the effectiveness of social innovations within the Belgium context.

Rita Žiauberytė-Jakštienė is a psychologist. Having HR consultancy experience, she is interested in the tandem of empowered personnel and successful businesses. Currently she is working at TNO, The Netherlands Organisation for Applied Scientific Research with projects related to workplace innovation and social innovation.

Chapter 19
Management Practices for Promoting Employee-Driven Innovation

Kåre Hansen, Oscar Amundsen, Tone Merethe Berg Aasen and Leif Jarle Gressgård

19.1 Introduction

Organizations increasingly recognize the innovation potential their employees represent. Consequently visionary organizations like Google or Apple allocate a significant amount of working time for their employees to present and develop own ideas, and a Swedish study shows that the organizations that are best at exploiting their employees' ideas in fact receive about 50 ideas per employee per year, in comparison to the average of one idea per employee (Östberg et al. 2010).

Involving employees in improvement and innovation is not a new invention (Tidd and Bessant 2009). Bessant and Caffyn, for instance, introduced the concept of 'High Involvement Innovation' in 1997, based on the idea that the general engagement of employees in innovation will lead to a higher degree of diversification and flexibility than can be seen in organizations where innovation processes are carried out by a limited group of people. Norway and the other Scandinavian countries have a long tradition of employee participation in working life

K. Hansen (✉) · L.J. Gressgård
Department of Social Sciences, International Research Institute of Stavanger, Stavanger, Norway
e-mail: khan@iris.no

L.J. Gressgård
e-mail: ljg@iris.no

O. Amundsen
Faculty of Social Sciences and Technology Management, Department of Education and Lifelong Learning, Norwegian University of Science and Technology, Trondheim, Norway
e-mail: oscar.amundsen@ntnu.no

T.M.B. Aasen
SINTEF Technology and Society, Trondheim, Norway
e-mail: tone.m.aasen@sintef.no

P.R.A. Oeij et al. (eds.), *Workplace Innovation*, Aligning Perspectives on Health, Safety and Well-Being, DOI 10.1007/978-3-319-56333-6_19

(Aasen et al. 2012). This tradition was expanded by the Norwegian government in the so-called White Paper on Innovation (Report no. 7, 2008–2009), where employee involvement was pointed out as an important measure in its policy of innovation.

The core idea of 'employee-driven innovation' (EDI) is that by inciting cooperation between employees and managers, companies' innovation performance will improve greatly. A similar idea, referred to as 'workplace innovation', has been promoted by the European Commission since 2013 (Pot et al. 2016). This can be seen as a paradigm shift presently penetrating the innovation policies in European countries. However, the Norwegian Confederation of Trade Unions took a leading initiative in promoting employee-driven innovation as early as 2006, and as indicated, EDI became part of the Norwegian Governments Innovation Policy two years later. To support the operationalization of the policy, a 'Manual of employee-driven innovation' was introduced by the Ministry of Trade and Industry in 2011 (Amundsen et al. 2011). The manual resulted from a research project seeking to identify a best practice for EDI in Norway. Cases were thus chosen based on expectations of their relevance (Flyvbjerg 2006). In addition to the executive research team, a reference group consisting of representatives from labour market parties actively participated in this process. The manual offers advice on how one may introduce EDI as part of the daily work practices in organizations, based on experiences from private and public enterprises in Norway. This approach now forms the basis for innovation processes in a number of Norwegian organizations.

This article is based on the above-mentioned study conducted for the Ministry of Trade and Industry in 2011. The paper investigates the impact of management practices on the general involvement of employees in innovation work. More specifically, the overall objective is to give answers to the following questions: What are the characteristics of management practices supportive of employee-driven innovation? How do these characteristics correspond with traditions of cooperation and employee participation in Norwegian working life?

In the first part of the article, we discuss the term employee-driven innovation (EDI) in the context of Norwegian working life traditions, followed by a brief review of relevant research on the topic. Further on, we explain the study's methodological approach. The results of the study are presented in the third part, emphasising the characteristics of management practices influencing employee-driven innovation. Finally, findings are discussed in light of the research questions, and then summarized.

19.2 The Context of EDI in Norway

19.2.1 Historical Background

The Norwegian EDI initiatives are grounded in a long research tradition in Norwegian working life, as far back as to the 1950s, where employees' participation in development activities has been important (for further details, see Gustavsen 2011). In a Nordic context, employee participation implies a modification of the owners' management prerogative and of the hierarchical decision-making structure that is found in corporate companies. Institutionally, this modification is achieved by legislation and/or agreements whereby employees are given certain rights regarding their participation in different decision-making processes in the workplace. In the Nordic countries, one has established a set of formalized arrangements intended to guarantee the employees a say in the development of the company. The right to representation on the Board and the establishment of working environment committees, works councils, and departmental councils are all examples of arrangements to ensure the employees' interests are looked after.

When discussing the Nordic model of employee participation it is necessary to distinguish between direct and indirect participation. *Direct participation* is understood as the type of involvement whereby individual employees are given the opportunity to influence conditions in their own workplace. In practice, this form of individual participation, which is often called *participative democracy* (Pateman 1970), is closely related to the question of the employees' self-determination in the work situation. *Indirect participation*, on the other hand, takes place through representatives. The employees elect people to represent them vis-à-vis the owners and the management, and their influence is thus exercised indirectly. This representation may be on behalf of the employees as a group or on behalf of those employees who are members of a trade union (Hansen 2002).

The philosophy of employee-driven innovation is typically based on *individual (direct) participation* and the assumption of a community of interests between the players in the company. However, the Norwegian Confederation of Trade Unions also sees employee driven innovation as a way to democratize innovation processes. They argue that EDI must be based on both direct and indirect participation and that this combination is essential to the way EDI should be performed in Norwegian companies. We will return to this question later on in our discussion of EDI and the role of management.

19.2.2 The Basic Agreement as a Backbone for EDI

In 1982, a supplement was added to the Basic Agreement between the private sector labour market parties in Norway. This new part of the major collective agreement has been and is an important legal backbone for the development of EDI practices in Norwegian companies. The supplement is named 'Agreement on enterprise development' and is a joint initiative encouraging collaboration between employees and managers. It states that 'The objective with regard to enterprise development is to increase the creation of values through broad participation by the employees at the enterprise'.

Today, this part of the Basic Agreement has implemented the term Employee-Driven Innovation as its main concept. The Norwegian Confederation of Trade Unions argues that 'this agreement represents a responsibility for all employees. A responsibility to contribute actively to financial and production-related improvements. However, it is a two-way obligation. The management has an obligation to create conditions for the employees to be able to fulfill their responsibility. The agreement has always been related to productivity of work, but must now also be seen in relation to EDI.' (Kallevig 2012).

Even though the agreement only applies to private enterprises that are members in a participating association, which is about half of the Norwegian enterprises, it is highly interesting when studying how and to what extent EDI is practiced in the working life.

19.3 Why EDI?

There has been no public debate about EDI in Norway. However, the arguments put forward in favor of EDI, mainly from the labour market parties, are very similar to the arguments that have been used for decades when debating employee participation in general. Looking at past debates on these issues in Norway, we find four main lines of arguments. The first line aims to present employee participation as a *means of increasing productivity*. It is believed that giving the employees the opportunity to influence their own work situation will lead to increased job satisfaction and thus provide a basis for improved effort and more efficient production. This type of argument is also associated with the humanistic belief that employees represent a major competence base in the company, and exploiting this is a means to innovate and improve productivity. The second line of arguments is found in the presentation of participatory schemes as *a means of achieving more influence on the decision-making process*, wherein the purpose of employee participation is to achieve a more even distribution of power between employers and employees. What is viewed as the ideal power balance will vary depending, among other things, on the political ideology behind it. The third line of arguments presents employee participation as a *means of subduing conflicts*. It is assumed that closer cooperation between employer and employees may lead to a higher degree of social consensus, which further on will be regarded as a positive state of affairs and a good in itself. Lastly, we have the line of arguments advocating the introduction of employee participation as part of the general *democratisation of society*. Here, the democratisation of areas outside the traditional political institutions is deemed important for the overall democracy in a society (Pateman 1970).

Normally, when debating employee participation, the different arguments will overlap or be related to each other. Looking at the period from 1970 until today, we see a clear tendency towards the ideological and value-related arguments for employee participation being used less and less in debates, while one makes increasingly more use of more market-related arguments of increased productivity, not least by parts of the trade union movement. This is also the case when the labour market parties argue for the use of EDI. The rationale behind the arguments appears to be primarily economic, pointing at a need to increase companies' competitive strength and their ability for innovation. Still, looking at the arguments put forward by the Norwegian Confederation of Trade Unions, we also find more value oriented claims such as the following:

> Employee Driven Innovation represents a democratization of the innovation concept and means that employees contribute actively and systematically in an organized innovation process. Unlike innovation that is determined and driven from above, EDI is a bottom-up process and a more experience/knowledge based practice tied to employees' daily challenges. This implies an active and systematic involvement of all employees in the innovation process. (Kallevig 2012)

19.4 Theory

Innovation processes based on input from sources of competence external to the organization, like user driven innovation (Von Hippel 1995, 2005) and open innovation (Chesbrough 2003), have received considerable attention over the last 10–15 years. In contrast, researchers have generally been less occupied with investigating the significance of new ideas, processes, and methods suggested by own employees (Byrne et al. 2009). However, as the ability to innovate becomes ever more important to all types of organizations, both researchers and managers gradually direct attention towards the inherent opportunities of broad employee involvement in innovation work and continuous improvement.

19.4.1 What is Employee-Driven Innovation?

The Norwegian Confederation of Trade Unions, which has been a driving force in efforts to put EDI on the national and organization policy and research agenda, defines EDI as:

'... Innovations (new products, processes or services) that are produced through an open and inclusive process of innovation, based on a systematic use of people's ideas, knowledge, and experience – that is evolving for the business's overall innovation capability.' (www.lo.no). In line with this, the White Paper on Innovation (Report no. 7, 2008–2009) explains EDI as employees' active participation in the development of goods, services and production processes, and in spin-offs from existing business. There are however divergent views on the characteristics of EDI and what distinguishes it from other types of innovation processes. As an example, Kesselring et al. (2014) emphasize EDI as an informal, bottom-up process. This is in line with the definition proposed by Smith et al.

(2008, p. 1), explaining EDI as: '… the generation and implementation of novel ideas, products and processes originated by a single employee or by joint efforts of two or more employees.' In contrast to this, Høyrup (2012) claims that EDI is also distinguished by qualities associated with top-down innovation processes as well as the exchanges of top-down/bottom up initiatives. Høyrup (2010) refers to EPOC—European Foundation for the Improvement of Living and Working Conditions (1997, p. 15), which describes this central aspect of EDI:

> It is about the scope for improving employment and competitiveness through a better organization of work at the workplace, based on high skill, high trust and high quality. It is about the will and ability of management and workers to take initiatives, to improve the quality of goods and services, to make innovations, and to develop the production process and consumer relations.

This definition of EDI includes initiatives from employees acknowledged and supported by the management. Such initiatives may also become an integrated part of the organizational strategy and structure. Furthermore, the management may adapt for employees' active participation in innovation, making EDI part of the formalized innovation strategy. Thus, in its broad definition, EDI is similar to the more recent concept of 'workplace innovation' (Oeij et al. 2012).

The view on EDI as the broad and structured involvement of employees in innovation forms the basis for this paper. Accordingly, we claim that important driving forces for innovation are joint internal challenges and the interaction between managers and employees in order to create and share resources in the form of knowledge, experience, ideas, creativity, and capacity to meet said challenges. Implementing EDI implies integration of the innovative activities as a natural part of the everyday working life.

19.4.2 EDI and the Role of Management

EDI is based on the fundamental belief that all employees have the potential to contribute to innovation and growth in a company. Systematically involving employees in innovative work implies, however, a close and constructive dialogue between employees and managers (Tierny et al. 1999). Surprisingly, researchers have so far given the managers' role limited attention regarding this type of innovation; studies have generally been focusing on structure, group interaction, and organizational climate (Byrne et al. 2009). To some extent, research focus has also been on management actions affecting employee behavior (De Jong and Den Hartog 2007; Jayasingham et al. 2010; Allen et al. 2015). The finding by Manz et al. (1989) supports the importance of the latter strand of research claiming that management is the most important determinant of innovation. Even several decades ago, Pelz (1956) documented that employees' work motivation, and subsequently the quality of task accomplishment and development, was positively related to their leaders' encouragement of their involvement in everyday work processes. The most important leadership strategy in this respect was to provide the employees with a certain degree of autonomy (i.e. ability to make decisions) while at the same time being available for discussion and consultations. In line with this, clarity concerning expectations and possibilities when it comes to employees seeking improved solutions as part of their daily work tasks is emphasized as important for the organizational innovation capacity (Gjelsvik 2007). A Danish study underlines the importance of leaders' informal visibility and availability. The objective is to enable employees in all departments to engage in discussions about new ideas and suggestions for improvements with managers, in an atmosphere characterized by mutual trust. Based on case studies in Danish organizations, Brandi and Hasse (2012, p. 144) suggest that the most important role of leaders is to turn 'creative actions initiated in local work place cultures into innovation'.

Varied and rich interaction between managers and employees is thus an important factor when it comes to succeeding with EDI. Wilkinson and Dundon (2010) emphasize that this is relevant to both the direct involvement of the individual employee and for the 'representative' involvement of employees. Concerning enterprises with union representatives, one has claimed they may play important roles in building relationships and ensuring well-functioning cooperation between managers and employees (DISKO 1999).

In a recent literature review, Smith et al. (2012) discuss four main factors found to be significant for the development of EDI. These are 'management support', 'employee autonomy', 'cooperation', and 'organizational norms for exploration'. The first factor (management support) is regarded as the most important one. However, the authors suggest that the influence of this factor is dependent on the phase of the ongoing innovation process. At an early stage (idea generation), management support appears decisive, while managers' allocation of resources is more important in later stages (implementation, diffusion) (Gjelsvik 2007).

How, then, do these and similar research findings relate to established management theories? Fundamentally, 'management' refers to the ability to influence others (Sims et al. 2009), and there exists a myriad of theories and typologies developed to discuss the impact of management behaviour on employees' performance. A common distinction is made between classical and contemporary views of management. The former includes *aversive management* (i.e. use of threats, intimidation, reprimand, and punishment), *directive management* (i.e. giving directions, instructions, and commands), and *transactional management* (i.e. dispensation of personal and material rewards in exchange for compliance, effort, performance, and loyalty). The contemporary view refers to *transformational management* (i.e. generation of commitment through formulating and communicating a vision, exhortation, inspiration and persuasion, and challenge to the status quo) (Sims et al. 2009, pp. 150–151). This strand of literature often highlights charisma as an important trait for transformational leaders, and as such the category often incorporates *charismatic management*. Common to both the classical and contemporary management types is that they involve a top-down perspective; that is, the management set the direction and make decisions. In contrast, a more recent perspective on management based on theories of participation and self-management is called *empowering management* (Pearce et al. 2003). This is a bottom-up view on management, where the empowering managers coach others to lead themselves. Accordingly, the focus is on responsibility instead of obedience (Sims et al. 2009). Prerequisite for this is that an effective and rich exchange of information between managers and employees and factors related to work autonomy (i.e. initiative, self-responsibility, self-confidence, self-goal setting, positive opportunity thinking, and self-problem solving) are emphasized (Sims et al. 2009; Hmieleski and Ensley 2007).

19.5 Method

The present study is based on interviews with managers, employees and union representatives in twenty Norwegian enterprises. Common to these enterprises is the fact that they have succeeded in involving employees in innovation and continuous improvement. The selection of informants was based on a list of approximately 40 enterprises prepared by two main sources; the Norwegian Confederation of Trade Unions (LO), and the Confederation of Norwegian Enterprises (NHO). These confederations possess in-depth knowledge concerning Norwegian working life, and they supplied us with suggestions for 'best practice' enterprises. Consequently our methodological strategy was similar to what Miles and Huberman (1994) characterise as 'intensity'. This implies the search for cases that are particularly evident or appropriate to describe a phenomenon (Marshall and Rossman 2006). It is important, however, to emphasize that the researchers made the final choice of whom to interview. As one of our selection criteria, we aimed at obtaining a heterogeneous collection of cases, thus potentially developing a more general insight into the topic of EDI (Schofield 2002). The selected cases comprised both small and large enterprises, private and public. The industries included were construction, process, mechanical, energy, food, medical equipment, electronics, ICT, private and public services, and public register. A more detailed overview of the cases is supplied in the appendix.

Forty-eight respondents were interviewed. We conducted two interviews with individual respondents, the remaining part of the data collection was carried out as group interviews. This is in accordance with Gaskell (2000), who argues that group interviews are applicable where one does not face sensitive topics. He also points out the advantages associated with a setting in which respondents can build on each other's input in the interview situation. The interview guide was developed based on a theoretical framework consisting of general phases of the innovation process, and adapted to fit the particular context of EDI (De Jong and Kemp 2003; Smith et al. 2008). In practice the interviews focused on the manner in which companies were carrying out employee-driven innovation, what they considered to be key factors for success, which 'tools' they used to support the EDI practices, and the advice they would offer to other companies seeking to implement EDI.

The interviews lasted for 1–1½ h and were conducted by two researchers. The informants were divided into three categories; managers, employees and union representatives. All interviews were recorded and subsequently transcribed. The two interviewing researchers coded the interviews according to a common thematic template. Results were then discussed and further analysed by the total research group (which consisted of the four authors).

19.6 Findings

In our study of Norwegian companies practicing employee-driven innovation, we cannot find a unified 'best practice' when it comes to the role and significance of management. Nevertheless, we find some common characteristics among the enterprises, not least concerning the middle management's role.

Several of the informants pointed out that managers who want to promote employee-driven innovation should be less concerned with asserting their own opinions, but all the more focused on direct, listening-oriented interaction with the employees. Managers acting according to EDI principles describe their role as more challenging than before in some aspects, because of their reduced control. At the same time, they experience that this type of leadership is more efficient and 'safe' than the more traditional ways of exercising management. Because letting go, at least in part, of influence and power, leads to a more committed and motivated workforce and a stronger community feeling. Further, managers described their role as being a combination of supporter and promoter:

> I think the key to it all is commitment. It is a very important management task to motivate and push people who have ideas. If there are employees showing that they want to achieve something, you need to support them all the way.

> Managers must be very aware that there are people in all organizations who have clear heads, clear thoughts, and clear opinions. The key lies in the company's ability to capture what is potentially innovative through their employees.

To manage EDI, leaders need to be self-confident enough to give their employees the freedom to develop ideas themselves without intervening too much and controlling the direction—nor 'take over' tasks that employees are responsible for:

> What you must not do when employees start to become a little more motivated, is to finish the last 20% of the job after everyone has gone home. Then you've made a mistake. It is the most difficult tasks that remain. What is fun is done. So a good advice is to let them take everything, do not finish the job for them.

> I do not participate directly in the process, because I know that if I participate, then I'm going to control it. If I do, then I'm not going to get the type of commitment to the process that I want, so instead I try to be an interlocutor.

Broad employee involvement means that managers must accept there may be many activities going on in the enterprise they do not take part in and may not even know about. It also means that the management must make it evident, both in their conduct and manner of speaking, that they really want contributions from employees. An important aspect of this is that managers need to understand what kind of expectations they stimulate among the employees.[1]

We found that managers at all levels are important for promoting EDI, but the middle managers' role seems to be of particular interest. Middle managers, whether they are titled production manager, team leader, discipline leader, or something else, are responsible for the daily operations. As such, they are also the closest to the important internal source for ideas and suggestions: the employees. The middle managers are also in the position to motivate the employees directly and encourage an active approach where the opportunities for improving their own work situation is concerned. In some cases it may also be the middle manager who documents ideas and suggestions from employees. This may be relevant e.g. if employees do not have access to a computer or a whiteboard on their work site, or if they come from other countries. Our respondents indicate that there are certain personal qualities concerning the middle manager that are of particular importance to the success of EDI:

> … a delicate capability to communicate and to be gentle. Being a strong leader, but a listening one as well. One who is involved, one that involves himself in the wellbeing of the employees.

Another important dimension of EDI management is 'collaborative attitude'. Our respondents are very consistent that this is a crucial factor for success:

> … so it has been a success factor that the management and employee representatives have been able to work together. That we 'have ploughed the soil together', and that we have had a plan for how we will work further along. It has, in a way, been a success factor for both parties, that we have been able to see the benefits we achieve through cooperation.

The essential reciprocity in the relationship is evident in this quote: 'It is rather useless to talk about cooperation if the managers do not see the advantage of involving the employees both in planning and business development'. This may be the reason why, when the employees talk about collaboration with managers, confidence is a recurrent theme:

> … management must be willing to give trust, but it goes both ways. In this collaboration, we talk a lot about the fact that if you do not have accountability and take responsibility on both sides, the cooperation will not function. One must rely on one another when working towards the same goal.

Adopting EDI-principles also means that you interfere with the existing hierarchy of the organisation. This was supported by our respondents, who emphasized

---

[1]It should be noted that bonus systems were not found to be important for employee-driven innovation in our study (only 10% of the companies had some kind of reward systems). Thus we do not follow-up on this factor.

that authority conditions in their enterprises were more 'muted' than in other businesses, and that the internal hierarchy was less visible:

> I think that, among our managers, there is no one who is like that really, what should I say … in a way there is no hierarchy […] Where I worked before, people had a strong respect for the boss, but you did not really talk very much with him. If he came into my office, it was almost as if: "What is wrong now?"

Some of the elements mentioned here are especially prominent in companies having adopted ICT tools for idea generation and innovation management. This kind of software has been increasingly adopted during the last decade, in particular by larger companies. One consequence of this technology is that the whole innovation process opens up and unfolds internally. This also influences the role of managers in these companies:

> The work with […] and focus on that everyone should not only be sellers in the company, but that absolutely all employees should be innovators, it has meant that we simply have gone from being a black box when it comes to product development, to being a completely open company. Very high transparency. The discussions are open for everybody, one can disagree with the VP […] in the comments, without anyone reacting on it […] I doubt there are many other companies doing it our way. Here, management must take part in an open dialogue, internally, so to speak.

Where companies have applied software of this type, it is also emphasized that technology affects the whole power balance internally, because more people will have the opportunity to influence the innovation process:

> It (the ICT-tool) affects everyone in the company in a way that makes them to a large extent feel obliged to be creative. They feel obliged to try to add value or help if we have a challenge or a problem. People are much more eager to take part in the development of products and services. I think in many ways this changes the balance of power in the company, especially regarding activating and decision-making, moving from being a matter for a few people to actually involving all the employees. It is perhaps the biggest change we have had, that we earlier had about 10 employees who took care of product development and innovation, and know we have 200, who can have different opinions, and that really affect what we develop.

To sum up, the five most important factors related to management behaviour influential on enterprises' success with EDI, are: (1) *work autonomy* (i.e. freedom of action combined with clear expectations regarding employees' contributions), (2) *diversity promotion* (i.e. encourage and highlight the significance of employees' ideas), (3) *informality* (i.e. informal presence and cooperation), (4) *de-emphasizing organizational structures and roles* (i.e. reduction of status differences and hierarchical relations), and (5) *support* (i.e. backup of employees). These factors are relevant to all management levels in an organization, but as already emphasized, the middle manager seems to play a particular important role where succeeding with EDI is concerned.

19.7 Discussion

With reference to the somewhat sparse research on the management's role in EDI, we find that our results are in tune with the ideas of 'empowering management'. A cooperative and less controlling management style is needed to incite employees' engagement in innovation. Furthermore, successful EDI implies the development of a fair and transparent process of stimulating, capturing, and following up ideas from employees. Our findings also confirm that EDI must be founded on a strong relationship of trust between management and employees, and that such trust can deteriorate if mutual expectations are not met. In large enterprises, EDI management seems particularly dependent on effective sharing of information about the idea development process (e.g. evaluation criteria and reasons for rejection), as well as on development of clear organizational objectives shared by all organizational members.

Interestingly, several studies show that management in Norway and the other Nordic countries share some particular characteristics separating them from management styles in continental Europe and USA (e.g. Smith et al. 2003). These characteristics are very similar to the management qualities we have identified as being supportive of EDI. As an example, Brodbeck et al. (2000) analyzed country-level variation in the European data from the 61-nation GLOBE survey (House et al. 2003) of management behaviour. They found that while employees from Northern and Western Europe preferred inspirational leaders of high integrity, the opposite was true for employees from Southern and Eastern Europe. Another example is Lindell and Arvonen (1996), who asked more than 2500 employees to characterize the management style of their superiors in Finland, Sweden, Norway and Denmark. Responses were scored on scales measuring orientation towards the task, the employees, and change and development. Compared to corresponding scores from other European nations, Lindell and Arvonen found that Nordic managers put less emphasis on the structuring of tasks and showed more consideration towards their employees. Norwegian managers were found to have the highest scores among the Nordic countries on both these parameters. Similarly, Hampden-Turner and Trompenaars (1993) found that a higher percentage of managers in Sweden, Denmark, Norway, and Finland reported delegating decisions to subordinates than in many other nations. Recent studies confirm that Scandinavian countries still get high scores on job autonomy, employee involvement, and workplace innovation (Eurofound 2012, 2013).

Vie (2012) has analyzed further the management traditions predominant in Norway. He finds that they can be categorized into two main groups. The first tradition regards management as an *individual characteristic*, while the second approaches *management as an organizational function*. Within the second tradition, a strong emphasis is placed on collaboration and on employee involvement and empowerment (Grønhaug and Hansen 2001). Vie (2012) argues that this tradition is the one dominant in Norwegian working life, reflected in the way employers and employees think about organization and management. Seen this way, one could argue that the Norwegian labour market represents an ideal arena for employee-driven innovation, and that EDI can be implemented in any business, regardless of the managers' personal qualities, as long as the managers facilitate the development of culture and work processes supportive of EDI in collaboration with the employees. However, our data indicate that the role of managers is crucial in succeeding with EDI and representatives from the companies we have studied, both managers and employees claim that this role is not easily learned because it requires certain personal qualities. For example, managers of EDI need to be collaborative, inclusive, and perceived as trustworthy. As such, managers play an important part in the development of an enterprise culture supporting EDI.

As outlined in Amundsen et al. (2014), we have already identified nine cultural characteristics distinctive of enterprises successfully aligned with principles of EDI, see Table 19.1.

These characteristics are interrelated, which implies that efforts to change the nature of one of them probably would affect one or more of the others. Most of these characteristics can be linked directly or indirectly to management practices, but there are four conditions in particular that stand out and have clear implications for the management role, namely 'Cooperative orientation', 'Trust', 'Openness', and 'Autonomy'. The feature we have termed cooperative orientation implies a widespread perception in the enterprise that there is 'agreement to cooperate' and that cooperation is the best basis for the innovation work. Generally, one can say that organizations with this feature have a cooperative climate where employees and management are concerned. Maintaining and contributing to develop this cultural feature implies that managers are willing to cooperate through concrete actions.

When it comes to 'trust', this is typically expressed as a mutual relationship: The managers show trust in employees, while employees show trust in their managers

Table 19.1 Cultural characteristics

Cultural characteristic	Description
Commitment	High commitment towards innovation among employees
Cooperative orientation	Basic assumption that there is 'agreement to cooperate' between management and employees
Pride	Pride in working for this particular enterprise
Trust	Enterprises characterized by trustful relationships
Tolerance	Assumption that 'diversity is valuable', and that errors must be tolerated
Feeling of security	Feeling that 'thinking aloud is permitted' and that suggested improvements will be followed up by education, training, or opportunities of new job situations
Development orientation	Employees see improvement and innovation as an *integral* part of their jobs
Openness	Openness in communication, internally and externally
Autonomy	Employees have a high degree of influence in relation to the execution of various tasks

and each other. Trust is built gradually through interaction, which for the practice of management implies exercising and showing trust in their everyday practice, to 'initiate a virtuous circle'.

The cultural feature 'openness' means openness in communication, internally and externally, e.g. by using various types of open forums and extensive access to documents relating to strategy and decisions. In practice, this also means that 'disagreeing with the managers is allowed' and that this will not have negative consequences for employees. We also found that leaders typically practiced an 'open-door policy', i.e. that it was easy for employees to make informal contact.

It is no surprise that 'autonomy' is a feature of the organizational culture in our study. To the employee, this means freedom in problem solving, while, for the managerial role, it implies that one actually delegates responsibility in practice. Delegating responsibility can be challenging for a manager, but if autonomy is to be carried out, it will be necessary for the manager to avoid acting too control-oriented in the wake of delegation.

In several of the enterprises included in the present study, respondents reported about managers having succeeded in involving employees in innovation efforts, but also about others having failed to develop cooperative relations and the commitment necessary for successful EDI. Several respondents referred to individual managers' personal qualities and abilities when explaining EDI practices. Our findings thus indicate that all enterprises aiming to implement principles of employee-driven innovation should do a thorough assessment of their management styles, and not take for granted that all managers have, nor are motivated to develop, the leadership abilities to succeed with this kind of innovation strategy.

19.8 Conclusion

Employee-driven innovation is an important strategy in order to improve the exploitation of a company's innovation potential. However, this makes demands on both employee conduct and management practices. Our empirical findings lead us to suggest that employees should be given some level of autonomy, combined with clear expectations regarding their contributions to innovation and improvement. This also implies that managers should accept that authority structures are weakened and that their own perspectives will and should be challenged by the employees. Another major finding is that managers should adopt an informal and cooperative style of leadership. Furthermore, managers should show understanding and respect towards the employees' ideas and give the employees support and backup, especially during the first phases of the innovation process.

Appendix

Industry (NACE-classification)	Company	Sector		Size (employees)	Informants		
		Private	Public		Leader	Employee	Union rep.
Administrative & support service activities	1		X	50–100	1	2	
Construction	2	X		>250	1	1	1
	3	X		>250	1	1	
	4	X		50–100	1	1	
	5	X		50–100	2		
Electricity, gas, steam & air condition supply	6		X	>250	1	1	
Financial & insurance activities	7	X		10–50	1	1	
Information & communication	8	X		>250	1	1	
	9	X		100–250		2	
	10	X		50–100	1	2	
Manufacturing	11	X		>250	1		1
	12	X		10–50	1	2	
	13	X		10–50	1	2	
	14	X		>250	1	1	1
	15	X		>250	1	2	1
Professional, scientific & technical activities	16	X		50–100	1		
	17	X		50–100	1	1	1
	18	X		>250	2	1	
Public administration & defence; compulsory social security	19		X	>250		2	
	20		X	>250	1		

References

Aasen, T. M., Amundsen, O., Gressgård, L. J., & Hansen, K. (2012). In search of best practices for employee driven innovation: Experiences from Norwegian work life. In K. Møller et al. (Eds.), *Employee-driven innovation: A new approach to innovation* (pp. 57–76). London: Palgrave Macmillan.

Amundsen, O., Gressgård, L. J., Hansen, K., & Aasen, T. M. (2011) *Medarbeiderdrevet Innovasjon: En Håndbok med Råd og Veiledning*. Nærings og Handelsdepartementet (in Norwegian).

Amundsen, O., Aasen, T. M., Gressgård, L. J., & Hansen, K. (2014). Preparing organisations for employee-driven open innovation. *International Jouranal Business Science and Applied Management, 9*(1), 24–35.
Allen, M. R., Adomdza, G. K., & Meyer, M. H. (2015). Managing for innovation: Managerial control and employee level outcomes. *Journal of Business Research, 68,* 371–379.
Bessant, J., & Caffyn, S. (1997). High-involvement innovation through continuous improvement. *International Journal of Technology Management, 14*(1), 7–28.
Brandi, U., & Hasse, C. (2012). Employee-driven innovation and practice-based learning in organizational cultures. In S. Høyrup, M. Bonnafous-Boucher, C. Hasse, M. Lotz, & K. Møller (Eds.), *Employee-driven innovation: A new approach to innovation* (pp. 127–148). London: Palgrave Macmillan.
Brodbeck, F., Frese, M., et al. (2000). Cultural variation of leadership prototypes across 22 European countries. *Journal of Occupational and Organisational Psychology, 73,* 1–29.
Byrne, C. L., Mumford, M. D., Barrett, J. D., & Vessey, W. B. (2009). Examining the leaders of creative efforts: What do they do, and what do they think about? *Creativity and Innovation Management, 18*(4), 256–268.
Chesbrough, H. (2003). *Open innovation. The new imperative for creating and profiting from technology*. Boston: Harvard Business School Press.
De Jong, J. P. J., & Kemp, R. (2003). Determinants of co-workers' innovative behaviour: An investigation into knowledge intensive services. *International Journal of Innovation Management, 7*(2), 189–212.
De Jong, J. P. J., & Den Hartog, D. N. (2007). How leaders influence employees' innovative behavior. *European Journal of Innovation Management, 10*(1), 41–64.
DISKO. (1999). *The Danish Innovation System*, Rapport Nr. 9.
European Foundation for the Improvement of Living and Working Conditions. (1997). *EPOC—Employee direct Participation in Organisational Change*. New Forms of Work Organisation: European Research Report.
Eurofound. (2012). *Fifth European working conditions survey*. Luxembourg: Publications Office of the European Union.
Eurofound. (2013). *Work organisation and employee involvement in Europe*. Luxembourg: Publications Office of the European Union.
Flyvbjerg, B. (2006). Five misunderstandings about case-study research. *Qualitative Inquiry, 7*(2), 189–212.
Gaskell, G. (2000). Individual and group interviewing. In M. W. Bauer & G. Gaskell (Eds.), *Qualitative researching—With text, image and sound* (pp. 38–56). London: Sage Publications.
Gjelsvik, M. (2007). *Innovasjonsledelse*. Bergen: Fagbokforlaget. (in Norwegian).
Grønhaug, K., & Hansen, K. (Eds.). (2001). *Medvirkning, læring og konkurranseevne*. Bergen: Fagbokforlaget. (in Norwegian).
Gustavsen, B. (2011). The Nordic model of work organization. *Journal Knowledge Economy, 2,* 463–480.
Hampter-Turner, C., & Trompenaars, F. (1993). *The seven cultures of capitalism*. New York: Doubleday.
Hansen, K. (2002). Postulater om deltagelse, demokrati og arbeidsmiljø – historien om forsøket på å praktisere en teori. *Tidsskrift for Arbejdsliv, 1,* 29–45. (in Norwegian).
House, R. J., Hanges, P. W., Javidan, M., Dorfman, P., & Gupta, V. (Eds.). (2003). *Culture, leadership, and organizations: The globe study of 62 societies*. Beverly Hills: Sage.
Hmieleski, K. M., & Ensley, M. D. (2007). A contextual examination of new venture performance: Entrepreneur leadership behavior, top management team heterogeneity, and environmental dynamism. *Journal of Organizational Behavior, 28,* 865–889.
Høyrup, S. (2010). Employee-driven innovation and workplace learning: Basic concepts, approaches and themes. *Transfer: European Review of Labour and Research, 16,* 143–154.

Høyrup, S. (2012). Employee-driven innovation: A new phenomenon, concept and mode of innovation. In S. Høyrup et al. (Eds.), *Employee-driven innovation: A new approach to innovation* (pp. 3–33). London: Palgrave Macmillan.
Jayasingam, S., Ansari, M. A., & Jantan, M. (2010). Influencing knowledge workers: The power of top management. *Industrial Management and Data Systems, 110*(1), 134–151.
Kallevig, A. (2012). *Medarbeiderdrevet innovasjon*. Retrieved Okt 15 2012 from http://www.arbeidslivet.no/Arbeid1/Naringspolitikk/Medarbeiderdrevet-innovasjon/. (in Norwegian).
Kesselring, A., Blasy, C., & Scoppetta, A. (2014). *Workplace innovation. Concepts and indicators*. Report as part of the European Innovation Scoreboard 2013–2014 project for Directorate-General Enterprise and Industry.
Lindell, M., & Arvonen, J. (1996). The Nordic management style in a European Context. *International Studies of Management and Organization, 26*(3), 73–92.
Manz, C. C., Bastien, D. T., Hostager, T. J., & Shapiro, G. L. (1989). Leadership and innovation: A longitudinal process view. In A. H. Van de Ven, H. L. Angle, & M. S. Poole (Eds.), *Research on the management of innovation* (pp. 613–636). New York: Oxford University Press.
Marshall, C., & Rossman, G. B. (2006). *Designing qualitative research*. Thousand Oaks: Sage.
Miles, M. B., & Huberman, A. M. (1994). *Qualitative data analysis: A methods source book*. London: Sage.
Oeij, P. R. A., Dhondt, S., Kraan, K., Vergeer, R. & Pot, F. (2012). Workplace innovation and its relations with organizational performance and employee commitment. *Lifelong Learning in Europe, 4* (article 10). www.lline.fi
Östberg, L., Robinson, A. G., & Schroeder, D. M. (2010). *Small ideas, great results (Små idéer: stora resultat)*. Stockholm: SIS Forlag.
Pateman, C. (1970). *Participation and democratic theory*. Cambridge, UK: Cambridge University Press.
Pearce, C. L., Sims, H. P., Cox, J. F., Ball, G., Schnell, E., Smith, K. A., et al. (2003). Transactors, transformers and beyond: A multi-method development of a theoretical typology of leadership. *Journal of Management Development, 22*(4), 273–307.
Pelz, D. C. (1956). Some social factors related to performance in research organizations. *Administrative Science Quarterly, 1,* 310–325.
Pot, F., Totterdill, P., & Dhondt, S. (2016). Workplace innovation: European policy and theoretical foundation. *World Review of Entrepreneurship, Management and Sustainable Development, 12*(1), 13–32.
Schofield, J. W. (2002). Increasing the generalizablity of qualitative research. In: A. M. Huberman, & M. B. Miles (Eds.), *The qualitative researcher's companion*. Thousands Oaks: Sage.
Sims, H. P., Jr., Faraj, S., & Yun, S. (2009). When should a leader be directive or empowering? How to develop your own situational theory of leadership. *Business Horizons, 52,* 149–158.
Smith, P. B., Andersen, J. A., Ekelund, B., Graversen, G., & Ropo, A. (2003). In search of Nordic management styles. *Scandinavian Journal of Management, 19,* 491–507.
Smith, P., Kesting, P., & Ulhøi, J. P. (2008). *What are the driving forces of employee-driven innovation?* Paper Presented at the 9th International CINet Conference, Valencia, Spain, September 2008.
Smith, P., Ulhöi, J. P., & Kesting, P. (2012). Mapping key antecedents of employee-driven innovations. *International Journal of Human Resources Development and Management, 12*(3), 224–236.
Tidd, J., & Bessant, J. (2009). *Managing innovation. Integrating technological, market, and organizational change* (4th ed.). Chichester: Wiley.
Tierney, P., Farmer, S. M., & Graen, G. B. (1999). An examination of leadership and employee creativity: The relevance of traits and relationships. *Personnel Psychology, 52,* 591–620.
Vie, O. E. (2012). Ledelse på norsk. *Magma, 4,* 60–67.

Von Hippel, E. (1995). *The sources of innovation*. New York: Oxford University Press.
Von Hippel, E. (2005). *Democratizing innovation*. Boston: MIT Press Books.
Wilkinson, A., & Dundon, T. (2010). Direct employee participation. In A. Wilkinson, P. J. Gollan, M. Marchington, & D. Lewin (Eds.), *The Oxford handbook of participation in organizations* (pp. 167–185). New York: Oxford University Press.

Author Biographies

Kåre Hansen is a Research Director, Department of Social Science, International Research Institute of Stavanger. Current research: Interorganizational complexity and cooperation, employee autonomy and participation, organizational safety.

Oscar Amundsen is a Professor, Department of Education and Lifelong Learning, Norwegian University of Science and Technology. Current research: Employee-driven innovation and symbiotic learning in municipal care work, organizational change, innovation and participation in work-life.

Tone Merethe Berg Aasen is a Senior research scientist, Department of Industrial Management Institute of Technology and Society, SINTEF. Current research: business strategy and innovation in established enterprises, complex models of service innovation, disruptive innovation, employee-driven innovation.

Leif Jarle Gressgård is a Senior Research Scientist, Department of Social Science, International Research Institute of Stavanger. Current research: Interorganizational complexity and cooperation, technology management, knowledge exchange and organizational learning, innovation management.

Chapter 20
How Can Work and Organisational Psychologists Fortify the Practice of Workplace Innovation?

Maria Karanika-Murray and Peter R.A. Oeij

20.1 Introduction

The increasing adoption and implementation of workplace innovation (WPI) practices in business organisations poses a number of challenges for the role of work and organisational (W&O) psychologists in WPI. Here, by work and organisational psychologists we refer to researchers and practitioners in the fields of occupational psychology, occupational health psychology, industrial and organisational psychology, and cognate areas, whereas we use the term WPI to refer to innovations in deploying human talent and organising work processes that should result in good work and better organisational performance. The implications of WPI practice for W&O psychologists are the need to find synergy in people and organisational issues on the one hand, and the need to communicate the value and potential of WPI to stakeholders with different backgrounds, on the other. Challenges that emerge from the meeting of W&O psychology and WPI practice include, for example, W&O psychologists being called to provide rigorous evidence for relevant practice, often having to move between increasingly varied roles as both reflective practitioners and action researchers, and being required to communicate with diverse groups of stakeholders with different agendas and understandings. Unless such challenges are successfully addressed, they can become barriers for the successful utilisation of W&O psychological knowledge in the implementation of WPI practice.

M. Karanika-Murray (✉)
Department of Psychology, Nottingham Trent University,
Nottingham, United Kingdom
e-mail: maria.karanika-murray@ntu.ac.uk

P.R.A. Oeij
TNO, The Netherlands Organisation for Applied Scientific Research,
Leiden, The Netherlands
e-mail: peter.oeij@tno.nl

P.R.A. Oeij et al. (eds.), *Workplace Innovation*, Aligning Perspectives on Health, Safety and Well-Being, DOI 10.1007/978-3-319-56333-6_20

These challenges are not unique to the field. Rather, they reflect a long-standing concern about a practitioner-researcher divide in W&O psychology and in business and management (e.g., Anderson 2005; Argyris 1996; Anderson et al. 2001).[1] The practitioner-researcher divide denotes the phenomenon of practitioners and researchers operating in isolation from each other: research advancements are often ignored by practitioners and practical problems are often ignored in research. More broadly, a practitioner-researcher divide is also afflicting a range of fields including personnel selection (Anderson 2005), nursing practice (Arber 2006), education practice (Fraser 1997), design (Wampler 2010), occupational health and safety (Zanko and Dawson 2012), and even foreign policy (Nye 2008). Others too have called for management scholars to place practice and the pragmatic concerns of practitioners on their agenda (Zanko and Dawson 2012). Nevertheless, a recent upsurge in the solutions proposed to bridge this divide encourages optimism about the chances of success for using W&O psychology to support WPI practice.

In this chapter, we discuss how the practitioner-researcher challenges for W&O psychologists are framed within WPI practice. Then we identify a range of ways in which W&O psychologists can demonstrate the value of the field to WPI and examine ways in which the role of W&O psychologists can be strengthened for successful WPI practice. By examining the transaction between W&O psychology and WPI practice, with this chapter, we address the question "*how can work and organisational psychologists fortify the practice of workplace innovation?*". To achieve this, we draw from a range of literatures, such as W&O psychology, WPI, Human Resource Management (HRM), and industrial relations, taking a necessarily integrative and critical approach.

20.2 What Challenges Is W&O Psychology Called to Deal with in WPI Practice?

WPI practice poses unique challenges for W&O psychology and, at the same time, W&O psychology can offer opportunities for bolstering WPI practice. In practice, there is a risk for the practitioner-researcher divide to be exacerbated unless we can identify ways for the two fields to converge. Here, we discuss the meaning and practice of WPI and what challenges this context poses for W&O psychology research and practice.

[1]In a more general sense a likewise divide can be observed between HR, on one hand, and business, finance and marketing and operations management & engineering, on the other. The point in this article is to stress that the role of W&O psychologists, who often work in HR functions in organisations, can be much bigger if they climb out of their silos.

The definition of workplace innovation (WPI) that we employ here is that of:

> developed and implemented practice or combination of practices that structurally (structure orientation or a focus on division of labour) and/or culturally (culture orientation or a focus on empowerment) enable employees to participate in organisational change and renewal to improve quality of working life and organisational performance (Oeij et al. 2015, pp. 8, 14).

Importantly, the *structure-* and *culture-oriented WPI practices* are part of a broader comprehensive organisational *strategy* that provides the framework for implementing WPI in the specific organisational context and with the available resources. The *structure orientation* contains practices that structure work organisation and job design (De Sitter et al. 1997; Oeij et al. 2015). Structure-oriented practices can stimulate employee control and autonomy (De Sitter et al. 1997). These practices concern the division of labour, the division of controlling (or managing) and executing tasks, and providing employees with decision latitude or capacity for control. For instance, do employers allow employees a genuine say in organisational change initiatives by providing them with task autonomy and voice in decisions; or do they only offer a token to employee empowerment and employability by inviting ideas but not acting on them (Herriot 2001)? Such an approach goes beyond HR-dominated streams of practices (such as high performance work practices and high involvement work practices), because it is rooted in the choices made on how to design the production system. Hence, it goes beyond HR practices by supporting and improving the underlying causes of engagement and not merely softening the possible negative effects of non-engagement. The *culture orientation*, on the other hand, includes practices that provide opportunities for employees to participate in various ways such as, for example, in organisational decision-making (Oeij et al. 2015). Participation is more than being listened to; rather, employees co-decide on the issues that concern them and affect their day-to-day work and well-being (Oeij et al. 2015). Participation is not limited to employees but also applies to employee representatives engaging in dialogue and collective bargaining. Culture-oriented practices can stimulate commitment and provide employees and employee representatives with voice (Totterdill and Exton 2014). As such, not only do they allow for voice in contract negotiations and pay for performance decisions, but also consist of psychological rewards, such as appreciation, recognition and professional acknowledgement. Genuine commitment and voice find expression in 'formal' rewards and in the psychological contract and employee relations.

The practice of WPI poses four challenges that the field of W&O psychology is in a very good position to address. First, in order to practice WPI successfully and reap the benefits associated with it, one needs to *look at the organisation as a whole and consider the reciprocal effects of strategy, structure, and culture* (Howaldt et al. 2016; see also in this volume Van Amelsvoort and Van Hootegem, Chap. 17 and Oeij and Dhondt, Chap. 5). Although not uncontested, it was Chandler (1962) who coined the adage that structure follows strategy, to which we add that culture follows structure (see Fig. 20.1). Strategy determines the design of the production of products or services, based on the central purpose of the organisation. The evolving production system reflects a design built on a certain division of labour, which can be characterised in terms of high or low job autonomy, i.e., decentralised versus centralised. From here follows the nature of operational employment relationships (in particular, dealing with the degree of the division of managing and executing tasks and the splitting up of responsibilities and decision latitude in the working process), which is mirrored in the design of departments, teams, jobs, and tasks. Meanwhile, the management philosophy (i.e., centralised vs. decentralised) determines not only the production system, but also the type of HR system applied to support the production system. As such, the HR system can focus on either control or commitment. Importantly, strategy and structure set the boundaries for the organisational behaviour exhibited by leaders/managers and employees. A preference for centralised or decentralised production systems breeds a type of leadership that is either task-oriented or people-oriented (i.e., transactional and transformational leadership, respectively), and lays foundations for employee engagement. People-oriented leadership associates with a bottom-up approach giving space to employees, whereas task-oriented leadership relates more to top-down management where employee voice is underutilised. Such behaviour is further stimulated or facilitated by the HR system. Ultimately, the established design, i.e., the combined result of the production system, leadership styles and HR system, defines the social and contractual elements of the employment relationships and the features of the economic and psychological contract, described as employee involvement. Finally, strategy, structure, and culture together lead to a number of outcomes including quality of working life (autonomy, stress, motivation etc.), organisational performance (efficiency, effectiveness, customer satisfaction, market share, etc.), and innovative capability (resilience, creativity, resourcefulness, right to play, future proofing, etc.).

Figure 20.1 below displays this reasoning. The absence of a direct arrow from strategy to culture does not imply absence of a relationship between the two. Rather, it highlights the fact that managers design structures that stimulate certain behaviours. In other words, managers design organisations and, in turn, organisational design largely determines people's behaviour. In turn, behaviour and structures define the culture of the workplace itself. For example, people tend to behave differently within a top-down/centralised structure, which reflects a control strategy, as opposed to a bottom-up/decentralised structure, which reflects a commitment strategy.

W&O psychologists are in a good position to help understand the causal links among strategy, structure, and culture, which are too often overlooked (De Sitter et al. 1997; MacDuffie 1997). For example, few managers may consider how strategy impacts structure and consequently employee behaviour, as described in the example above. Few are also able to understand the multi-causal nature of several of these elements. For example, organisations that are run top-down can turn more democratic when stakeholders become more powerful to initiate bottom-up renewal, or when external powers force an organisation to be redesigned. Unfortunately, in practice, W&O psychologists tend to be marginalised, and viewed as peripheral, even juxtaposed to the primary purpose of the organisation, and this

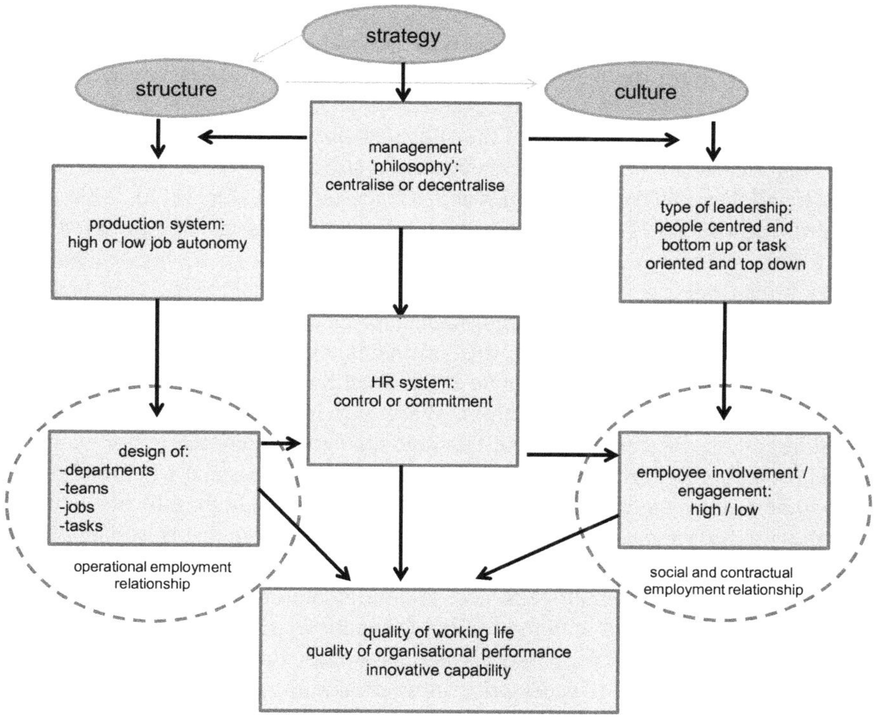

Fig. 20.1 Structure follows strategy, and culture follows structure

tends to limit their opportunities for access to board level decision-making (Karanika-Murray and Weyman 2013). In many organisations, W&O psychologists, especially those who are more practice-focused, are often too much of an island and, for various reasons, also unable to link their role to broader human resource management issues. For example, Alvesson and Sveningsson (2015) warn that when primarily HR-related staff lead change processes, the whole process of change might be considered as just "another HR-thing" (p. 177) among ordinary employees and line management. Shifting the focus away from HR may increase trust and bring ownership of the process to the participants and recipients of change, thus enhancing the potential success of the change process. In the next section we explore how W&O psychologists can position themselves differently and add value.

Second, *WPI is by nature multidisciplinary*: it brings together a range of stakeholders and draws from a range of knowledge and practice domains. WPI is not solely about worker engagement, workplace health, job design, or human resource management. Rather, it is about integrating a range of perspectives such as business and operations management. Too often, however, WPI seems to be approached solely as a human resource management topic, walking into Alvesson and Sveningson's "another HR-thing"-trap (2008). As a consequence, many underestimate the potential of W&O psychology to contribute to WPI, which may result in underusing the potential of WPI (Howaldt et al. 2016). Well-known examples come from the work-related stress literature. For instance, many practitioners and researchers tend to limit themselves to the application of stress management programmes that deal with the effects of stress, but overlook the causes of stress that are deeper within the organisation's structure (Cox et al. 2000, 2007; Karasek and Theorell 1992; Kompier and Kristensen 2001; Oeij et al. 2006).

Third, because WPI practice necessarily involves the organisation in its entirety, it also poses *communication challenges for those involved in its implementation*, including managers, researchers, practitioners, and other stakeholders. In practice, human resource, line, and operational managers seem to function within separate silos within organisations. Indeed, this communication issue is known (Petronio et al. 1998; Roehling et al. 2005; Stone 2004; Sutcliffe et al. 2004). By appreciating the stakeholders' different perspectives, W&O psychologists can help to identify and address their different needs and facilitate dialogue among them. For example, by understanding both research and the needs of the business and its commitments to customers, they are able to better translate research findings into practice and align these to business priorities. By understanding leadership theory and employee motivation, they are able to appreciate the challenges that managers have, identify the motivation needs of employees, and smooth communication between the two. And by getting acquainted with the basics of operations management, they are able to become better partners for engineers and shop floor managers.

Fourth, although WPI is necessarily an affair among multiple stakeholders, the hierarchical nature of organisations often means that *power rather than relevance or expertise determine the influence of specific stakeholders* and this is especially the case in WPI practice. Power in most organisations is asymmetrically distributed (Buchanan and Badham 2008), which means that owners and managers have higher decision-making power than those carrying out the work. Often, management fads, opinions, and desires feed change, rather than rigorous evidence and evidence-based good practice. How managers think largely influences how the organisation is or should be run. A management philosophy, for instance, to centralise or decentralise, may strongly affect whether an organisation is led more top-down or bottom-up, respectively. Convincing examples stem from the literature on lean management. Originally, lean management saw high quality of working life and genuine team autonomy as key drivers for enhancing the quality of performance (Suzaki 1987; Womack and Jones 1996; Womack et al. 1990). However, the practical application of lean thinking has been dominated by a drive to improve cost-efficiency at the detriment of the quality of jobs, essentially increasing workload (Oeij et al. 2013). In this case, the potential of W&O psychology to take a whole-systems approach can be beneficial. The context of WPI makes collaboration between practitioners and researchers and between W&O psychologists and other professionals extremely important. In the next section, we make the potential value of W&O Psychology (WOP) more tangible.

20.3 How Can WPI Practice Recognize the Untapped Potential of W&O Psychology?

Achieving a more substantive use of W&O psychology in WPI practice relies on two conditions: that WPI recognises the potential that W&O psychology can offer and, at the same time, that W&O psychologists broaden their role in WPI practice. For WPI to recognise the potential of W&O psychology, two recommendations can be made.

First, it is necessary that all WPI stakeholders develop a recognition that WPI practice is multidisciplinary and involves a strategic focus on the whole organisation. Power and influence is important only to the extent that it is functional and can help to achieve an agreed common goal. In this case, the common goal is to successfully implement WPI, which can only be achieved if all the elements of WPI are met and if all stakeholders and WPI practitioners (psychologists, HR specialists, and social science practitioners) collaborate.

In addition, because of their training, W&O psychologists are in a good position to deliver the evidence in evidence-based management practice. Chartered W&O psychologists are trained intensively in all EU countries, but this training rarely includes a focus on organisational strategy and structure. Integrating this focus in W&O psychologists' training would help to contextualise their knowledge, make it more easily applicable in practice, and strengthen its transferability in a range of settings. This is all the more relevant in the context of WPI, given that, WPI research falls into the realm of applied science and involves offering solutions to problems (mode 2 of research) rather than developing 'scientific inquiry' (mode 1 of research) (cf. Anderson et al. 2001; Gibbons et al. 1994). This is in line with Argyris's (1996) notion of a need for actionable knowledge, that is, knowledge that can be used practically to improve the functioning of organisations. For instance, he points out that, whereas there is much work in the empirical literature on the relevance of trust in managing, little attention has been paid to how managers can create trust. Mobilising, translating, adapting, and applying research findings in order to develop relevant practice that is based on solid evidence is a strength that W&O psychologists bring.

All WPI practitioners could consider the fact that in practicing WPI, *culture* is dependent on both *structure* and *strategy*, and that these are determined by management, marketing, business and sales, and (technical/operational) engineers. This requires adopting a more pluralistic approach to collaboration. Indeed, for instance, team innovativeness is dependent on both team climate and team structure (Anderson et al. 2014). Adopting a pluralistic approach is a matter of self-reflection for all those involved in WPI practice in order to make the most of everyone's skills, knowledge, and expertise.

20.4 How Can W&O Psychologists Strengthen Their Contribution to WPI Practice?

W&O psychologists too can implement some changes in order to claim a place or develop a stronger foothold in WPI practice. Here we present our recommendations on how this can be achieved.

First, for W&O psychologists to influence WPI, they must surpass HR management and *become acquainted with production systems design*. This means that they should understand the relationship between operational systems and job tasks and how these job tasks relate to human resource issues. Adopting such a role would enable them to partake in improving both performance and the quality of working life. It is thus possible to broaden the immediate focus of W&O psychology (from human resource management issues, individual health, and job design, for example) and become acquainted with organisational strategy, structure, production systems design, marketing, and IT systems. As Fig. 20.2 below indicates, the role of W&O psychologists can be expanded beyond human resource staff or 'general' managers (such as engineers, marketers, and technical managers) to that of consultancy partners or interlocutors of functionaries. Engineers, IT designers, and operational managers design the (technological) production system, which defines whether job autonomy will be centralised or decentralised. Marketers develop products in conjunction with manufacturing that determines how production orders flow through the organisation, namely with or without a say of internal production experts. Human resource staff design human resource systems as 'supporting' or 'advising the business', which has consequences for workers in becoming docile or proactive task executors. Finally, managers and team leaders may wish their employees to follow what markets demand or to absorb market knowledge themselves from customers. Consequentially, employees may become trivial task executors or co-innovators of the firm's products or services. Building this expertise also means and perhaps even requires that W&O psychologists build additional theoretical knowledge about innovation as a phenomenon and process. Being fully conversant in both WPI and HR issues would place them in an excellent position as 'the spider in the web' that is linking the conversation about strategy, structure, and culture for the benefit of WPI. The W&O psychologist is purposefully not depicted in Fig. 20.2, as his or her role is to offer advice to the change leader who is more central to the process and to enable the stakeholders to engage with the issues when WPI interventions are being developed and implemented. Not depicted either in this scheme for reasons of simplicity, are employees/employee representatives and top management, who of course play either a direct or an indirect role (via representatives). Whether W&O psychologists embrace their role as active consultants or accept a secondary dependent role largely determines how their expertise is used and developed. If W&O psychologists choose the first avenue, W&O psychology can become more 'organisational' in relation to WPI practice. This is a matter of self-learning and expanding the W&O psychology knowledge base.

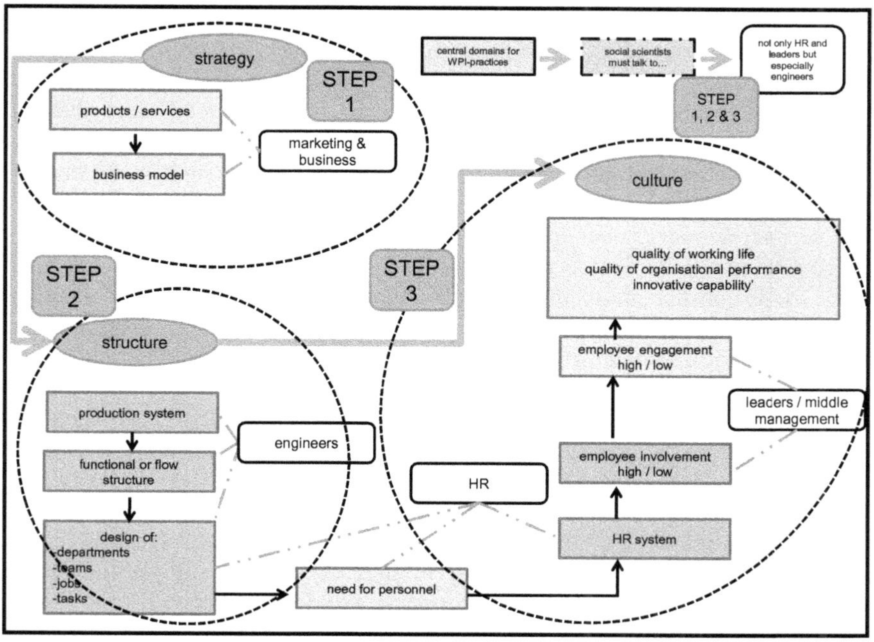

Fig. 20.2 Flowchart to conversations about the design of strategy, structure, and culture from the W&O psychologist's (or social scientist's) perspective. Note: *White boxes* the interlocutors of the W&O psychologist/consultant; *Grey* central domains for the implementation of WPI-practices; *Blue* domains less central to the design of WPI-practices, but with consequences for WPI-practices or how WPI-practices play out (e.g. related to technical interventions in *Step 1*, team design and staffing in *Step 2*; reward systems in *Step 3*); *Orange* W&O psychologists are not allowed to ignore that they must talk to *White box* interlocutors about *Grey* WPI issues if they want to steer on causes, and not just on effects ('symptoms'). *Step 1* At the strategic level: talk to marketing and business people who are responsible for products/services and the business model. *Step 2* at the structure level talk to engineers who are responsible for designing the production system into smaller segments like departments and tasks; align the talking to engineers with the talking to HR people (links with *Step 3*), who are responsible for staff, and the co-design of departments, teams, jobs and tasks, and the HR system. *Step 3* concerning culture, continue to talk to HR people and leaders and managers about involving and engaging organisational members. Leadership styles and mature ways of communication with bottom-up inputs are options for choice. Notice that the order of *Steps 1–3* suggest linearity in a change process; this is never the case. The steps will likely iterate (color figure online)

In addition, it is important for an organisation's management to understand that *W&O psychologists' expertise can contribute to both better jobs and better organisational performance* (De Sitter et al. 1997; Pot 2011; Ramstadt 2014). The two are inseparable. W&O psychologists are also able to help achieve a balance between a focus of WPI at the organisational level with a focus of WPI at the individual level. This implies balancing business values and corporate economic objectives with humanistic and societal values (Lefkowitz 2008). This is a matter of

W&O psychologists adopting a new role and becoming allies with top management, decision-makers, and business owners, rather than simply acting as researchers or consultants in the process of WPI implementation. Those who make the decisions need expert input on matters on which they are not as knowledgeable. A combination of knowledge and decision-making authority can lead to more responsive practice and this can only be achieved by delegating a more strategic role to W&O psychologists in organisations practicing WPI.

W&O psychologists also *have a catalytic role for evidence-based management practice* (Cascio 2007; Rousseau and McCarthy 2007). By sharing their expertise, they can demonstrate how research can provide solutions to broader strategic challenges. By communicating and translating research findings they can help practitioners solve problems (Shapiro et al. 2007). For example, Aguinis et al. (2010) described how psychologists can demonstrate rigour and relevance of research for specific groups in specific contexts by collecting additional quantitative data of more localised qualitative data to supplement existing knowledge. This can be achieved by striving for a balance between the particular (relevance) and the general (rigour) and for strong research that is relevant for the aims and practices of business organisations. Neither overemphasising relevance at the expense of rigour (Aram and Salipante 2003) nor pushing for rigorous research, whose findings are not readily applicable to organisational practice (Anderson et al. 2001), is useful.

Furthermore, where the evidence is scarce, W&O psychologists can apply their research skills to investigate specific practitioner-oriented research issues (Shapiro et al. 2007). The generation of such evidence has to be problem-initiated rather than a purely intellectual activity, transcend epistemological doctrinaire views, and geared at testing the validity of research as "utilization of the knowledge in the world of practice" (Aram and Salipante 2003, p. 203) thus as being actionable (Argyris 1996). The essential question is: can this research evidence or new knowledge be immediately applied into practice? In line with this, Hirschkorn and Geelan (2008) suggested that creating research translation roles is one of the four essential solutions for closing the research-practice gap (the others being: fixing the researcher, fixing the practitioner, and fixing the research). Creating a role for the 'research translator' who "would be adept at speaking the language of both practitioners and researchers and would be able to translate research findings into a form that is comprehensible, plausible, and appears potentially fruitful to practitioners, as well as to convey the interests and concerns of practitioners to researchers" (Hirschkorn and Geelan 2008, p. 11) would also be useful. To paraphrase Lewin, there is nothing as practical as applicable research findings.

Of course, meeting these challenges and redefining these roles can only be achieved by no other than W&O psychologists themselves who ought to be equipped with specific tools. We use 'tools' rather than 'skills' to emphasize practical immediacy and application in organisational practice. *One of the most important tools in this respect is political acumen.* Indeed, "evidence-based management is an inherently political project" which masks "underlying fundamental differences of interpretation, purpose, and power among the various stakeholders situated on both sides of the academic practitioner/policy divide" (Hodgkinson

2012; p. 404). W&O psychologists need to "engage in political activity in order to reduce or redirect the influence of the key stakeholders" (Anderson et al. 2001). As Anderson et al. (2001) observe, the push and pull between two groups of stakeholders, powerful academics and organisational clients, drives practitioners towards either pedantic or populist science and away from the ideal of pragmatic science. By exercising political acumen and taking a more strategic approach to collaboration, W&O psychologists can help to balance practical relevance with methodological rigour (Anderson et al. 2001; Buchanan and Badham 2008; Cascio 2007).

Furthermore, redistribution of power and influence necessarily involves the *development of communities of practice* who can be crucial for translating and adopting research into practice and for highlighting practical problems to guide research. If participatory action research is essential for WPI, communities of practice can offer the bridges by which W&O psychologists can produce knowledge *for* WPI practice. As Bartunek (2007) notes, "the most frequent means of creating academic practitioner relationships is through engaged scholarship, or collaborative research", which implies "relationships between researchers and practitioners that jointly produce knowledge that can both advance the scientific enterprise and enlighten a community of practitioners" (p. 1328). Thus, 'engaged scholarship' as a mode of linking research and practice can both boost the relevance of research to practice and also contribute to enhanced domain knowledge (Van de Ven 2007; Van de Ven and Johnson 2006; also see McKelvey 2006). Developing communities of practice may be difficult, but it is possible. It may necessitate aligning researchers' and practitioners' disparate beliefs about science and the relevance of the scientific method for the workplace (McIntyre 1990). Because W&O psychologists in academic and applied settings tend to differ in their work values, (Brooks et al. 2003), developing communities of practice may also necessitate acknowledging and being more tolerant of these differences. For example, Brooks et al. (2003) showed that autonomy and scientific research were more important for academics, whereas affiliation, money, and a structured work environment were more important for practitioners. By applying his or her specialised analytical background into real-world practical settings, the experienced academic practitioner is in a position to appreciate differences in values and priorities, and align the needs of practice with the values of research.

20.5 Concluding Thoughts

There has been increasing concern in W&O psychology about the divide between research and practice, which is clearly evident in the context of WPI. In this essay we have highlighted a range of ways to achieve a meaningful and productive engagement between the two. Although a small minority believe that the researcher practitioner divide is too challenging to bridge (e.g., Kieser and Leiner 2009) or that the scientist-practitioner model too challenging to adopt (e.g., Brooks et al. 2003; Murphy and Saal 1990), we have highlighted many reasons to be optimistic. As some scholars note, researchers and practitioners are more alike than different (e.g., Bartunek and Rynes 2014) and bridging the gap "is already happening" (Hodgkinson and Rousseau 2009). Appreciating the underused potential of W&O psychology is essential for enabling psychologists to make a unique contribution to WPI practice. Bridging the gap requires W&O psychologists to further expand their knowledge by learning from other fields such as business and operations management. Only by embracing an 'integral perspective' (De Sitter et al. 1997; also MacDuffie 1997; Van Amelsvoort and Van Hootegem, Chap. 17 in this volume) can W&O psychologists become suitable interlocutors for management, and suitable service providers for both employees and managers. Both these key organisational stakeholders can benefit from the W&O psychologists' input in order to perform productively in their jobs and at the same time enable healthy and challenging workplaces. Moreover, by offering such input, W&O psychologists can bring together their natural focus on people and behaviour (i.e., culture and leadership) and their developing understanding of systems and institutions (i.e., strategy, structure, and power).

Acknowledgements The work by Maria Karanika-Murray is based on a project supported by the European Union Programme for Employment and Social Solidarity - PROGRESS (2007–2013) which is implemented by the European Commission. For more information see: http://ec.europa.eu/inprogress. "The information contained in this publication does not necessarily reflect the position or opinion of the European Commission".

The authors and publisher gratefully acknowledge the following permission to use the material in this book: Maria Karanika-Murray & Peter Oeij, The role of work and organisational psychology for workplace innovation practice: From short-sightedness to eagle view. In: European Work and Organisational Psychology in Practice. Special issue on Workplace Innovation, 2017, Volume 1, 19–30.

References

Aguinis, H., Werner, S., Abbott, J. L., Angert, C., Park, J. H., & Kohlhausen, D. (2010). Customer-centric science: Reporting significant research results with rigor, relevance, and practical impact in mind. *Organizational Research Methods, 13*(3), 515–539.

Alvesson, M., & Sveningsson, S. (2015). *Changing organizational culture: Cultural change work in progress* (1st ed. 2008). New York: Routledge.

Anderson, N. (2005). Relationships between practice and research in personnel selection: Does the left hand know what the right is doing. In A. Evers, N. Anderson, & O. Smit-Voskuijl (Eds.), *The Blackwell handbook of personnel selection* (pp. 1–24). New York: Wiley.

Anderson, N., Herriot, P., & Hodgkinson, G. P. (2001). The practitioner-researcher divide in industrial, work and organizational (IWO) psychology: Where are we now, and where do we go from here? *Journal of Occupational and Organizational Psychology, 74*(4), 391–411.

Anderson, N., Potočnik, K., & Zhou, J. (2014). Innovation and creativity in organizations a state-of-the-science review, prospective commentary, and guiding framework. *Journal of Management, 40*(5), 1297–1333.

Aram, J. D., & Salipante, P. F. (2003). Bridging scholarship in management: Epistemological reflections. *British Journal of Management, 14*(3), 189–205.

Arber, A. (2006). Reflexivity: A challenge for the researcher as practitioner? *Journal of Research in Nursing, 11*(2), 147–157.

Argyris, C. (1996). Actionable knowledge: Design causality in the service of consequential theory. *The Journal of Applied Behavioral Science, 32*(4), 390–406.

Bartunek, J. M. (2007). Academic-practitioner collaboration need not require joint or relevant research: Toward a relational scholarship of integration. *Academy of Management Journal, 50* (6), 1323–1333.

Bartunek, J. M., & Rynes, S. L. (2014). Academics and practitioners are alike and unlike: The paradoxes of academic-practitioner relationships. *Journal of Management, 40*(5), 1181–1201.

Brooks, M. E., Grauer, E., Thornbury, E. E., & Highhouse, S. (2003). Value differences between scientists and practitioners: A survey of SIOP members. *The Industrial-Organizational Psychologist, 40*(4), 17–23.

Buchanan, D., & Badham, R. (2008). *Power, politics, and organizational change: Winning the turf game*. London: Sage.

Cascio, W. F. (2007). Evidence-based management and the marketplace for ideas. *Academy of Management Journal, 50*(5), 1009–1012.

Chandler, A. D., Jr. (1962). *Strategy and structure: Chapters in the history of the American industrial enterprise*. Cambridge. Boston, MA: MIT Press.

Cox, T., Griffiths, A., & Rial-González, E. (2000). *Research on work-related stress*. Bilbao: European Agency for Safety and Health at Work.

Cox, T., Karanika, M., Griffiths, A., & Houdmont, J. (2007). Evaluating organizational-level work stress interventions: Beyond traditional methods. *Work & Stress, 21*(4), 348–362.

De Sitter, L. U., Den Hertog, J. F., & Dankbaar, B. (1997). From complex organizations with simple jobs to simple organizations with complex jobs. *Human Relations, 50*(5), 497–534.

Fraser, D. M. (1997). Ethical dilemmas and practical problems for the practitioner researcher. *Educational Action Research, 5*(1), 161–171.

Gibbons, M., Limoges, C., Nowotny, H., Schwartzman, S., Scott, P., & Trow, M. (1994). *The new production of knowledge: The dynamics of science and research in contemporary societies*. London: Sage.

Herriot, P. (2001). *The employment relationship: A psychological perspective*. Hove, East Sussex: Routledge.

Hirschkorn, M., & Geelan, D. (2008). Bridging the research-practice gap: Research translation and/or research transformation. *The Alberta Journal of Educational Research, 54*(1), 1–13.

Hodgkinson, G. P. (2012). The politics of evidence-based decision making. In D. M. Rousseau (Ed.) *The Oxford handbook of evidence based management* (pp. 404–419). Oxford: Oxford University Press.

Hodgkinson, G. P., & Rousseau, D. M. (2009). Bridging the rigour–relevance gap in management research: It's already happening! *Journal of Management Studies, 46*(3), 534–546.

Howaldt, J., Oeij, P. R. A., Dhondt, S., & Fruytier, B. (2016). Workplace innovation and social innovation: an introduction. *World Review of Entrepreneurship, Management and Sustainable Development, 12*(1), 1–12.

Karanika-Murray, M., & Weyman, A. K. (2013). Optimising workplace interventions for health and well-being: A commentary on the limitations of the public health perspective within the workplace health arena. *International Journal of Workplace Health Management, 6*(2), 104–117.

Karasek, R., & Theorell, T. (1992). *Healthy work: Stress, productivity, and the reconstruction of working life*. New York: Basic Books.

Kieser, A., & Leiner, L. (2009). Why the rigour–relevance gap in management research is unbridgeable. *Journal of Management Studies, 46*(3), 516–533.

Kompier, M. A. J., & Kristensen, T. S. (2001). Organizational work stress interventions in a theoretical, methodological and practical context. In J. Dunham (Ed.), *Stress in the workplace: Past, present and future* (pp. 164–190). Philadelphia, PA, US: Whurr Publishers.

Lefkowitz, J. (2008). To prosper, organizational psychology should… expand the values of organizational psychology to match the quality of its ethics. *Journal of Organizational Behavior, 29*(4), 439–453.

MacDuffie, J. P. (1997). The road to "root cause": Shop-floor problem-solving at three auto assembly plants. *Management Science, 43*(4), 479–502.

McKelvey, B. (2006). Van De Ven and Johnson's "engaged scholarship": Nice try, but…. *Academy of Management Review, 31*(4), 822–829.

McIntyre, R. M. (1990). Our science-practice: The ghost of industrial-organizational psychology yet to come. In K. R. Murphy & F. E. Saal (Eds.), *Psychology in organizations: Integrating science and practice* (pp. 25–48). Hillsdale, NJ: Erlbaum.

Murphy, K. R., & Saal, F. E. (1990). What should we expect from scientist-practitioners? In K. R. Murphy & F. E. Saal (Eds.), *Psychology in organizations: Integrating science and practice* (pp. 49–66). Hillsdale, NJ: Erlbaum.

Nye, J. S. (2008). Bridging the gap between theory and policy. *Political Psychology, 29*(4), 593–603.

Oeij, P. R. A., Kraan, K., & Dhondt, S. (2013). *Work teams and psychosocial risks and work stress*. OSH Wiki. https://oshwiki.eu/wiki/Work_teams_and_psychosocial_risks_and_work_stress. Retrieved December 31, 2015.

Oeij, P. R. A., Wiezer, N. M., Elo, A.-L., Nielsen, K., Vega, S., Wetzstein, A., et al. (2006). Combating psychosocial risks in work organizations: Some European practices. In S. McIntyre & J. Houdmont (Eds.), *Occupational health psychology: European perspectives on research, education and practice* (Vol. 1, pp. 233–263). Maia, Portugal: ISMAI Publishing.

Oeij, P. R. A., Žiauberytė-Jakštienė, R., Dhondt, S., Corral, A., Totterdill, P., & Preenen, P. (2015). *Workplace Innovation in European companies.* Study commissioned by Eurofound. Luxemburg: Office for Official Publications of the European Communities.

Petronio, S., Ellemers, N., Giles, H., & Gallois, C. (1998). (Mis) communicating across boundaries: Interpersonal and intergroup considerations. *Communication Research, 25*(6), 571–595.

Pot, F. D. (2011). Workplace innovation for better jobs and performance. *International Journal of Productivity and Performance Management, 60*(4), 404–415.

Ramstad, E. (2014). Can high-involvement innovation practices improve productivity and quality of working-life simultaneously? Management and employee views on comparisons. *Nordic Journal of Working Life Studies, 4*(4), 25–45.

Roehling, M. V., Boswell, W. R., Caligiuri, P., Feldman, D., Graham, M. E., Guthrie, J. P., et al. (2005). The future of HR management: Research needs and directions. *Human Resource Management, 44*(2), 207–216.

Rousseau, D. M., & McCarthy, S. (2007). Educating managers from an evidence-based perspective. *Academy of Management Learning & Education, 6*(1), 84–101.

Shapiro, D. L., Kirkman, B. L., & Courtney, H. G. (2007). Perceived causes and solutions of the translation problem in management research. *Academy of Management Journal, 50*, 249–266.

Stone, F. (2004). Deconstructing silos and supporting collaboration. *Employment Relations Today, 31*(1), 11–18.

Sutcliffe, K. M., Lewton, E., & Rosenthal, M. M. (2004). Communication failures: an insidious contributor to medical mishaps. *Academic Medicine, 79*(2), 186–194.

Suzaki, K. (1987). *The new manufacturing challenge: Techniques for continuous improvement.* New York: The Free Press.

Totterdill, P., & Exton, R. (2014). Defining workplace innovation: The fifth element. *Strategic Direction, 30*(9), 12–16.

Van de Ven, A. H. (2007). *Engaged scholarship: A guide for organizational and research knowledge*. NY: Oxford UP.

Van de Ven, A. H., & Johnson, P. E. (2006). Knowledge for theory and practice. *Academy of Management Review, 31*(4), 802–821.

Wampler, J. S. (2010). *Methods and strategies for briding the design practitioner-researcher gap.* Unpublished Masters' thesis. Columbia: University of Missouri.

Womack, J. P., Jones, D., & Roos, D. (1990). *The machine that changed the world: The story of lean production*. New York: Rawson/Harper Perennial.
Womack, J. P., & Jones, D. T. (1996). *Lean thinking: Banish waste and create wealth in your organization*. New York, NY: Simon and Shuster.
Zanko, M., & Dawson, P. (2012). Occupational health and safety management in organizations: A review. *International Journal of Management Reviews, 14*(3), 328–344.

Author Biographies

Maria Karanika-Murray is associate professor in occupational health psychology. Her research seeks to understand the determinants and conditions that can support work-related well-being and performance, with a focus on workplace design, intervention evaluation, presenteeism (attending work while sick), employability, and older workers.

Peter R.A. Oeij is a senior research scientist and consultant affiliated with TNO, The Netherlands Organisation for Applied Scientific Research in Leiden, The Netherlands. His field of work is innovation management, workplace innovation, social innovation and team dynamics.

Chapter 21
Case Studies Can Support Definitions of Workplace Innovation in Practice

Fietje Vaas and Rita Žiauberytė-Jakštienė

21.1 Definitions of Workplace Innovation Are Not Very Actionable

Practitioners interested in implementing Workplace Innovation (WPI) practices are in dire need of clear definitions of the concept. A recent study on WPI and its indicators conducted by the European Commission (2014), in the frame of the European Innovation Scoreboard (EIS) 2013–2014 project, identified 8 different definitions of WPI (pp. 13–14). Whereas they differ to a certain extent, all of these definitions conceptualize WPI as being a field or fields of action, a participative process and a win-win outcome for the organisation and employees. Moreover, some provide specific examples of WPI practices, such as: participative job design, self-organised teams and continuous improvement.

In Table 21.1, we provide a selection of four of these definitions that showcase the elements mentioned above.

Whereas these definitions are clear, most remain relatively abstract. This can be helpful given that implementing WPI successfully is highly dependent on the situation, the context and the history of any given organisation. In this sense, the definition should leave room for local interpretation since the concept of WPI itself implies that management should not impose a blueprint of an action plan upon

F. Vaas (✉) · R. Žiauberytė-Jakštienė
TNO, The Netherlands Organisation for Applied Scientific Research, Leiden, The Netherlands
e-mail: fietje.vaas@tno.nl

R. Žiauberytė-Jakštienė
e-mail: rita.ziauberytejakstiene@tno.nl

P.R.A. Oeij et al. (eds.), *Workplace Innovation*, Aligning Perspectives on Health, Safety and Well-Being, DOI 10.1007/978-3-319-56333-6_21

Table 21.1 Workplace innovation definitions

Source	Definition
European Commission (2014)	"*Those innovations aim at improving staff motivation and working conditions, thereby enhancing labour productivity, organisational performance, innovation capability, reactivity to market changes and consequently business competitiveness.*" (p. 13)
Dortmund Brussels position paper on WPI (2012)	"*Workplace Innovation is defined as a social process which shapes work organisation and working life, combining their human, organisational and technological dimensions. Examples include participative job design, self-organised teams, continuous improvement, high involvement innovation and employee involvement in corporate decision making. Such interventions are highly participatory, integrating the knowledge, experience and creativity of management and employees at all levels of the organisation in a process of cocreation and co-design. This simultaneously results in improved organisational performance and enhanced quality of working life ...*" (p. 2)
Eurofound (2013)	"*High performance work practices or innovations in work organisation are defined as deliberate changes that can affect how employees undertake their job and/or their broader experience of work and refer to any element of people management*" (p. 11)
Pot et al. (2012)	"*Workplace innovations are strategically induced and participatory adopted changes in an organisation's practice of managing, organising and deploying human and non-human resources that lead to simultaneously improved organisational performance and improved quality of working life.*" (p. 262)

those involved. Employee participation in the design of WPI measures and in the implementation process is a necessary condition for real WPI (Oeij et al. 2015a; Pot 2011; Ramstad 2009; Totterdill and Exton 2014).

However, while it is inevitable from a theoretical point of view for definitions to be abstract, from a practical standpoint, more actionable (Argyris 1996) information would be helpful. It is possible that this lack of actionable information on WPI could explain why relatively few organisations have implemented WPI practices, although they promise win-win outcomes for all involved. Indeed, we have often encountered the following questions cropping up in companies preparing to implement WPI: "What are we talking about? What changes in job design, work organisation and management are we supposed to implement? Can't you give us examples of good WPI practices, so that we can understand?"

We conclude that, next to clear theoretical definitions of WPI, practitioners could benefit from examples of cases that embody good WPI practice. Therefore, the study, documentation and dissemination of cases representing best practices are very important for practitioners.[1] Moreover, for researchers, it is important to know how to produce and report case studies that can lead to learning in practical situations.

In this article we focus on the following question: what information should a case study provide to help practitioners, managers and employees in their search for concrete WPI interventions? In other words: what makes a WPI case a good example for practical purposes?

First, based on our practical experience, we will reflect on the criteria that would render a case a good example of WPI. Second, we will use the case studies analysed and described in mini cases[2] in the Eurofound project: *Workplace innovation in European companies* (Oeij et al. 2015a, b) to more thoroughly explain these criteria. Note that we are not developing a case study methodology. We simply want to share our practical experience with WPI and provide a simple framework that can be used by practitioners discussing an intervention with stakeholders and by researchers when reporting a case study.

21.2 What Makes a Case a Good Example of WPI Practice?

First, a case, that is a good example of WPI practice, should exhibit real, substantial WPI practices. Whereas this is a common sense argument that might even sound trivial, we deem it important, given that, in some cases, lip service has been paid to WPI without any structural changes (see e.g. Ramstad 2009). In other words, the case should show that workplace innovation practices have been implemented that have led to (or theoretically promise to lead to) improved company performance and improved quality of work. However, determining whether a WPI case does indeed showcase substantial WPI practices is tricky, given that WPI definitions are neither extremely detailed nor actionable in themselves. This issue also arose in the Eurofound study and was solved by asking WPI experts to assess the cases based on their theoretical knowledge and experience with WPI interventions, and to subsequently, categorise them as exhibiting high, medium or less substantial WPI practices (see the paragraph about the Eurofound study below).

Second, for a case to be a good example of WPI practice, a thorough description of the case needs to be available. Without a thorough, detailed description, the value added for practitioners in other organisations is significantly diminished. In this respect, a case description should outline the specific practices/measures/actions/interventions initially planned to be implemented and discuss the ones that

[1]There are two knowledge banks where many WPI-cases can be found: www.workplaceinnovation.org and EUWIN's knowledge bank: http://portal.ukwon.eu/euwin-knowledge-bank-menu-new.

[2]The Eurofound website presents all the 51 mini cases: http://www.eurofound.europa.eu/surveys/2013/european-company-survey-2013-00.

have actually been implemented. Preferably, the case description should be broken down into clear steps that could be replicated in other organisations, although they might have to be adapted to fit these other contexts. In other words, the information presented should be *actionable*. As Argyris (1996) suggested, attention should be paid to informing the reader on how to create—in this case—Workplace Innovation. That is, it should inform people in other organisational contexts what has been changed in the structure (job design, work organisation, hierarchical structure) and the culture (empowerment, management, trust, HR practices) of the organisation to achieve WPI.

Third, the case has to be inspiring. This means that the description really gets across the story of the change in this organisation. Profound storytelling and high-quality narratives have the capacity to reveal important, real-life facts of a case (Gabriel 2013). Preferably, the description is a narrative that gets the reader to be enthusiastic about the changes in jobs and roles, organisational structure, corporate climate, behaviour, labour relations and trust, etc. It shows the reader that the bundle of WPI practices that is, or was, implemented promise to result or have resulted in the improvement of the company's performance, sustainability and innovative capabilities as well as in providing challenging jobs and good work prospects for all employees.

One strong mechanism to provoke inspiration is identification. Positive collective narratives that people identify with can influence behaviour, cognition and emotions in a favourable way (Rappaport 1995). Therefore, the narrative should include the possibility for various groups of readers to identify with, in the sense that they can recognize the kind of work, their position (manager, employee, employee representative, trade-unionist, user), the sector, the size of the organisation and/or the cultural context. For instance, for employees, it is important that the narrative showcases employees in the case-organisation who speak positively about the change process and/or the (predicted) results. Another factor that can lead to identification or, for that matter, to non-identification, is the cultural context of the case as well as national or regional labour relations. For instance, the results of a Danish case could seem so unattainable to Greek managers and employees that the case turns out to be scary instead of inspiring. The same may count for cases from different sectors. Employees of a chemical factory might think (rightly or wrongly) when hearing about practices successfully implemented in a school that: '*this will not work for us*'. Therefore, it is important to embed a case in its cultural or sector-related context, so that a practitioner can decide whether he or she can use it in a specific organisation.

Figure 21.1 summarises the criteria that preferably should be present in a case description of WPI practices.

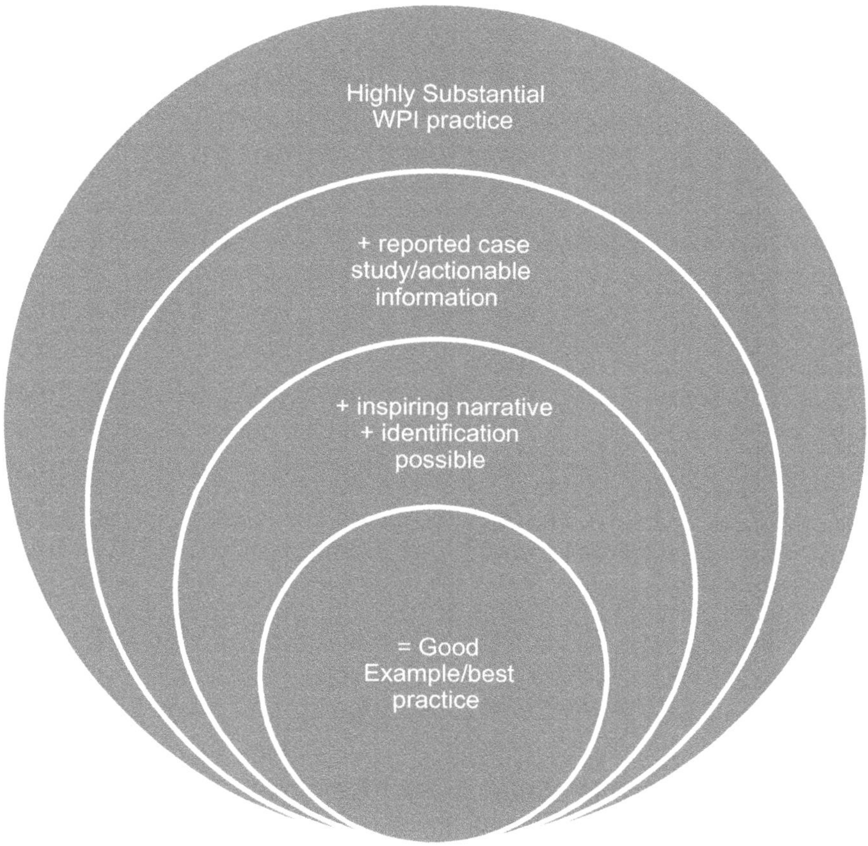

Fig. 21.1 The criteria for a case study to be a good example of WPI practice

In the next paragraphs we will elaborate on these criteria and show how they can be applied to existing case studies. To this end, we will use the case studies described and analysed in the Eurofound study[3]. We will introduce that study first.

[3]Also see the website: http://www.eurofound.europa.eu/surveys/2013/european-company-survey-2013-00.

21.3 The Eurofound Study: A Source for WPI Cases

The Eurofound study was commissioned by the European Foundation for the improvement of Living and Working Conditions (Eurofound) (Oeij et al. 2015a, b). The purpose of the study was to investigate why and how companies in the European Union apply WPI. The authors of this article were involved in this study as part of a larger international research team. During 2014–2015, interviews were held in 51 selected companies covering ten EU member states. These companies represent the 5% best scoring companies out of 30,000 companies on a workplace innovation-index in the European Company survey 2013 (Dhondt et al. 2014). Interviews were held with managers, employees and employee representatives and mini-case study reports were written.

As part of the Eurofound study, each case was assessed by three experts.[4] They based their assessment on theoretical notions about the types of interventions that may lead to the desired outcomes of improving the quality of working life and organisational performance. They assessed whether in each case, structural and/or cultural WPI measures, supported by HR measures were indeed implemented. All 51 cases were subsequently assigned a score from 0 to 10 based on a number of criteria representing structural (0–4 points), cultural (0–4 points), bundled or HR supportive interventions (0–2 points). Based on the final-scores, the cases were classified as being exemplary of highly (16), medium (21) or less (14) substantial WPI practices (Oeij et al. 2015b).

21.4 Cases That Distinguish Between Substantial and Less Substantial WPI Practices

For a better understanding of what highly substantial and less substantial WPI practices mean, below, we will provide a description of a case with a high score and one with a low score. We chose to present the Danish manufacturing case because it scored the highest (9.67 out of 10) out of all 51 cases. It received such a high score because it embodies a promising combination of structural and cultural bundles of measures integrated by an explicit vision on WPI. Moreover, the case study exhibits positive outcomes for both the organisation and the workers. The Polish school was selected as an example of a case with the lowest score on WPI (2.33 out of 10). Although good practices were implemented, they were not considered to be Workplace Innovations, since they were merely IT-related measures.

[4] The three experts discussed their individual ratings and the final score was reached via consensus.

The cases with the highest and the lowest WPI score

The case with the highest WPI score in the Eurofound study

DK-MANU-FABRIC-S[5] [Denmark-manufacturing-making fabrics] is an international company that produces fabrics for different customer groups (retail, hotels, furniture manufacturers etc.). The headquarter of the company is in Denmark and it counts one hundred and forty-five employees there who support the manufacturing process in different European countries.

The company operates from a vision that workplace innovation is *"a way to ensure renewal and the ability to offer a service that is so good that customers will chose this company over others*" (Denmark_Minicase_Kvadrat, p. 1/3). This company has implemented the following WPI practices: (1) customer segment teams, (2) autonomous and semi-autonomous team work, (3) a flat hierarchy, *(structural improvements according to the experts)* (4) room for proposals to improve processes, products or services, (5) participative formulating of performance goals and targets, (6) innovation meetings for product development (*cultural or mixed improvements, according to the experts*).

The customer segment teams are multi-disciplinary teams that serve segments of clients, such as retail or hospitals. Those employees, who have direct or indirect contact with a group of customers, work together in a team. In supporting departments, such as the Order Expedition, employees work in autonomous or semi-autonomous teams. They distribute their work themselves at a daily morning meeting. There are only three hierarchical layers: the board of managers, middle managers and employees. Every employee can make a proposal directly to the management or colleagues and, if possible, these ideas are directly implemented.

Performance goals and scores are formulated together with the employees, which leads to mutual trust building. The department for product development organises continuous innovation meetings, where creative thinking is promoted.

According to the manager and the employees that were interviewed, these interventions resulted in very positive outcomes. The manager stressed that an innovative culture, knowledge sharing and autonomous, interdisciplinary teams give the company competitive advantage. Employees described working in multi-disciplinary teams as highly motivating and said they feel in control over the incremental innovation process that enables them to bring in their ideas.

The case with the lowest WPI score in the Eurofound study

This school PL-EDUC-INFO-SCHOOL-S [Poland-education-informatics-school organisation] was established in 1996 in Poland. The school developed and implemented two IT systems. The first integrates information processing and communication procedures and facilitates communication

[5]The indicators S and L stand for Small = 50–250 and Large $\geq$ 250 employees.

among the institution, students and employees. The second is an access control system with integrated monitoring. Administrative staff has been trained to work with these systems. Many of the functionalities in these systems were suggested by employees.

The implementation of the two systems has brought benefits for employees, students and for the school: transparency has improved and there is a large reduction of administrative burden. Students can have direct and remote contact with the school and access to administrative processes and educational resources. The image of the school improved.

In the rest of this article we will exclusively focus on the 16 cases that were qualified by the experts as having implemented highly substantial WPI practices, given that is a core element of a case to be a good example of WPI practice. To this end, we will select cases that best embody the remaining criteria.

21.5 Cases that Distinguish Between Actionable and Less Actionable Information for WPI

In order to expand on the criterion of providing actionable information for WPI, we will present a case that provides information that is actionable and one that could do better in that respect. As an example of a case that presents highly actionable information, we chose the case of a Danish school. Firstly, the narrative describes the stakeholders involved and the actions that have been taken in terms of co-opting them: joint training sessions, setting common goals, etc. Secondly, the organisational principle for the design of autonomous teams is made explicit, namely, to create autonomous teams, each entirely responsible for educating one group of students. Of course this is specific for a school, but on a higher level of abstraction, one can state that there was a change from a disciplinary- or functional-oriented to a client-oriented work organisation. This principle can be transferred to any other work situations, therefore, it can be called actionable.

As a contrast, we briefly present the case of a Lithuanian manufacturer that has implemented highly substantial WPI practices, but could have done better in providing actionable information about them. From the description, it is not entirely clear how the work organisation has been reformed to allow the teams to function as relatively independent self-managing teams. In addition, a more thorough explanation regarding the implemented procedures and the manner in which flex-time helps self-management would have improved the actionability of this information.

Two cases showing more or less actionable information about WPI practices

A case study with actionable information

The Danish municipal primary and lower secondary school [DK-EDUC-SCHOOL-S] has 90 employees and 700 students. Following a merger between different schools, the implementation of several educational reforms, a conflict with the unions about the payment of hours for preparation, and given the lowest influx of pupils in the municipality, it was deemed important to create a more exciting and innovative school. Workplace innovation was expected to help. To this end, the following WPI practices were implemented: (1) partnerships with stakeholders were established, (2) autonomous and interdisciplinary teams were introduced, (3) discussion forums were established.

At the start of the change project, partnerships between management, employees and (union) representatives were established to support dialogue, to enhance trust in management, to formulate common goals, and to ensure that all stakeholders were involved as partners in the change process. The creation of autonomous teams responsible for the education of an entire group of students was a structural and radical intervention. Teachers who used to be responsible only for teaching a specific subject were now made responsible for the education of a group of pupils, together with colleagues from other disciplines. Changes were discussed in a number of forums with both direct and indirect (representatives from the union) employee participation.

Many teachers appreciate the changes, although some still struggle with the fact that they cannot just teach 'their own' subject. All stakeholders have learned that, in the beginning, autonomous teams might need a framework that clarifies and delimits their areas of responsibility for action and decisions. Union representatives appreciate that it is easier for employees to get a fair share of the exciting subjects; hence, jobs have become more challenging for all employees. In the period these practices were implemented the school has also increased the student influx. The school expects to become more sustainable given that employee competences and ideas are utilized better in coping with changes.

A case study that could do better in providing actionable information

LT-MANU-RESPIR-L [Lithuania-Manufacturing-making respiratory products] is Europe's leading designer, manufacturer and supplier of a wide range of medical devices for respiratory support. The factory in Lithuania is the main production and scientific research site, part of an international group headquartered in the UK. The company employs over 1700 people in Lithuania.

The company has implemented the following WPI practices: (1) self-managing teams, (2) flexible working times, (3) (electronic) post-boxes to submit ideas, (4) several concrete measures to improve organisational and personal learning and development.

Regarding the self-managing teams the mini case states that '*Now, the teams at the production lines are more responsible for planning (especially production time management), quality and other aspects of the production process.*' (Lithuania_Minicase_Intersurgical, p. 1/3). The reader can imagine that there is some impact on the way tasks and responsibilities in the primary process are divided, but it is not clear how the new work organisation is designed.

The case also states that working times can be arranged in a flexible way, so that employees have the opportunity to balance their work and private duties. However, the narrative does not clarify what "flexible ways" means, therefore, the information is not actionable.

Regarding the implemented idea-management practices, the case states that there is a review and a feasibility assessment of the submitted ideas. In addition, to develop the promising ideas '*responsible people are assigned to organise the implementation process*' (Ibidem p. 2/3). More information regarding the participation of the involved employees would have improved the actionability of this information.

Regarding training and development opportunities, collective training sessions take place each year. Individual training plans—supported by an electronic Learning Management Information System—are drawn up and implemented. Internal career opportunities are offered and supported, and much is done to stimulate learning that helps to adopt new technology. The learning and development practices are very well described and, therefore, can be transferred.

21.6 Inspiring WPI Cases

For a case to serve as good example the case study should not only provide actionable information, but it should also be inspiring for various groups who want to learn from it.

In the box below we present three cases that inspire by their narrative. We chose the narrative about the German pet food company because it gets across the positive feeling, or 'the buzz' (Totterdill and Exton 2014, p. 16), deriving from a complete renewal of old fashioned production processes and shows that employees and managers are excited about the future and progress. The Irish pharmaceutical company was selected because it provides us with an example that many stakeholders can identify with, especially employees and union representatives. All the workplace innovation practices are embedded in a workplace partnership between management, employees and trade union representatives. In the case of the British bathroom equipment producer many elements of Lean production theory are recognizable, especially the notion of continuous improvement (KAIZEN). Practices from this management theory are popular in Western European manufacturing. However, Lean Production is applied in very different ways. Oftentimes it is merely a top down measure to realize operational excellence and employee involvement only goes as far as needed to reach that goal. In this case, however, elements of Lean Production are used to structurally and culturally empower and enable employees to participate in organisational change and renewal in order to improve the quality of working life and organisational performance, that is, WPI. Therefore, this highly substantial case can be a very inspiring example for many Western European manufacturing companies.

Three WPI cases that inspire

A case that inspires by its narrative

The German company DE-AGRO-PETFOOD-S [Germany-Agriculture-food for pets] is part of a holding and produces pet food. A few years ago the company had to go through a complete upheaval in order to stay competitive. The whole production process was modernised by introducing machines instead of craft-based production. The new production lines required new employee competencies. Employees were educated and trained to take over every task in the manufacturing process. Moreover, some unskilled workers were trained to do production jobs—every employee had the chance to upskill and take over a skilled worker's task. Thus former unskilled workers were trained in robotics and are now machine operators. At the request of employee representatives and the works council, every employee was given the opportunity to develop and to implement a training and career plan, which led to more qualified employees taking over more demanding jobs.

In addition, workers at the production site have leeway to organise their work as long as they meet the production deadlines given by the customers. For example, they can decide themselves to take a break, to have a meeting, or to watch a football championship game as long as they fulfil the production targets.

All these measures helped the company stay competitive. Employees like the production flexibility and the chance to improve their skills and competences as well as their new jobs.

A case many different stakeholders can identify with

IE-MANU-PHARMA-S [Ireland-Manufacturing-Pharmaceutic] is a leading manufacturer of pharmacy products mainly for the treatment of angina pectoris and related conditions. The site in Ireland was bought by a Belgian company in 2007, and there are 124 people working here.

Since 2012 the emphasis has been on integrating this plant within the larger group. This was supported by the arrival of the new HR Director who focused on strengthening employee engagement, collaboration amongst employees, creating a partnership with the principal trade union and introducing a culture of continuous improvement.

The WPI practices that were implemented are: (1) systematic follow up of the engagement survey, (2) effective management culture, (3) workplace partnership and (4) embedding quality and compliance.

Of these practices, the workplace partnership is the one that illustrates the important role employees and union representatives have played in the change process and the one that employees and unions can identify with. In this respect, the mini case states: "*During the period of restructuring and its aftermath issues were brought formally to management by union representatives on a regular basis. In time a sense of mutual trust and understanding developed [...] Management is increasingly proactive, finding opportunities to discuss emerging issues with the union official at an early stage.*" (Minicase-IE-UCB (Schwarz), p. 2/3).

The HR manager ascribes results such as time efficiency, productivity, improved communication, improved decision making to this partnership and to the proactive discussion with union representatives about emerging issues, thus avoiding conflicts and promoting engagement at all levels.

A case that can provide identification via region and sector

This British company UK-MANU-BATH-L [United Kingdom-Manufacturing-Sanitary products] is a good example for both the North West European region and the manufacturing sector.

The company was founded in 1977 in Birmingham (UK) as a family-owned business. It is now a leading supplier of showers, taps and bathrooms in the UK with exports to Europe and Russia.

Continuous improvement (CI or Kaizen) is at the heart of the company's strategy. The company's strategy is very much directed at operational excellence and innovation to meet customer's needs. This is practised by engaging all employees and stimulating them to participate in the continuous improvement. The Chief Executive points out that '*all employees have two duties: to develop themselves and to change and develop their roles*' (Minicase-UK-Bristan, p. 1/3). Teams are empowered to address issues from customers directly, working within boundaries set to provide room for entrepreneurial behaviour. Moreover, everyone in the warehouse is involved in the integration of technology for 'Just in time' with their team working practices.

The company does the best it can to use the talents and ideas of all employees in cross-functional project teams, in self-managing teams, in an Employee Forum and 4–6 Kaizen events per year.

21.7 Cases that Are Good Examples of WPI

As we have argued before, cases that are good examples of WPI, have implemented highly substantial WPI practices, provide actionable information and are inspiring. Out of the 16 Eurofound cases that were considered to have implemented highly substantial WPI practices, we have selected three cases that fulfil all of these criteria and that can be used as good examples in a discussion about how to realize WPI in practice. Specifically, we chose to focus on these three because of the following reasons.

The Spanish paper product manufacturer implemented a wide range of structural and cultural measures, supported by HR practices. Most practices in this case are described in an actionable way. The case is inspiring since the story of this company makes the reader believe that it will succeed in surviving a very complicated and threatening market development by mobilizing and unifying the competences, ideas and engagement of many stakeholders in the region. Moreover, we deem this a case with which different stakeholders can identify. In short, the case shows that the whole change process is carefully orchestrated by taking into account the interests of a diverse group of stakeholders.

The case of the Danish museum was seen as an interesting bundle of structural, cultural and 'traditional' HR practices. The information about the WPI practices is actionable as it shows how seeking for a common goal and trusting social dialogue can lead to WPI. The reader can learn to trust that teams—supported by training—can find the best way to organise their work to contribute to a shared goal. Given that the museum did this and succeeded can be inspiring for others and it can lead to identification for managers, employees and union representatives.

The case of the Dutch publishing company provides actionable information. It describes how salesmen's and journalists' jobs were enlarged and enriched in a more client- or theme-oriented organisational structure. This case inspires because one feels that local management and employees have fought hard for the survival of an independent regional newspaper within the framework of being owned by a foreign company with different business ideas. Their success can be attributed to giving employees more leeway and more autonomy. People working in companies that need to search for new earning opportunities and a (partly) new business model will certainly be able to identify with this case. And many readers will recognize and identify with the employees of the media group struggling with foreign investment companies executing their own agenda.

Three cases that are good examples of WPI practices

ES-MANU-PAPER-S [Spain-manufacturing-paper products]

This company is a paper product producer with a total workforce of about 200 employees. The majority are men working on production-related tasks. This company has implemented a number of WPI practices over the last few years. (1) The frame for these practices is provided by good social dialogue between management, employees and three active regional trade unions. Management provides transparent information and consults employees regarding important decisions early. (2) The common goals that facilitate this dialogue are a shared concern about the sustainability of the company (an important employer in the region) and the quality of work. (3) In terms of HR measures, there is a collective agreement that guarantees high salaries, a 35 working-hour and no overtime working is allowed. (4) The production is done in self-steering teams "*that enjoy a certain degree of autonomy in the way their human resources are organized*" (*ES Minicase, Company X, p.* 2/2). (5) The multi-skilling of the workers is stimulated by a "*sliding work organization system*" (ibidem, *p.* 2/2). Employees are encouraged to take up activities that are part of a job higher in rank and in different production lines. (6) Employee representatives have suggested using *biological calendars* to schedule production in a way that allows satisfying the needs for rest and sleep. (7) Employee representatives participate in the process of internal reflection on new business opportunities. (8) In the sphere of human resources management, the company is paying much attention to health and safety risk prevention, by providing information and training courses.

The stakeholders show enthusiasm about the (expected) results. According to management, the WPI practices have led to a good employment climate, skilled, well-motivated and committed staff, and a good use of the innovative ideas of the employees. This in turn, has resulted in improved economic and financial company performance. The employees think this company is a good employer and that the employment conditions are excellent. The unions especially appreciate the actions that contribute to the net creation of employment in the region, such as the financial results and the 35 h-working week with no overtime.

DK-ART-MUSEUM-S [Denmark-Art and Leisure sector-Museum]

This organisation is a museum with 140 employees divided over nine worksites in a larger Danish municipality. During the last years, the pressure to compete with other attractions (cinemas, amusement parks) has increased, and the museum realises the importance of employee involvement to create appealing guest experiences.

This case study focussed on WPI practices that concerned two professional groups: hosts and craftsmen in the open-air museum. The implemented WPI practices were: (1) autonomous team work for the craftsmen, (2) a good hosting project, and (3) partnership with unions.

Funen Village is a living workplace, where guests watch craftsmen work on the antique buildings. Previously, the craftsmens' work centred around maintenance with eighty percent of their working day being spent on following a mandated plan. Now, the teams autonomously organise the task sequence, the planning, work mode, duty roster and budget. They have the opportunity to plan their work in such a way that they can perform jobs that are interesting for the guests to see, during business hours. They also have more time to talk with visitors about their craft.

The museum hosts sell tickets and make sure that the exhibitions function. The hosting project was based on experiences from other types of businesses (e.g., Legoland, the famous Danish theme park) on how to create an exciting guest experience. The museum hosts were trained and, subsequently, formulated new work codes on, for instance, how to greet the guests and how to inspire the guests to explore the exhibitions. They received a book about good hosting experiences providing many tools which changed their jobs. The partnership took shape in a joint committee with union representatives, Occupational Health and Safety representatives and management representatives (based on a regional framework).

Management, representatives and employees agree that the new practices provide greater job satisfaction for employees while simultaneously enabling the company to provide better service. The focus, in general, has shifted from a narrow focus on specific disciplines to a customer focus.

NL-INFO-NEWS-L [Netherlands-Information-Newspaper]

This company is the publisher of a regional newspaper in the Netherlands with about 300 employees. For more than a decade, the company's existence is being threatened by diminishing earnings from selling subscriptions and advertisements and by the development of new information technologies. In addition, there was a stream of changing international owners and management with different strategies.

By engaging in coalitions supported or initiated by the Works Council and employees, the local management succeeded in implementing WPI practices with the aim of mobilizing employee talents and finding new earning models. These practices were: (1) upskilling and redesigning the jobs of the advertising salespeople into account managers, (2) restructuring the editorial department and redesigning the journalists' jobs, (3) dialogue between employees and CEO regarding ideas for business improvement, (4) four cross-functional teams that each develop a new idea, (5) cooperation with external partners, such as a university for applied science and a broadcasting company.

The role and the targets of the salespeople have changed completely: instead of selling advertisements by phone, they now have to build a relationship with the clients and to consult them on how to best reach their intended customers. The editorial work used to be organized along regions in the province, producing daily news messages. Given that nowadays this information is more easily distributed by social media, now the value added of a news outlet is to produce more background stories and in-depth studies. Therefore, most of the journalists were placed in the Research department and were organized into specific Theme groups (e.g., Health, Sports, Education). This new way of organizing in multifunctional theme groups provides opportunities for journalists to cooperate with account managers in writing targeted articles.

Management-employee dialogue resulting in cross-functional teams developing promising ideas are familiar practices in other companies as well. However, what is striking here is the thoughtful way in which they organised this. For instance, team members work full-time in these cross-functional teams for a certain period of time and are thoroughly trained and coached to do the job.

The results for the company seem positive. The newspaper is still on the market and is investing in new earning opportunities. The employees have more challenging jobs, have better career prospects and the feeling of being heard and taken seriously.

Table 21.2 shows that these three cases meet all the criteria and can be used as good examples.

Table 21.2 Summarizing the three cases that are good examples of WPI

	ES-MANU-PAPER-S	DK-ART-MUSEUM-S	NL-INFO-NEWS-L
Highly substantial case	Yes, score: 7.5[a]	Yes, score: 7.7	Yes, score: 8
+ Case study/actionable information	Mini-case information could be more actionable	Mini-case, including actionable information about redesign of 2 jobs and team working	Mini-case, including actionable information about redesign of jobs and organisational structure, employee groups for business development
+ Inspiring + Possibilities for identification	Yes, by mobilizing many stakeholders in the region to retain employment	Yes, the image of the museum modernized; managers and unions can identify because of partnership model	Yes, for companies and stakeholders that need to mobilize all talents to develop new business opportunities. Employee representatives can be inspired by the role of the works council here

[a]Scores given by the Eurofound experts

21.8 Conclusion

Many practitioners find it problematic to define Workplace Innovation. The well-known definitions hardly provide any actionable information. In this contribution, we have argued that, for practitioners, examples of cases describing best WPI practices can provide value added to purely theoretical definitions of WPI. Therefore, we tried to answer the question: *What makes a case a good example of WPI practice?*

In this article we proposed the following criteria. First, the company that serves as a case should have implemented good, highly substantial WPI practices in a participative way. Those practices must have led to or promise to lead to improved company performance and improved quality of work for the employees involved. Second, the case description needs to provide actionable information about the WPI practices implemented. That is, it provides information on what has been changed in the structure, the culture and in supporting HR practices and how that has been done. Third, the case has to be inspiring; that means that the narrative really gets across the story of the change in the organisational structure, culture, climate and behaviour of management and employees, as well as, their enthusiasm about the (expected) results. An inspiring case also provides the opportunity for diverse groups in other organisations to identify with it. That is, others can recognize and identify with the kind of work, the sector and/or the cultural context.

To illustrate these criteria, we used the cases from the recent Eurofound study. First, we focused on cases that were identified by experts in the Eurofound study as having implemented highly substantial WPI practices as well as less substantial WPI practices. Subsequently, we focused our attention on the sample of 16 highly substantial cases in the Eurofound study, given that this is a sine-qua-non for a case to be able to provide any useful information on WPI best practices. Within these 16 mini-cases we found and described some that provided actionable information and some that could do better on that criterion.

Moreover, we found cases that are inspiring because the narrative not only presents what has been changed in the company and how it has been done, but also gets across 'the buzz' generated. Finally, we described three cases that embody all three criteria and can be used by practitioners as examples of best practices for WPI.

Some take-away messages for practitioners are the following: A case that is a good example of WPI practice can help clarify what managers and employees might change in the structure and culture of an organisation and how to do that. Therefore, practitioners should consider finding cases that have been rated by experts as having implemented effective WPI practices. Moreover, these cases should have an inspiring narrative that describes the practices and the change process in an actionable way. Practitioners might also want to consider in how far people in their organisation would be able to identify with that case. Depending on the situation, that might be the sector, the region or a specific approach, such as a dialogue.

Surely, this is not a full-proof recipe and we advise practitioners to be flexible in their approach. Surprisingly, a case from a service sector might inspire employees from a manufacturing company. Practitioners should also keep in mind that a case description is not a handbook for how to implement workplace innovation. The best- practices-approach has its own general limitations, and, practitioners should be prepared to be flexible and adapt any best practices to their own context. In this sense, a good case example can be a source of information for finding practical solutions for specific problems in an organisation as well as provide inspiration for discovering new perspectives.

Acknowledgements The authors and publisher gratefully acknowledge the following permission to use the material in this book: Fietje Vaas, Rita Žiauberytė-Jakštienė & Peter Oeij, Case studies can support definitions of workplace innovation in practice. In: European Work and Organisational Psychology in Practice. Special issue on Workplace Innovation, 2017, Volume 1, 61–72.

References

Argyris, C. (1996). Actionable knowledge: Design causality in the service of consequential theory. *The Journal of Applied Behavioral Science, 32*(4), 390–406.

Dhondt, S., Preenen, P., Oeij, P., Corral, A., Isusi, I., Totterdill, P., et al. (2014). *European company survey: Construction of the workplace innovation index and selection of companies*. Leiden: TNO.

Dortmund/Brussels Position Paper on Workplace Innovation. June 26, 2012. http://www.kennisbanksocialeinnovatie.nl/nl/kennis/kennisbank/workplace-innovation-as-social-innovation/1009/download/dortmund-brussels-position-paper-12th-june-2012/1980

Eurofound. (2013). *Work organisation and innovation*. Luxembourg: Publications Office of the European Union.

European Agency for Safety and Health at Work. (2012). *Review of workplace innovation and its relation with occupational safety and health*. Luxembourg. https://osha.europa.eu/en/publications/reports/workplace-innovation-review

European Commission. (2014). Kesselring, A., Blasy, C., & Scoppetta, A. *Workplace innovation: Concepts and indicators*. Brussels: European Commission, DG for Enterprise and Industry.

Gabriel, Y. (2013). Researchers as storytellers: Storytelling in organizational research. In M. Gotti & C. S. Guinda (Eds.), *Narratives in academic and professional genres* (pp. 105–122). Bern: Peter Lang.

Oeij, P., Žiauberytė-Jakštienė, R., Dhondt, S., Corral, A., Totterdill, P., & Preenen, P. (2015a). *Third European company survey: Workplace innovation in European companies*. Study commissioned by Eurofound. Luxemburg: Publication Office of the European Union.

Oeij, P., Žiauberytė-Jakštienė, R., Dhondt, S., Corral, A., Totterdill, P., & Preenen, P. (2015b). *Workplace innovation in European companies. Technical annex*. Dublin: Eurofound. https://www.eurofound.europa.eu/sites/default/files/ef1540_-_technical_annex_third_european_company_survey_workplace_innovation_in_european_companies.pdf

Pot, F., Dhondt, S., & Oeij, P. (2012). Social innovation of work and employment. In H.-W. Franz, J. Hochgerner, & J. Howaldt (Eds.), *Challenge social innovation. Potentials for business, social entrepreneurship, welfare and civil society* (pp. 261–274). Berlin, Heidelberg: Springer.

Pot, F. D. (2011). Workplace innovation for better jobs and performance. *International Journal of Productivity and Performance Management, 64*(4), 404–415.

Ramstad, E. (2009). Promoting performance and the quality of working life simultaneously. *International Journal of Productivity and Performance Management, 58*(5), 423–436.
Rappaport, J. (1995). Empowerment meets narrative: Listening to stories and creating settings. *American Journal of Community Psychology, 23*(5), 795.
Totterdill, P., & Exton, R. (2014). Defining workplace innovation: The fifth element. *Strategic Direction, 30*(9), 12–16.

Author Biographies

Fietje Vaas is a psychologist. Until her retirement she was employed by and is now temporarily affiliated with TNO, The Netherlands Organisation for Applied Scientific Research Innovation for Life. Currently she is working on topics such as innovation of work organisation and the management and employment relations of private and public organisations.

Rita Žiauberytė-Jakštienė is a psychologist. Rita Žiauberytė-Jakštienė is a junior researcher at TNO, The Netherlands Organisation for Applied Scientific Research Innovation for Life. Having HR consultancy experience, she is interested in the tandem of empowered personnel and successful businesses. Formerly she has been a recruitment consultant in personnel selection agencies in Lithuania. Currently she is working at TNO on projects related to workplace innovation and social innovation. Her research interests include social innovation, workplace innovation, and goal-setting topics.

Chapter 22
From Automated Defensive Behaviour to Innovation Resilience Behaviour: A Tool for Resilient Teamwork as an Example of Workplace Innovation

Peter R.A. Oeij

22.1 Introduction

Workplace innovation entails renewing ways of working and organising that lead to an improvement in the quality of work and organisational performance, and, as a consequence, in the innovative capability of the organisation as a whole (Chap. 5). This chapter introduces the Innovation Resilience Behaviour tool (IRB-tool) aimed at improving teamwork as an example of a workplace innovation intervention. Teamwork refers to a group of employees that work on a common goal and who have interdependent tasks. The type of teamwork we focus on is that of teams working on innovation projects. Innovations are, for instance, implemented renewals like products, processes, methods and organisational forms.

Characteristic for this type of teamwork is that it concerns complex work, because innovation confronts people with new elements in their work. The complexity can be related to the content of inventing and innovating—intellectually difficult issues—, or to the context, such as the presence of many stakeholders, differing interests, and intensive interactions inside and outside the team. However, most people dislike complexity. Indeed, people in general, and more so in complex

This chapter is partly based on Oeij, P.R.A. (2016). From automated defensive behaviour to innovation resilience behaviour: Improving the management of R&D and innovation projects. Invited paper for *IOSH 2016 Annual Conference "Influential leadership: delivering impact—sustaining change"*. IOSH, Institution of Occupational Safety and Health, ICC ExCel London, London, United Kingdom and Oeij, P.R.A. Oeij, Preenen, P.T.Y., & Van der Meulen, F.A. (December 2014). *From unnatural behaviour to Innovation Resilience Behaviour: Prototype of a change tool.* [ETP Behaviour and Innovation]. Leiden: TNO Healthy Living. Both sources will be forwarded free of charge upon request to the author.

P.R.A. Oeij (✉)
TNO, The Netherlands Organisation for Applied Scientific Research, Leiden, The Netherlands
e-mail: peter.oeij@tno.nl

P.R.A. Oeij et al. (eds.), *Workplace Innovation*, Aligning Perspectives on Health, Safety and Well-Being, DOI 10.1007/978-3-319-56333-6_22

projects, have an innate tendency to strive for control and structure. Moreover, they seek out certainty regarding the process, the project resources and the expected results in order to feel competent and decrease feelings of threat and insecurity (Kahneman 2011). This need for certainty is inherently at odds with the complex nature of innovation projects. Given that much is uncertain and unplannable, the processes involved in these projects are largely difficult to predict (Mulder 2012).

Interestingly, the literature reports relatively high failure rates (about 70%) of team innovation projects (Mulder 2012). Therefore, a crucial question is how can team processes be improved to render teams less prone to failure and/or better able to deal with failure? One possible answer could be to make innovation teams more 'innovation resilient' (Oeij et al. 2016b). Resilience is the ability to bounce back after a setback. Complex teamwork, however, requires more than just being able to bounce back. Teams must also be able to foresee and prevent problems, and they must be capable of staying on track once they have bounced back successfully (Weick and Sutcliffe 2007).

This contribution will discuss the IRB-tool or Team IRB-tool, that companies can implement in strengthening team resilience. First we will discuss the development of the tool, then summarise its content. Finally, we provide an outline of the tool in the Annex. Given that the tool is still a prototype, we are actively inviting feedback for improving it.

22.2 Development of the IRB Tool

The purpose of developing the IRB tool was to render innovation teams, more aware of and better able to deal with unexpected events and setbacks during an innovation project. To enable such behaviour, called 'innovation resilience behaviour', teams need both certain internal competencies as well as certain organisational features in place (Oeij et al. 2014).

For this purpose, we turned to the literature on crisis management and safety science, because the HRO-teams in these fields operate under risky situations and hardly ever fail (Weick and Sutcliffe 2007). HROs are, for example, nuclear plants, operating theatres, fire brigades and aircraft carriers. From this literature, we gleaned that mindful infrastructures enable innovation resilience behaviour (Oeij et al. 2016a). Mindful infrastructure is a team environment built upon team psychological safety (making mistakes is not punished), team learning (experimenting is stimulated), team voice and constructive organisational politics (commitment through (joint) decision making) and complexity leadership (overcome controversies and ambiguity through seeking synergy). Hence, mindful infrastructures enable a team to be extremely alert for mistakes and motivated to act on them, even if it costs a lot of energy.

In these studies, mindful infrastructures have been shown to be helpful for teams in overcoming critical incidents (Flanagan 1954), that is, events that can significantly get an innovation project off track, for example, as a consequence of financial decisions, changing customer demands, technical setback, and conflicts with stakeholders. Specifically, mindful infrastructures have been found to enable a resilient type of problem solving behaviour in these teams called 'innovation resilience behaviour' (Team IRB) (Oeij et al. 2016b). Team IRB are team behaviours such as being alert for weak signals about failure, resisting overly simplistic reasoning, ensuring that shop-floor work is unambiguously derived from the organisation's strategy, being able to change course when needed, and valuing expertise higher than rank (Oeij et al. 2016b; Weick and Sutcliffe 2007). Our previous research has shown that innovation teams, who often operate in organisations facing economic turmoil and technological change, can benefit from these mindful infrastructures and IRB when it comes to being innovative and staying competitive or cost-efficient (Oeij et al. 2016a, c).

To develop the IRB tool, we took a two-pronged approach, whereby we relied on previous research as well as on feedback from practitioners working in innovation teams. Our primary goal was to transform research findings into a practical tool that reflects the complexity of teamwork and yet is simple—but not simplistic (Pacanowsky 1995). To this end, we relied on instruments we had used in previous research (Oeij et al. 2014, 2016a, c) as well as on instruments developed by others which we made domain-specific for innovation teams. Next, we discussed our ideas with a number of innovation practitioners in different organisations in order to receive feedback and refine our thinking. Moreover, we tested parts of the tool within our own project team (five persons, during a team meeting).

Below, we provide a birds-eye view summary of the rest of the chapter, which we will expand upon in the following. First, organisations that wish to improve team processes and innovative team behaviour need to carefully consider whether they are prepared to engage in IRB, given that it requires a psychologically safe environment and the teams' motivation and cognitive effort to analyse their own behaviour. Second, once teams have decided that they want to improve their teamwork, they need to address the following issues: recognize defensive behaviours in their team and make these discussable; assess the degree to which mindful infrastructure and innovation resilience behaviour are present and assess which elements should be improved; and develop simple practical tools to be applied during team meetings.

In the rest of this section, we first summarise the reasoning behind the IRB-tool, based on the literature. Specifically, we discuss how defensive behaviour and team members' preference for psychological comfort can hamper innovative performance and how IRB can help overcome these hurdles. Next, we discuss the methodological steps we took to check the soundness of the theoretical model underpinning the Team IRB tool.

22.2.1 Theoretical Underpinnings

Innovation projects require people to step out of their comfort zones and engage in exploration and a certain degree of controlled risk-taking. However, defensive behaviour favouring the status quo often occurs and tends to lead to risk-avoidance. This risk-avoidance, in turn, can lead to subpar performance in innovation teams. To help people step out of their comfort zones and prevent risk-avoidance, natural, effortless behaviour would need to be replaced by unnatural or effortful behaviour (Kahneman 2011). Natural behaviour is characterised by high efficiency, and is performed in a cognitively effortless or automatic manner, such as our natural fight or flight response when faced with danger. In this sense, natural behaviour does not demand much psychological effort or motivation. Indeed, we almost engage in it instinctively when we experience danger, discomfort or social threat (Argyris 1990, 2002). In situations that can be dealt with by simple, in routiniz reactions, natural behaviour is useful given that it helps individuals react quickly and efficiently without effort expenditure. However, in situations in which one needs to perform and learn new behaviour, such as in innovation projects, effortful behaviour is needed. In these situations, individuals have to suppress their inclination to engage in (automatic) natural behaviour and replace it by (effortful) unnatural behaviour. It is, for example, natural to not consider all thinkable alternatives when you are in a hurry, and, hence, jump to conclusions instead of validating your decisions. However, to engage in unnatural behaviour, people need to experience a sense of urgency that motivates them to process information in a more effortful manner (Kahneman 2011).

Yet, even in situations that should trigger a high sense of urgency to engage in effortful behaviour people often tend to engage in natural, in routinized behaviours (Weick and Sutcliffe 2007). One reason for this could be that individuals might be purposefully ignoring certain important weak signals in the environment, because addressing these signals would require effort. An example could be that a critical remark from someone in an innovation team is interpreted as a personal attack (automatic, first natural response) instead of as an invitation to look into the issue raised (effortful, unnatural response). However, these weak signals need attention. If left unaddressed, they often hinder the success of innovation projects because people just move on with their natural behaviour and ignore issues that require attention.

Research from the field of crisis management and safety science shows that HROs have found a way to effectively deal with weak signals and the unexpected. For instance, even if a fire brigade has experienced a certain situation before, for example, how an open wildfire develops, they will suppress their inclination to think that they have seen it all before, because they know that not every fire unfolds in similar ways (Weick and Sutcliffe 2007). Indeed, inspired by insights from research on HROs, our research (Oeij et al. 2016a, c) indicates that enhancing the alertness to weak signals in the innovation process improves team innovative capability, because it can on the one hand, reduce certain natural behaviours (i.e., risk-avoidance) and on the other hand it can induce certain unnatural behaviours (e.g., controlled risk-taking behaviour). The more general behavioural pattern associated with controlled risk-taking is 'innovation resilience behaviour' (IRB).

In sum, based on our previous research (Oeij et al. 2014), we use the model (Fig. 22.1) below as the basis for the IRB-tool.

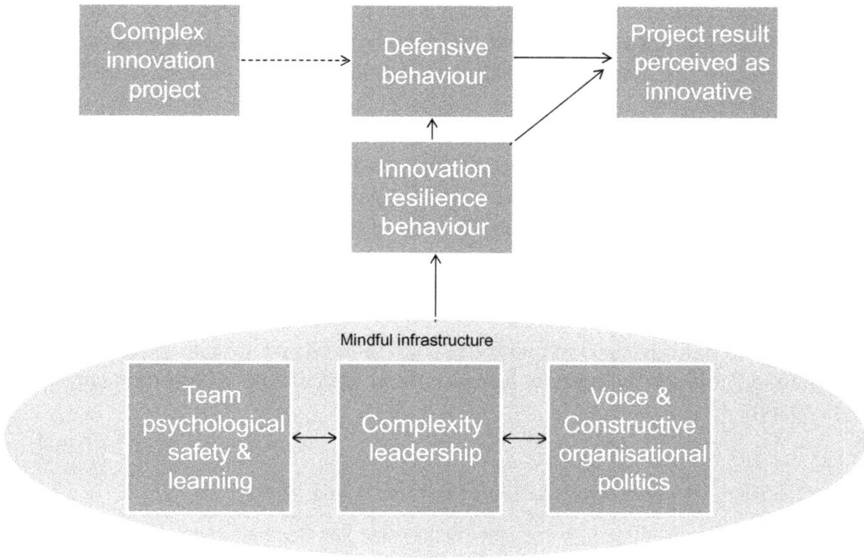

Fig. 22.1 Mindful infrastructure and IRB related to complexity, defensiveness and project results (adapted from Oeij et al. 2014)

This model assumes that complex innovation projects may induce defensive behaviours, such as risk-avoidance, which could result from the experience of threat, feelings of incompetence and shame (Argyris 1990). These defensive behaviours, in turn, may affect how the project results are perceived in terms of degree of innovativeness (i.e., self-perceptions by team members that their project is achieving the desired results). We argue that mindful infrastructure, consisting of team psychological safety, team learning, complexity leadership and team voice, constitute the team environment that enables team IRB (Oeij et al. 2016a, c). Furthermore, we argue that team IRB could help suppress defensive behaviour and help teams deal more effectively with complex aspects of the project, which, in turn, should affect how the project results are perceived in terms of degree of innovativeness. Specifically, IRB should lead to more positive perceptions of project results.

22.2.2 *Methodology of Investigating the Basis of the Model for the IRB Tool*

We have undertaken four steps to investigate the soundness of our model. First, results of case studies and a survey underline that the model in Fig. 22.1 is helpful in describing and understanding team dynamics in innovation teams (Oeij et al. 2014, 2016a).

By discussing 'critical incidents' in innovation teams we identified to what extent elements of mindful infrastructure were present or absent and how they, in turn, enabled IRB. Specifically, we found that mindful infrastructure was positively related to IRB (Oeij et al. 2016a). For instance, when a team leader, actively involved team members (participative decision making) and talked extensively to stakeholders to meet diverging needs and expectations (complexity leadership), he or she was able to avoid obstacles by being attentive to specific needs (alert to weak signals) and, at the same time, steer the project into directions that were backed by most of them (remain sensitive to the overall picture and eradicate ambiguity). This created enough transparency to deal with difficulties (non-defensive processing of complex tasks), which ensured better outcomes for all (project results).

Second, we held discussions with the studied teams on how to assess defensive behaviours (Table 22.1) and how to make them discussable within the teams (Oeij et al. 2016b, c).

Table 22.1 List of 14 defensive behaviours (based on Ardon 2009)

1. Compliance strategy: if your superior tries to persuade you to show commitment, you say that you comply regardless of whether you really do or not
2. Undergo strategy: if your superior initiates a change process, you undergo the interventions passively and do not clarify that you don't think that this is going to work
3. Plan strategy: you agree to make a plan and act as if you comply with the plan; this way you contribute to change and stay in your comfort zone
4. Blame strategy: if change does not succeed, you blame others and attribute negative intentions to them (scapegoating)
5. Assume strategy: you keep your negative assumptions about other individuals' intentions and situations private.
6. Withdraw strategy: if there are communication difficulties, you do not raise the issue with the persons involved; rather, you withdraw and think up a new initiative or discuss the difficulties with peers
7. Ignore strategy: if you observe patterns that are difficult to deal with (e.g., that your employees are not committed, do not ask questions), instead of dealing with the issue, you increase pressure on them to comply
8. Reduction strategy: if things become threatening or embarrassing, reduce the problem until it is controllable again
9. Denial strategy: if things become threatening or embarrassing, deny the problem until it is controllable again
10. Distance strategy: if the discussion gets too close for comfort, change the subject to discuss 'other' parties or to make general observations, such as employees, middle management, or 'the organisation'
11. 'We' strategy: talk in terms of 'our responsibility' and 'what we should do'; as a consequence, nobody has to feel personally responsible
12. Non-intervention strategy: not speaking up, otherwise one might be criticized by the others
13. Joke strategy: if things become threatening or embarrassing, make a joke and change the subject
14. Shirk strategy: shift the responsibility to an 'outsider' and avoid sharing your own opinion about the process or colleagues

Many of the defensive behaviours in Table 22.1 were recognized as being present in team interactions and they were mostly acknowledged as possibly harming the innovation process, for instance, by hindering open communication and collaboration. Hence, we concluded that improving transparency by making defensiveness discussible could be beneficial to the process of collaboration and the innovation process as a whole. When asked, the teams confirmed this assumption.

Third, a few presentations and workshops were held on the topic of IRB (Oeij et al. 2014), during which participants were invited to assess defensiveness within their teams and projects and to discuss how they could more effectively deal with it.

Fourth, and finally, we tested how we could make defensiveness discussable during an internal team meeting (within our own project team). First, participants were invited to closely read a list of 14 defensive behaviours (Table 22.1), which may lead to risk-avoiding behaviour. Second, they were instructed to observe whether they experienced any of these defensive behaviours within themselves or in others during the meeting. Third, if so, they were requested to press a virtual red alert button (which was simply a piece of paper with a red alarm button printed on it; individuals had to pretend pressing this button by slamming the table, indicating they had something to say). In such instances, the meeting was stopped to enable a meta-discussion about the observations of one of the participants. It proved to be difficult in the sense that it was hard to assess defensiveness in action (reflection in action). The participants concluded that one should first have a good understanding of the 14 defensive behaviours. Then, one should practice to identify and find solutions for these behaviours. Nonetheless, the team concluded that being skilled in assessing defensiveness on the spot could improve collaboration during meetings.

In sum, we developed the Team IRB-tool based on our theoretical model (Fig. 22.1) as well as on feedback from practitioners involved in team innovation projects. It is important to mention that the tool is still a prototype and requires further application, refinement and validation in a larger representative sample of teams.

22.3 The IRB-Tool

The instruments that comprise the Team IRB-tool are based on the work of renowned experts Argyris (1990, 1996, 2002), Donald Schön (Argyris and Schön 1974, 1996), Edmondson (2012), Robert Quinn (Lawrence et al. 2009), Karl Weick (Weick and Roberts 1993), and Kathleen Sutcliffe (Weick and Sutcliffe 2007).

The tool aims to improve innovation resilience behaviour in innovation project teams and it can be applied at the organisational or at the team level. An extended summary of the tool is attached as an appendix. The tool taps first into defensive behaviour, and second into mindful infrastructure and Team IRB and provides insights into weak and strong points on these dimensions. Subsequently, teams can discuss the 'gaps' between what is and what is desirable and what kind of follow-up action will be most useful. Top management and the innovation leader can provide support to make the follow-up effective.

The tool thus is intended to support the transition from automatic defensive routines to automatic learning routines. Hence, it addresses the key question of how you can stimulate automatic learning routines, where people ask for feedback, learn from it, reflect together, and search for systemic solutions, creating a learning stance that reinforces behaviours that result in structured, consistent, and persistent organisational inquiry (Sales et al. 2013).

The IRB tool is mainly a diagnostic tool to assess the present situation regarding three aspects:

1. the presence of defensiveness, and thus insight into possible causes for risk-avoidance;
2. the presence of mindful infrastructure, that is, characteristics that facilitate IRB;
3. the presence of IRB, that is, the behaviours and competences to keep an innovation team on track and to get it back on track.

The IRB tool consists of three steps and the above aspects are part of step 1 (aspect 1) and step 2 (aspect 2 and 3).

Step 1

Step 1 consists of 'Assessing your present state of defensiveness'. The purpose of this step is to assess the presence of defensiveness, which can simply be understood as risk-averse behaviour, that can negatively affect the innovation process. Making risk-averse behaviour discussable should help the team become transparent and better prepared to identify relevant bottlenecks in communication and collaboration. The exercise (Step 1 Exercise 1, see the Annex) is meant to 'Assess defensiveness'. Argyris and Schön (1974) developed the so called 'two-column model' which allows participants to make an inventory of things that have been said in difficult conversations, and defensive thoughts that were present but not uttered. The exercise helps to assess patterns of communication in teams, such as circular mechanisms (Smith 2008). The text box below offers an explanation of defensive behaviour for users of the tool.

Explanation of Defensive Behaviour
Perceived complexity in projects, e.g., innovation projects, renders them difficult and fallible endeavours, which may invoke feelings of anxiety, scepticism, moral duty and commitment, which, can be both encouraging and inhibiting. People have a preference for control because it makes them feel comfortable. If they perceive a situation as complicated and as difficult to master, they may feel anxiety and are inclined to restore the situation to a state of control. Argyris (1990) refers to this kind of behaviour as organisational defensive routines, which are any action, policy, or practice that prevents organisational participants from experiencing embarrassment or threat and, at the same time, prevents them from discovering the causes of the

embarrassment or threat. Examples of defensive behaviour are shifting responsibility, blaming others, avoiding conflicts, which all prevent feelings of embarrassment, threat and anxiety (see Table 22.1). Complexity is a source for one of the most frequently occurring defensive behaviours, namely sending mixed messages. Argyris (1990) illustrates how this works: "'Mary, you run the department, but check with Bill', or 'John, be innovative, but careful'. The logic is:

a. send a message that is inconsistent;
b. act as if it is not inconsistent;
c. make steps (a) and (b) undiscussable; and
d. make the undiscussability undiscussable.

The theory of action helps explain why humans perform defensive behaviours (Argyris 1990). Human beings hold two types of theories of action. The one they espouse is usually expressed in the form of stated beliefs and values (the 'espoused theory'); the other is their 'theory-in-use' which can only be inferred from observing their actions, their actual behaviour. In practice, this means that individuals craft their positions, evaluations, and attributions in ways that inhibit both inquiry and their testing with the use of independent logic. Consequently, the corresponding action strategies are likely to be defensiveness which, in turn, result in misunderstandings, self-fulfilling and self-sealing processes. Figure 22.2 depicts the theory-in-use.

Our theory-in-use, leads us to manifest our governing values in our action strategies. Our purpose is to feel good, stay in control, win and avoid losing

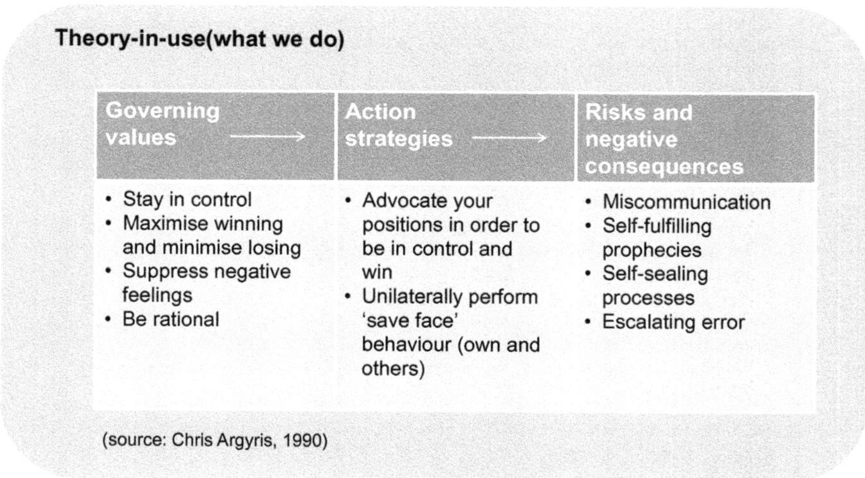

Fig. 22.2 Theory of action model I: theory-in-use

(face). The risks and consequences of this type of behaviour are mixed messaging, embracing 'tame' problems, covering up what happens. You could look again at Table 22.1 and re-read the defensive behaviours that may be at play and try to assess which ones occur in your project team. You may not clearly 'see' them, but you might sense that some of them are or were at work. The next step would be to reflect on these questions: How can defensive behaviours explain risk-avoidance? How could risk-avoidance possibly prevent us from achieving the (innovation) goal of the project? How can we assess defensiveness?

The second theory of action from Argyris is the espoused theory (Fig. 22.3). It consists of governing values that are highly rewarded, but that most of us underuse. The assumption of this model is that its application could rule out most defensive behaviours. Consequently, teamwork would improve and increase the chances for successful innovations. The most important governing value is to validate the information which forms the basis of your decisions. The most important characteristic of the action strategies is that you show the courage to speak up when needed and deliberately take/accept the risk that you may hurt yourself or others. The trick is, of course, how to do that in a constructive manner. Argyris' experience is that, to date, most people are unable to do this. However, if the team succeeds to act like this, the results are promising: no more defensiveness, no mixed messages, no ambiguity, better problem-solving, better learning, and no repeating mistakes. In short: controlled risk-taking or entrepreneurial behaviour in a controlled way.

Espoused theory (what we think how we should do things - ***but what we most often don't do***

Governing values →	Action strategies →	Opportunities and positive consequences
• Get valid (validatable information • Free and informed choice (fact based decisions) • Internal commitment to the choice • Monitor (the actions') effectiveness	• Advocate your positions and combine with inquiry and public testing • Minimise unilateral face-saving • If being critical be constructive	• Reduction of self-fulfilling, self-sealing, error-escalating processes • No mixed messages • Effective problem-solving • Better learning

(source: Chris Argyris, 1990)

Fig. 22.3 Theory of action model II: espoused theory

An important bottleneck is, as already mentioned, to be frank, open, and confronting when needed. Most individuals avoid such conflicts, to the detriment of clarity and progress.

After Step 1 the team has an overview of its defensive behaviour and can decide whether it wants to do something about it or not. If the team decides to do something, it can progress to Step 2 (see Annex).

Step 2

Step 2 is 'Move and go about it' which means the team has decided to do something about the defensiveness that threatens project progress. The purpose of this step is to assess whether a mindful infrastructure is present within the team. With this insight the team can determine whether the preconditions for being resilient in the innovation process are in place. There are two exercises. The first is to 'Assess the mindful infrastructure' (Step 2, Exercise 2 in the Annex) by completing three checklists of team (psychological) safety and team learning, (complexity) leadership, and team voice (and team participative decision making). Based on these completed checklists, team discussions can be organised to discuss the state of the mindful infrastructure. If the team decides to improve the mindful infrastructure, it is advisable to create an action list. Tip: think of what is needed to improve trust, exploration, dealing with ambiguity and commitment. The second exercise is to 'Assess Innovation Resilience Behaviour (IRB)' (Step 2, Exercise 3 in the Annex) of the team, by completing a checklist and subsequently, discussing the findings within the team. This step provides insight into whether the team operates mindfully and in an innovation resilient manner. If the team is not alert to weak signals, simplifies explanations, is unable to change course when needed, values rank higher than expertise, and shop-floor and organisational goals are not aligned, then the process is at risk and chances of failure are high. This insight thus helps establish whether the team is, currently, well equipped to act resiliently in the innovation process.

Step 3

After Steps 1 and 2, which have assessed defensiveness, the presence of mindful infrastructure and IRB, the third step is called 'Wrap Up' and contains 3 more exercises. The first exercise is: 'Assess if you are going to do it' (Step 3, Exercise 4, in the Annex). This step is important, because IRB requires quite some effort; it namely implies unnatural behaviour and deviating from the norm. In short, the team is advised to validate information, take fact-based decisions, ensure commitment to choices, and monitor effects. The second exercise is to 'Assess which competencies to improve' (Step 3, Exercise 5, in the Annex). The team should make a one-time inventory of tasks in the process of an innovation project and assess the team's competency in carrying out each of these tasks. If competencies related to IRB, such as acting proactively, are lacking, one can decide which competencies must be improved at the team or the personal level. The third exercise 'Develop your own tools' (Step 3, Exercise 6, in the Annex) is of a different nature, as it demands creativity. HRO-teams try to automate unnatural behaviour or organisational learning by creating procedures such as briefing and debriefing, and continuously improving processes and behaviours. Innovation teams could do this as well and develop their own briefing and debriefing tools.

The tools that the teams develop are owned by the teams, which in turn, enhances the likelihood of their implementation and should ensure team commitment. A golden rule for tool development, is that they should be simple but not simplistic (Pacanowsky 1995). If the tools are simple, they are more likely to be used, and the more they are used, the more they improve performance as well as individual and team competencies and skills. In other words, over time, through their usage they become automated organisational learning. For example, if the team sums up the decisions that have been taken during a meeting, they could apply a checklist from the clients' perspective (see below) to assess the quality of every major decision:

- Alertness: Are we aware of the wishes of our clients, and could this decision harm their interests?
- Avoid simplification: Did we consider all possible alternatives and is our decision based on facts?
- Sensitivity: Have we checked the effect of the decision for the rest of the organisation, other teams, other aspects of the innovation, other projects for the same client?
- Resilience: Do we know all consequences of our decision and do we have alternative/restoring actions in place?
- Expertise: Do we know who to turn to in the event of every thinkable unwanted effect, and is this person/expertise available when we need it?

If project teams apply this repeatedly during meetings the checklist eventually becomes automated behaviour that no longer requires much time or separate learning. The advantages of such tools are their low thresholds to create a mindful organisational culture with limited effort.

22.4 Conclusion and Discussion

WPI aims at improving the quality of performance, the quality of jobs, and in addition, the innovative capability of the organisation as a whole. Perhaps the most salient feature of WPI is the engagement and involvement of employees when it comes to organisational change and renewal. The IRB Tool pinpoints the role of employees in relation to team performance. By focusing on team defensiveness, a team's mindful infrastructure and Team IRB the tool helps uncover a number of issues that may hamper the success of innovation projects and provides ways to overcome these hurdles. Specifically, it focuses on (1) the issue of risk-avoidance and social safety, (2) the absence of the right organisational conditions that are conducive to desired risk-taking or entrepreneurship, and (3) team behaviours that are required to overcome critical incidents and prevent problems from escalating. Applying the IRB-tool is in itself a workplace innovation intervention, because employees participate in the process.

Certainly, there are some limitations to this contribution. First, although the IRB-tool seems to be promising at face value, it needs to be tested, validated and potentially refined. Second, the IRB-tool can only be applied if there is a minimum level of trust within the team and/or the organisation. If there is obvious conflict, applying the tool will undoubtedly lead to invalid information as people may turn out to be defensive without realising it. Informed consent of each participant is a precondition for use.

The primary advantage of applying the IRB-tool is the opportunity to boost entrepreneurship, intrapreneurship and innovation within the organisation. The tool and its theoretical underpinnings reflect the vision that innovation requires employee participation. Another advantage is that the application of the tool is not necessarily restricted to innovation teams. The tool basically deals with improving problem-solving behaviour, and this could be relevant to other types of teams, to project management in general and to any project-based organisation.

Annex Innovation Resilience Behaviour Tool/IRB Tool

Source: Oeij et al. (2014) (Extended summary).

By applying this tool, a user will obtain:

- Insight into the presence of defensiveness, and, thus, insight into possible causes for risk-avoidance;
- Insight into the degree of the presence of mindful infrastructure, that is, into the presence of characteristics that facilitate IRB;
- Insight into the presence of IRB, that is, into the presence of behaviours and competences to keep an innovation team on track and to get an innovation team back on track;
- A guide with which to develop simple and applicable team meeting tools for own use.

The tool contains 3 steps:

Steps:

Step 1: Assess your present state
Exercise 1: Assess defensiveness (two-column model)

Step 2: Move and go about it
Exercise 2: Assess mindful infrastructure

- Team safety and team learning
- Leadership
- Team voice and influence

Exercise 3: Assess Innovation Resilience Behaviour

- Acting mindfully and alertly

Step 3: Wrap Up
Exercise 4: Assess whether you are going to do it
Exercise 5: Assess which competencies to improve
Exercise 6: Develop your own tools

Step 1: Assess your present state
Explain the step
The purpose is to assess the presence of defensiveness. Defensiveness can be understood as risk-averse behaviour, and risk-averse behaviour can negatively affect the innovation process. Making risk-averse behaviour discussable will help the team to become transparent and better able to identify bottlenecks in communication and collaboration.

Exercise 1: Assess defensiveness
Execute

First: Write down what was said in the right column as literally (verbatim) as you can remember. Write down everything said by you and your conversation partner in the sequence of that talk. Take a good look at what you have written, reread it, and assess whether it is complete according to your own memory.

Second: Now for each part of the dialogue, write in the left column what you were thinking, but did not say (exactly at the level of your own turns). These were probably thoughts that were emotionally-laden and it is likely that these thoughts could have had a strong impact on the conversation if you had spoken them.

Third: Now please look at the following questions:

- Why did you not say what you were thinking?
- What might the consequences have been if you had spoken your thoughts?

- What is the deeper reason behind why you did not say what you were thinking?

Fourth: Look at the defence mechanisms in Table 22.1.

- In hindsight, did you, your conversation partners, or others who were present, undertake any of these defensive practices? If so, which one(s)?
- If so, can you explain why this/these defensive behaviours(s) were used? What was the effect of their usage?

Fifth: Take a step back from this concrete example, and reflect on the following question: to what extent could applying this/these defensive behaviours(s) affect the effectiveness of team work, especially with regard to performing an innovation project?

The same procedure can be used with your team as a whole, if it is safe to do so.

It could be helpful to collect all the team's experiences with defensive behaviours. Then you could discuss questions like these:

- Do we see a pattern in how we communicate or miscommunicate?
- Are we moving in circles from which we do not seem able to escape?
- Is defensiveness related to certain issues, problems, persons, situations?
- What does it say about our own ability to critically but constructively reflect on what happens?
- Are we self-critical or are we scapegoating our environment or 'others'?
- Are we addressing issues we can influence ourselves or are we addressing issues that lie outside our sphere of influence?
- Is there a group bias to favour conformity which excludes deviant thinking and thinking out-of-the box?
- Is this a way to keep mixed messages unresolved?

Result

The result of this exercise should be personal or team awareness regarding your own defensiveness and how that possibly affects the process of the innovation project and/or the team. The fundamental question is: are you prepared to do something about it? If so, please continue reading.

Step 2: Move and go for it

Innovative behaviour is not only a matter of characteristics of the behaviour of individuals but also a matter of organisational design and the design of jobs (i.e., active jobs; Karasek & Theorell, 1990). At the team level, a facilitating factor is the presence of a mindful infrastructure (Fig. 22.1). Once this is in place, there is greater potential for the emergence of IRB (Fig. 22.1).

Explain the step

The purpose of this step is to assess whether a mindful infrastructure is present in your team. With this insight you can determine whether your team is well situated to be resilient in the innovation process. If that is not the case, you can decide what to improve.

Exercise 2: Assess mindful infrastructure

Execute

- Team psychological safety & learning.

Answer the following questions for your team/department/organisation **Do you agree or disagree with the following statements about team safety and team learning?**	Agree = 1 Disagree = 0
If you make a mistake on this team, it is never held against you	
Members of this team are able to bring up problems and tough issues	
People on this team never reject others for being different	
It is safe to take a risk in this team	
It is easy to ask other members of this team for help	
No one on this team would deliberately act in a way that undermines my efforts	
Working with members of this team, my unique skills and talents are valued and utilised	
We regularly take time to work out ways to improve our team's work processes	
This team tends to handle differences of opinion privately or off-line, rather than addressing them directly as a group	
Team members go out and get all the information they possibly can from others, such as customers, or other parts of the organisation	
This team frequently seeks new information that leads us to make important changes	
In this team, someone always makes sure that we stop to reflect on the team's work process	
People in this team often speak up to test assumptions about issues under discussion	
We invite people from outside the team to present information or have discussions with us	
Team safety and team learning Score: 10–14 = Present; 1–5 = Absent; 6–9 = Present to a limited degree	

If your score is 6 or lower, consider improvement.

Results

- Draw your conclusions, based on data and discussion.
- Define future actions to take.

Execute

- Leadership

Answer the following questions for your team/department/organisation **Do you agree or disagree with the following statements about leadership?** (*Remember: leadership can be performed by one individual but also by professionals and by the group as a whole*)	Agree = 1 Disagree = 0
Leadership concerning *collaboration* is:	
Making it legitimate to contribute opinions	
Maintaining an open climate for discussion	
Employing participative decision making	
Leadership concerning *creativity* is …	
Launching important new efforts	
Getting unit members to exceed traditional performance patterns	
Encouraging direct reports to try new things	
Leadership concerning *control* is …	
Keeping projects under control	
Ensuring that corporate procedures are understood	
Expecting people to get the details of their work right	
Leadership concerning *competition* is …	
Demonstrating full efforts on the job	
Getting work done quicker in the unit	
Providing fast responses to emerging issues	
Leadership concerning *tough issues* is…	
Ability to provide clear directions	
Ability to serve compatible needs in the organisation	
Ability to rule out ambiguity	
Leadership of the leader and the team: Score: 10–15 = Present; 1–5 = Absent; 6–9 = Present to a limited degree	

If your score is 6 or lower, consider improvement.

Results

- Draw your conclusions based on data and discussion.
- Define future actions to take.

Execute

- Team voice and influence.

Answer the following questions for your team/department/organisation **Do you agree or disagree with the following statements about team voice and influence?**	Agree = 1 Disagree = 0
Here each (team) member:	
Develops and makes recommendations concerning issues that affect this work group	
Speaks up and encourages others in this group to get involved in issues that affect the group	
Communicates their opinions about work issues to others in this group even if their opinion is different and others in the group disagree with them	
Keeps well informed about issues where their opinion might be useful to this work group	
Gets involved in issues that affect the quality of work life here in this group	
Speaks up in this group with ideas for new projects or changes in procedures	
Here in this organisation:	
We have a "we are together" attitude	
There are real attempts to share information throughout the project team	
We decide many issues together, or at least have influence on matters that concern us	
Team voice and team participative decision making: Score: 7–9 = present; 1–4 = absent; 5–6 = present to a limited degree	

If your score is 5 or lower, consider improvement.

Results

- Draw your conclusions based on data and discussion.
- Define future actions to take.

Based on these completed checklists and questions you should have a fair picture of the mindful infrastructure of your team. If you decide to improve the mindful infrastructure: make a list of your actions.

Explain the step

The purpose of this step is to assess whether Team IRB is present in your team. This step helps you gain insight into whether the team operates in a mindful and alert manner.

Exercise 3: Assess Innovation Resilience Behaviour (Team IRB)
Execute

- Team IRB.

Answer the following questions for your team/department/organisation **Do you agree or disagree with the following statements about acting mindfully and alertly?**	Agree = 1 Disagree = 0
Preoccupation with failure	
We actively look for risks and try to understand them	
We are keen for cues to understand why our expectations are not met	
When members spot potential risks we discuss them extensively	
Reluctance to simplify	
Members of this team never take things for granted	
Team members listen carefully, and it is rare that someone's view goes unheard	
We actively seek more explanations and viewpoints before taking a decision	
Sensitivity to operations	
Team members put effort into building a clear picture of the current situation of the project	
We constantly monitor the progress of the project in a profound manner	
The team has discretion to resolve unexpected problems as they arise	
Commitment to resilience	
We always learn from every mistake made	
Most members have the skills to act on unexpected problems that arise	
This team is extremely resourceful	
Deference to expertise	
Team members typically "own" a problem until it is resolved	
In this organisation expertise is valued over hierarchical rank in most decisions	
Instead of muddling through, the team quickly obtains any external expertise if needed	
Our innovation resilience behaviour: Score: 11–15 = present; 1–5 = absent; 6–9 = present to a limited degree	

If your score is 6 or lower, consider improvement.

Results

- Draw your conclusions based on data and discussion.
- Define future actions to take.

The challenge here is to make a list of IRB-behaviours that could be improved by discussing them within teams or departments. The purpose of the discussion is to create a common awareness of what is needed. This could lead to an action list.

Step 3: Wrap up
Explain the step
By now you should have a good idea of the defensiveness, mindful infrastructure and Team IRB of your team/department or organisation. You may also have some suggestions about how to move forward. This step links back to the 'espoused model' and will help you design your own tools.

The 'espoused model' will help you combat defensiveness in your team. It tries to make defensiveness discussible. There is no easy guide for this: you and the team must be prepared to do this and find your own way in how to do it.

Exercise 4: Assess whether you are going to do it
Execute
Are you prepared to apply the values of the 'espoused theory''? For example, in the situation of taking decisions during team meetings:

- Will you make the effort to gather valid information (evidence)?
- Will you make your decisions fact-based?
- Will you seek internal commitment?
- Will you monitor the effectiveness of actions?

Results

- Draw your conclusions based on your answers and discussion.
- Define future actions to take.

This exercise should help you to come to a conclusion about what to do next.

Exercise 5: Assess the competences that need to be improved
Results

- Draw your conclusions.
- Define future actions to take (think of what competencies are needed to fulfil tasks and how these competencies can be acquired).

Exercise 6: Develop your own tools
Explanation
HRO teams try to automate unnatural behaviour by creating procedures such as briefing and debriefing, and continuously improving processes and behaviours. Teams working on innovation could develop such tools as well (see the example of the checklist for decisions from the client's perspective in the text; see page 386).

Prepare

- Choose domains of teamwork for which IRB-tools are helpful. For example: decision-making, stakeholder management, requirements for end-result, future market opportunities, development of a pilot to test the results.

Execute

- Apply the five IRB-competencies to the tasks of the team/team members in relation to the selected domain(s): make a list that you can consult/walk through during a team meeting.

References

Ardon, A. J. (2009). *Moving moments. Leadership and interventions in dynamically complex change processes*. Ph.D. Dissertation. Amsterdam: Free University of Amsterdam.

Argyris, C. (1990). *Overcoming organizational defenses. Facilitating organizational learning*. Upper Saddle River, NJ: Prentice Hall.

Argyris, C. (1996). Actionable knowledge: Design causality in the service of consequential theory. *The Journal of Applied Behavioral Science, 32*(4), 390–406.

Argyris, C. (2002). Double-loop learning, teaching, and research. *Academy of Management Learning & Education, 1*(2), 206–219.

Argyris, C., & Schön, D. A. (1974). *Theory in practice: Increasing professional effectiveness*. Oxford: Jossey-Bass.

Argyris, C., & Schön, D. A. (1996). *Organizational learning II. Theory, method, and practice* (2nd ed., 1st ed. 1978). Reading (MA): Addison-Wesley.

Edmondson, A. C. (2012). *Teaming. How organizations learn, innovate, and compete in the knowledge economy*. San Francisco, CA: Jossey-Bass.

Flanagan, J. C. (1954). The critical incident technique. *Psychological Bulletin, 51*(4), 327–358.

Kahneman, D. (2011). *Thinking, fast and slow*. New York: Farrar, Straus and Giroux.

Lawrence, K. A., Lenk, P., & Quinn, R. E. (2009). Behavioral complexity in leadership: The psychometric properties of a new instrument to measure behavioral repertoire. *The Leadership Quarterly, 20*(2), 87–102.

Mulder, N. T. (2012). *Value-based project management. A design approach to develop a project management approach for chaordic projects from the perspective of complexity thinking*. Ph. D. Dissertation. Eindhoven University of Technology (in Dutch).

Oeij, P. R. A., Dhondt, S., & Gaspersz, J. B. R. (2016a). Mindful infrastructure as an enabler of innovation resilience behavior in innovation teams. *Team Performance Management: An International Journal, 22*(7/8), 334–353.

Oeij, P. R. A., Dhondt, S., Gaspersz, J. B. R., & de Vroome, E. M. M. (2016c). Can teams benefit from using a mindful infrastructure when defensive behaviour threatens complex innovation projects? *International Journal of Project Organisation and Management, 8*(3), 241–258.

Oeij, P. R. A., Dhondt, S., Gaspersz, J. B. R., & Van Vuuren, T. (2016b). Innovation resilience behaviour and critical incidents: The relevance for the management of R&D and innovation projects. Paper presented at *EURAM 2016 'Manageable Cooperation?'*, Paris, France, Université Paris-Est Créteil (UPEC). June 1–4, 2016.

Oeij, P. R. A., Preenen, P. T. Y., & van der Meulen, F. A. (2014). *From Unnatural Behaviour to Innovation Resilience Behaviour: Prototype of a Change Tool*. [ETP Behaviour and Innovation]. Leiden: TNO Healthy Living. (December 2014).

Pacanowsky, M. (1995). Team tools for wicked problems. *Organizational Dynamics, 23*(3), 36–51.

Sales, M., Vogt, J. W., Singer, S. J., & Cooper, J. B. (2013). From automatic defensive routines to automatic learning routines. *Reflections: The SoL Journal on Knowledge, Learning, and Change, 13*(1), 31–42.

Smith, D. M. (2008). *Divide or conquer: How great teams turn conflict into strength*. New York: Portfolio, Penguin Group.

Weick, K. E., & Roberts, K. H. (1993). Collective mind in organizations: Heedful interrelating on flight decks. *Administrative Science Quarterly, 38*(3), 357–381.
Weick, K. E., & Sutcliffe, K. M. (2007). *Managing the unexpected. Resilient performance in an age of uncertainty* (2nd ed.; 1st ed. 2001). San Francisco: Jossey-Bass.

Author Biography

Peter R.A. Oeij is a senior research scientist & consultant affiliated with TNO, The Netherlands Organisation for Applied Scientific Research in Leiden, The Netherlands. His field of work is innovation management, workplace innovation, social innovation and team dynamics.

Part V
Conclusion

Chapter 23
Conclusion: The Way Forward with Workplace Innovation

Peter R.A. Oeij, Diana Rus and Frank D. Pot

23.1 Introduction

The past two decades have seen a proliferation of the concept of Workplace Innovation in policy circles, as well as among academics and practitioners. Workplace innovation has come to be seen as key to ensuring that organisations and the people within them can adapt to and engage in healthy, sustainable change in a turbulent world. The wider recognition and adoption of workplace innovation across countries and industries has yielded promising new opportunities for companies to generate value while simultaneously creating the conditions whereby people are centrally involved in improving the quality of working life.

In this book, '*Workplace Innovation: Theory, Research and Practice*', we have featured a compilation of chapters on workplace innovation taking a multi-level, multi-disciplinary perspective that combines policy, theory, research, and practical perspectives. In this respect, the book presents the current state of the art on European and national policies on workplace innovation, novel theoretical approaches to workplace innovation, as well as international workplace innovation case studies, and practical tools for the implementation of workplace innovation.

Recognising a need within societies and workplaces for more integrated approaches to problem solving, this book outlines how issues of wellbeing at work

P.R.A. Oeij (✉)
TNO, The Netherlands Organisation for Applied Scientific Research, Leiden, The Netherlands
e-mail: peter.oeij@tno.nl

D. Rus
Creative Peas, Amsterdam, The Netherlands
e-mail: d.rus@creative-peas.com

F.D. Pot
Radboud University, Nijmegen, The Netherlands
e-mail: frank.pot@ardan.demon.nl

P.R.A. Oeij et al. (eds.), *Workplace Innovation*, Aligning Perspectives on Health, Safety and Well-Being, DOI 10.1007/978-3-319-56333-6_23

and organisational performance can not only be better understood, but also solved simultaneously. It does so by integrating a number of research perspectives, in particular occupational safety and health and organisation theory, as well as by suggesting a combination of policies on work organisation, human resource management and employment relations. In this sense, the book contributes to the objectives of the Springer book series 'Aligning Perspectives on Health, Safety and Well-being'.

Whereas the field of workplace innovation has come a long way from its humble beginnings, tremendous amounts of work still lie ahead of us. In this chapter, we aim to (1) summarise the state of the art on workplace innovation policy in Europe; (2) discuss some of the similarities and dissimilarities across approaches presented in this book; (3) identify some points of convergence and mutual reinforcement within and across policy, theory, empirical, and practical approaches; and (4) offer some pointers for the future of workplace innovation in the areas of policy, theory, research, and practice.

23.2 State of the Art on Workplace Innovation Policy in Europe

European policymaking stresses the importance of innovation. Both at the societal and at the company level, technological and business model innovation alone are no longer deemed sufficient to realise the full potential of innovation, renewal and change. Indeed, the crucial role of human talent and human deployment has been acknowledged as a sine qua non for innovation success. At the societal level, this newfound understanding of the role of human talent for innovation has been conceptualised as social innovation (i.e., addressing social needs by social means), whereas at the company level, it has been conceptualised as workplace innovation (WPI). According to DG GROW of the European Commission "*Workplace innovation can mean many things such as a change in business structure, Human Resources management, relationships with clients and suppliers, or in the work environment itself. It improves motivation and working conditions for employees, which leads to increased labour productivity, innovation capability, market resilience, and overall business competitiveness*".[1]

Although workplace innovation has been gaining visibility as an emerging European policy, namely by being promoted as part of a broader EU economic and social policy to support organisational change in companies, the EU has only partially embraced workplace innovation, leaving most of the support to disparate national policies (see Chap. 2). These national programs and initiatives show that, across European countries, workplace innovation programs tend to exclusively

[1]https://ec.europa.eu/growth/industry/innovation/policy/workplace_en (Accessed June 1, 2017).

utilise soft regulation in its various forms. Apart from that, the activities are still unevenly distributed by geographical area, with the highest proliferation of programs in Northern and Western Europe (see Chap. 3).

On the one hand, we can conclude that European policy making is supportive of WPI to a certain extent. In this respect, the best policy initiative to date is perhaps the set-up of EUWIN,[2] the European Workplace Innovation Network by DG GROW in 2012 around the time of the publication of the Dortmund/Brussels position paper 'Workplace innovation as social innovation'.[3] By 2016, EUWIN has managed to connect more than 10,000 companies and other local, regional and national practitioners, policymakers and researchers that share know-how and expertise on workplace innovation via workshops and conferences as well as via a Knowledge Bank[4] that hosts numerous case studies, films, articles and other learning resources. On the other hand, EU policy making does not show evidence of firm and systemic support for WPI, thus making Europe's love for WPI only half-hearted. At the national level, national programs differ markedly across countries, leaving adoption and support of WPI rather scattered with several blank spots on the European map (e.g., the newer EU member states). That being said, new and promising developments are currently underway in many countries, such as Estonia, Lithuania and Poland, with no history of such programs.

23.3 Alignment of Research Perspectives: Similarities and Dissimilarities

Whereas the policy landscape provides a patchwork picture of WPI, so do theory building and research. As evidenced in the different chapters, policy makers, practitioners and researchers do tend to approach WPI from different angles, reflecting their own assumptions, viewpoints and interests. Therefore, we can conclude, that, to date, no overarching, commonly accepted theory of WPI has emerged. However, we do see calls for and movement towards a more unified approach to WPI.

One such development is noticeable in the evolution of the definition of WPI over time (see Chap. 5 for a more thorough discussion) and across stakeholders. For instance, the definition of WPI by DG GROW provided above, suggests that WPI can mean numerous things ranging from changes in structures to changes in the workplace itself as well as to improved outcomes at the organisational and individual level. This definitional variety is also reflected in the different chapters in this book. Nonetheless, at a more abstract level, the various definitions of WPI can be

---

[2]http://portal.ukwon.eu/.

[3]http://workplaceinnovation.org/nl/kennis/kennisbank/dortmund-brussels-position-paper-12th-June-2012/1009.

[4]A link to the Knowledge Bank: http://portal.ukwon.eu/euwin-knowledge-bank-menu-new.

classified into three distinct categories. Some focus primarily on the goals of WPI, namely that it should benefit both organisations and the people within them. Others tend to see WPI as an activity and focus primarily on the process of innovative change and adaptation. Finally, others focus on WPI as a way of framing and understanding and, therefore, focus on WPI as a theoretical conceptualisation of what is happening in organisations that apply WPI-like measures and interventions. These three different ways of defining WPI are, however, not mutually exclusive and, across chapters (see Chaps. 5, 9 and 17) there are calls and proposals for integrating these perspectives.

Another commonality across the chapters in the book is that most of them deal with goals, drivers, and leverage factors for WPI, albeit with different focal points and from different perspectives. The goals of WPI are deemed to be rather straightforward and there seems to be little disagreement in this respect, except maybe on the relative emphasis that is placed on these goals. WPI is aimed either (1) at achieving better organisational performance, such as competitiveness, cost-effectiveness, productivity, and innovative capabilities, or (2) at achieving a better quality of working life, including healthy, productive and meaningful work, or (3) the combination of both. One distinctly noticeable trend across the chapters, that we wholeheartedly endorse, is to insist on conceptualising the goals of WPI as *simultaneously* focusing on improved organisational performance and quality of working life.

The drivers of WPI that have been put forward are manifold. Yet, they can all be classified under a limited number of categories, namely: people, organisational design, and change processes. The first driver-category—people—focuses on better deploying human talents through HR measures, leadership, social dialogue, participation and involvement, and by creating engagement and commitment. The aim here is to influence and shift behaviour within organisations towards increased trust, cooperation, creativity, and motivation, with the ultimate goal of engendering improvements in organisational performance and/or quality of working life (see Chaps. 8, 10–13 and 16). The second driver-category focuses on redesigning structural elements of the organisation. Examples of such structural redesigns are: diminishing organisational layers, teamwork redesign that enables teams to execute 'complete' tasks, enriching jobs with autonomy and learning possibilities to create 'active jobs', and providing decentralised decision latitude wherever that is possible and functional (see Chaps. 8, 10, 16 and 17). The third driver-category refers to the process of change, the role of the different actors, and to the implementation of interventions and measures. Characteristic for workplace innovation is a bottom-up approach in defining and developing interventions. This implies serious dialogue with and/or involvement of employees, employee representatives and unions when it comes to change (see Chaps. 10–12, 19 and 20). This, however, does not imply that economic goals are made subservient in the process. On the contrary, economic motives often represent the starting point for WPI-interventions, as was shown in a Eurofound study across 10 EU member states (see Chap. 10). Nonetheless, most approaches, be they theoretical or practical, emphasise the crucial role played by employees for successful WPI implementation (see Chaps. 5, 9– 14, 18, 19 and 22).

This implies either engaging them, that is, make them feel heard, or involving them, that is, provide them with decision latitude.

Leverage factors for successful WPI implementation centre around sound relationships and trust-based partnerships within organisations. The above-mentioned Eurofound study shows that employee involvement, top management commitment and supportive leadership were the main leverage factors in 51 companies that had developed mature industrial relations between management, employees and employee representatives (see Chaps. 10– 12). Indeed, in practice, we observe that employees play a significant role when it comes to developing and implementing workplace innovation. When genuine cooperation between management and employees exists, both organisational and employee goals tend to be realised.

Our summary of this wide variety of goals, drivers and leverage factors could lead one to conclude that any combination of goals, drivers and leverage factors (or—to put it differently—any intervention whether it be focused on HRM, work organisation or labour relations), could be called 'workplace innovation'.[5] However, if that were the case, the relatively new concept of WPI would be indistinguishable from any of the other concepts and would, therefore, be meaningless. The workplace innovation community, which is represented in this book, stresses the need for an integral approach (see also Chaps. 5 and 17) that takes into account (1) organisational performance and quality of working life simultaneously (as goals), (2) people, organisational design and change processes (as drivers); and (3) sound relationships and partnerships within organisations (as leverage factors). This integral approach is also important for research and evaluation. Workplace innovation claims to improve organisational performance and quality of working life simultaneously. However, this can only be the case, if drivers and leverage factors have also been approached in an integral fashion. Calling separate interventions or top-down interventions workplace innovation and concluding that they did not lead to the intended goals does not do justice to the notion of workplace innovation.

In terms of theorising, the field has been marred by quite some divergence during the past years. This has largely been the case due to the fact that different theorists have focused mainly on one of the three driver-categories for WPI (see above), that is, people (cultural perspective), organisation design (structural perspective), and process-approaches (process and change perspective). We strongly believe that the time has come to start with the convergence and integration of these theoretical viewpoints. A theoretical chapter (see Chap. 5) as well as a design chapter (Chap. 17) tackle this issue by trying to align these various approaches and present an integral perspective. Both chapters conceptualise WPI as aiming to improve organisational performance and quality of working life simultaneously, and both identify a firm relationship between organisational design and

[5]The mentioned Eurofound study assessed that there are a limited number of configurations (i.e. combinations of factors) that indeed lead to WPI, whereas the majority of other configurations do not (see Eurofound 2015; Howaldt et al. 2016 in Chap. 10). Hence, not anything goes!

organisational behaviour. These authors suggest that WPI should integrate these various perspectives in measures and interventions, and they identify several approaches in the literature that could strengthen each other in achieving this purpose. For example, the ideas and starting points of Modern Sociotechnology, High Performance Work Systems, Dynamic Capabilities, Relational Coordination, Employee-Driven Innovation, and Democratic Dialogue, to mention just a few, share a common foundation in that they see the human factor as crucial to business success. In line with this integral approach, Chap. 7 argues that, while much research as well as EU-policies emphasise improving WPI via individual coping behaviour, attention to organisational level improvements (that is, a conditional approach) would be a more promising strategy to realise the promises of WPI and reduce well-being risks.

Whereas we do strongly support the notion that the field of WPI is in dire need of theoretical convergence this does not imply that we advocate a blind 'assimilation of viewpoints'. Rather it implies that we believe that researchers can benefit from taking a systemic perspective, leverage commonalities to be found within theoretical diversity and thereby, create a common shared base. The practical implication of this would be that actors trying to implement WPI have a common base to start from and, yet, can choose from a plurality of perspectives to customise their practice.

23.4 Towards Convergence and Mutual Reinforcement

As the field of WPI is maturing, some convergence within and between policymaking, theory, evidence-based research, and practice is needed. From a European policy perspective, a more systematic and committed approach to WPI would be more than welcomed. In addition, recent developments in digitisation and robotisation offer promising opportunities for the integration of technological innovation and skills, occupational health and workplace innovation. This would be the logical consequence of the historical development of the innovation of workplaces. This development can be traced back to the 1990s, when policymaking started placing an increasing emphasis on work organisation instead of on technology and economics alone. Indeed, at the turn of the century, EU policymaking shifted its focus towards more employment, productivity and the improvement of working life (healthy work). During the past decade, we have observed a strong focus on innovation and its social dimension, marked, for instance, by the EUWIN project. The EU could set the tone and provide support to WPI initiatives within and especially across member states. Chapter 3 describes successful national and regional initiatives which have developed more intense collaboration and the exchange of experiences under the umbrella of EUWIN.

At the national level, the main challenges lie in overcoming a technological or economical deterministic type of innovation support that still emphasises technological and business model innovation without sufficient attention to social aspects.

For instance, the German Industry 4.0 movement has a very optimistic view of technological innovation but may underestimate its social dimension. Drawing from earlier experience with "Halle 54"—a Volkswagen plant that became famous for its failed vision of a factory without people that relied on computer-integrated manufacturing—the authors of Chap. 4 illustrate the dysfunctionalities and contradictions of technology-centred approaches, whose excessive automation ambitions, despite having failed repeatedly, are currently experiencing a resurgence in "Industrie 4.0". Halle 54 ultimately failed because of its radically contra-anthropocentric rationalisation strategy. The lesson here would be that organisational development is more likely to be successful if it follows social demand rather than a technological push. This implies activating experience-based knowledge and inviting participation in order to clarify employee needs. "Industrie 4.0" would thus be better served if it followed a more comprehensive innovation strategy with social innovation and workplace innovation at the core. This implies paying attention to the development of sustainable and integrated business models, and the enhancement of companies' abilities to innovate through the comprehensive utilisation of the potential of their employees and of society. In Germany, this debate has been labelled "Work 4.0" (Arbeiten 4.0). It is important to note that this focus on technological innovation alone is not a uniquely German phenomenon, but rather prevalent across European countries.

Surely, policy makers will want evidence that the alignment of technological and economic approaches to innovation with workplace innovation and social innovation lead to superior outcomes as compared to a technological focus alone. In this respect, Chap. 6 presents quite some positive evidence for organisational performance and Chap. 8 shows strong positive evidence for effects on wellbeing (see also Chap. 16). To date, there is some research that shows a positive association between WPI, company performance and quality of working life simultaneously. Nonetheless, we need to be cautious in terms of drawing firm conclusions and instead, should encourage the proliferation of more and better quality research. Currently, the employed theoretical research models tend to be rather complex, which in turn, hinders the understanding and the diffusion of WPI by practitioners. Given the complexity of the models, organisations cannot just implement WPI based on others' experiences. In addition, the applied research methodologies and methods vary in quality in terms of their validity, representativeness and generalisability. In this respect, more high-quality research, improved data collection, and data analysis methods are needed to provide policymakers and organisations with a more solid framework for understanding which combinations of WPI practices increase organisational performance and quality of working life (see Chaps. 6 and 16). In his foreword Karasek adds to this that a connection with the level of society, in particular the economy, is necessary to make sure that workplace innovation is contributing to the development from a 'commodity-based economy' to a 'conducive economy'. This means creating value for citizens in what he calls Bargain #2:

producing the psychosocial wellbeing and social integration for healthy work as well as new ideas and skills for innovative economic development.

Two points of convergence can be seen across the majority of theoretical, empirical and practice-oriented chapters: (1) most of them highlight that employees play a crucial role for successful WPI implementation, and (2) most of them strongly recommend taking a systemic approach to WPI.

Indeed, some of the empirical chapters show evidence for the critical role played by employees in successful WPI implementation:

- successful WPI implementation results from an interplay between management-driven business goals and employee-driven quality of work goals (Chap. 10);
- a consistent approach to shared leadership can stimulate employee empowerment and bottom-up initiative, which, in turn, lead to successful WPI interventions (Chap. 12);
- lean management methods can only be a successful tool for WPI if employees are actively involved in the process (Chaps. 13 and 16);
- workplace innovation should really include the aspect of quality of working life otherwise low employee engagement will be the consequence (Chaps. 14 and 16);
- institutional alliances are relevant for the sustainability of WPI activities within companies (Chap. 11).

Mirroring the empirical findings of the empirical chapters, a number of the practice-oriented chapters also stress the critical need for employees to participate in organisational change, and the development and implementation of WPI-measures and activities and provide tools or recommendations on how this can be achieved. For instance:

- a step-by-step plan for implementing workplace innovation is offered that places a strong emphasis on empowering jobs and self-managed teams; people-centred management practices; systematic opportunities for employee-driven innovation; and co-created and distributed leadership combined with employee voice in strategic decision-making (Chap. 18);
- management practices for developing employee-driven innovation are proposed such as providing work autonomy, promoting diversity, acting informally, de-emphasising organisational structures and roles and providing support (Chap. 19);
- a tool is offered to help teams boost their innovation success. The tool focuses on team processes, such as resilience, psychological safety, learning, voice, and leadership. By uncovering team defensiveness and a team's mindful infrastructure as well as innovation resilience behaviour, the tool helps teams become more aware of issues that could hamper the success of innovation projects and provides ways to overcome these hurdles (Chap. 22).

Given the multi-level, multi-disciplinary nature of WPI, the focus on taking a systemic, holistic view found across chapters is not surprising and more than welcome. Indeed, the field would also benefit from a tighter integration of theoretical approaches as well as feedback loops between theory, practice and policy. In this respect, all of the theoretical chapters in the book argue for a systemic perspective on WPI (see Chaps. 5–8). In addition, some of the empirical chapters provide strong evidence that a systemic view would be beneficial. For instance, a review of existing WPI studies from a work and organisational psychology perspective, finds that job autonomy, job flexibility and participation in organisational life are generally positively related to both improved organisational performance and quality of working life. However, these relationships are highly contingent on taking into account both structural and cultural elements within the organisation. Therefore, given the intertwined nature of organisational factors in promoting higher organisational performance and quality of working life, there is a need to take a systemic perspective in the study and implementation of WPI (see Chap. 9). Moreover, an analysis of the new sharing and platform economies, for instance Uber and 'Uberisation', concludes that they offer both risks and opportunities for the quality of work, which depend on whether these renewals will be regulated by markets, regulations or organised labour (see Chap. 15). This would indeed suggest that a tighter integration between policy, research and practice would be needed. This systemic approach is also mirrored in the majority of the practice-oriented chapters. For instance, Chap. 17 provides a systemic design or analysis tool for organisations interested in implementing WPI. Moreover, Chap. 21 aims at distilling a set of criteria for identifying and creating case studies of WPI that could be useful for practice. In doing so, the authors strongly argue for a more thorough integration of applied research and practice. Finally, Chap. 20 presents tips for work and organisational psychologists on how to position themselves to be able to provide more value to WPI, given their understanding of both human behaviour and organisational structures. This is, of course, not only an assignment for psychologists. The holistic, systemic approach presumes an integration of psychology, sociology, economics, business administration, technological knowledge, and occupational safety and health expertise, to a larger extent than has already been achieved in this book.

All these streams—policy making, theory and research, and practice—have the potential to mutually reinforce each other, provided that the accumulated knowledge is shared across 'silos' and that there is increased interaction and cross-pollination of ideas. One of the best examples in this respect is Norway where companies, government, social partners and researchers have closely worked together to develop and implement employee-driven innovation (Chap. 19). One way to increase knowledge access would be to streamline and integrate all of the different resources on WPI on one platform. The Workplace Innovation Knowledge bank[6] is instrumental in this regard, but not enough. The next step would be to synthesise the current stream of diverging views into one of converging views.

23.5 The Future

To converge, workplace innovation policy, research and practice activities would require the joint effort on the part of the different stakeholders and better connections among them. For instance, public authorities might be tempted to say: "If the benefits of workplace innovation are beyond doubt, organisations will apply WPI by themselves and no government activity is needed." Unfortunately, the issue is more complicated than that. Sometimes there is a lack of knowledge regarding WPI or the existing labour relations may not allow for co-creation. Sometimes managers might be hesitant to share power by involving employees, in the mistaken assumption that they are weakening their own position or simply because they do not see the possibilities that come with that. Sometimes it is the employees who do not want to change and do not see the value added of being more involved and proactive. Sometimes managers may wait to see what the competition is doing or focus exclusively on short-term results instead of considering long-term innovation and sustainability. Indeed, as the case studies illustrated in this book explicitly show (see Chap. 12 for instance) changing leadership style and engaging in co-creation takes courage and time. Individual organisations are unlikely to do this as a matter of course unless there is a strong internal vision regarding WPI. Therefore, as the chapters in the policy section have shown, government campaigns are needed, in concert with the support of employers' associations and trade unions.

An excellent starting point for bringing together various WPI stakeholders such as policymakers, practitioners and researchers is the previously mentioned EUWIN. Apart from the Workplace Innovation Knowledge bank that helps disseminate case studies and practical know-how on WPI, recently, new publication channels have been established that would allow for a wider dissemination of research and practice, such as the *European Journal of Workplace Innovation*[7] and the *European Public & Social Innovation Review.*[8] In addition, research activities on WPI could be better streamlined by intensive exchanges between academic and more applied researchers via conferences, workshops, and by creating 'special interest groups' (SIGs) in networks such as EURAM, EGOS, DRUID and ISPIM, to mention just a few. Such networks could continue to encourage the cross-pollination of ideas and

---

[6]See http://portal.ukwon.eu/euwin-knowledge-bank-menu-new and http://www.workplaceinnovation.org/nl/kennis/kennisbank.

[7]http://journal.uia.no/index.php/EJWI.

[8]http://pub.sinnergiak.org/index.php/esir/index.

knowledge dissemination between policy makers, practitioners and researchers in the manner that EUWIN has done so far. Given that EUWIN has been closed as a EU-project in 2016, it is all the more important that networks are built and maintained that would allow for these exchanges to continue.[9] In addition, one could think of encouraging the creation of national workplace innovation associations that could support practitioners and maybe even create certification programs for WPI practitioners,[10] while at the same time, keeping abreast of current research and maintaining close ties with policymakers.

In terms of the future, this book has made it clear that, whereas we have booked significant progress in terms of WPI, there is still a lot of work ahead of us whether we are thinking about policy, research or practice. In terms of policymaking, the following could be some action points to be considered:

- Define a European agenda on how to better integrate technological and economic innovation with workplace and social innovation; this requires the collaboration of different DGs, such as GROW, EMPL, REGIO and RTD.
- Stimulate raising awareness about the opportunities that WPI could offer through business channels and across industrial sectors;
- Ensure that WPI implementation works by developing a combination of business assistance programs (in particular for SMEs) and practical research programs that companies can learn from;
- Incorporate WPI systemically into technological innovation programs in cooperation with business.

In terms of research, in the future we would benefit from:

- Reaching conceptual clarity and shared agreement on a definition of WPI;
- Developing an integrative systemic theory of WPI that accounts for the concept's multi-disciplinary, multi-level nature;
- Improving the overall quality of the applied research methods as well as employing a wider variety of both quantitative and qualitative research methods;
- Disseminating empirical research results more widely to a practitioner and policymaking audience;
- Making better use of available databases in Europe (Chap. 16).

In terms of practice, companies might want to consider:

- Taking a systemic approach to the implementation of WPI;
- Developing a 'participation and trust' management regime, to be distinguished from a 'command and control' management regime;

[9]EUWIN will be continued by the current network of six partners (TNO–NL, Workplace Innovation Ltd.–UK, DLR–GE, ZSI–AU, ARC Fund—BU, and Sinnergiak—SP).

[10]Flanders Synergy has certification programme for consultants who are allowed to assist in WPI projects.

- Developing employee participation, both direct participation (job autonomy and shop floor consultation) and formal (representative) participation;
- Developing innovation skills on the job as well as through external training to develop the organisation's innovative capability.

The nature of work and the ways in which it is performed have changed multiple times throughout human history. However, the pace of change has accelerated significantly, largely due to digital technologies. New technologies have already started to impact standard job definitions and patterns, transform the relationship between employers and employees, change work organisation and business models. Many of today's jobs and most sought-after skills did not exist a decade ago, whereas routine and even some non-routine occupations are vulnerable to automation.

Success in the new industrial revolution, of course, requires that companies, hospitals, government services etc. make use of the best available technologies. However, they also have to increase their focus on the human factor. This is the main resource in Europe and it is not used enough. People need to find a new place in smart factories, hospitals, government services etc. and organisations need to anticipate these changes in order to ensure that both the workforce and the workplace are fit for the future.

European organisations need to adapt. Advances in automation, digitisation, ICT and manufacturing represent enormous opportunities for employers and employees. Nonetheless, too few companies are actually rethinking the way people work and collaborate. Too few companies are re-modelling their internal organisation to tap into the capacities of all their employees—not only in their R&D departments. To be a leader or a proficient follower of the new industrial revolution means looking beyond technology alone. It requires having workplace innovation as the very DNA of the organisation. To achieve this, support from governments and social partners makes a difference.

Author Biographies

Peter R.A. Oeij Ph.D. is a senior research scientist & consultant affiliated with TNO, The Netherlands Organisation for Applied Scientific Research in Leiden, The Netherlands. His field of work is innovation management, workplace innovation, social innovation and team dynamics.

Diana Rus is a managing director at Creative Peas and a senior lecturer at the University of Groningen in The Netherlands. Her field of work is innovation management, open innovation and leadership development.

Frank D. Pot is an Emeritus professor of Social Innovation of Work and Employment, Radboud University Nijmegen (NL) and former director of TNO, The Netherlands Organisation for Applied Scientific Research Work and Employment.

Index

P.R.A. Oeij et al. (eds.), *Workplace Innovation*, Aligning Perspectives on Health, Safety and Well-Being, DOI 10.1007/978-3-319-56333-6

www.ingramcontent.com/pod-product-compliance
Ingram Content Group UK Ltd.
Pitfield, Milton Keynes, MK11 3LW, UK
UKHW020131290926
15471UKWH00003B/415
* 9 7 8 3 3 1 9 8 5 8 8 7 6 *